Technology Convergence and System Divergence

Hikari Nohara · Lars Medbo
Editors

Technology Convergence and System Divergence

Toyota and Uddevalla, Volvo

palgrave
macmillan

Editors
Hikari Nohara
Nagoya, Japan

Lars Medbo
Department of Technology
Management and Economics
Chalmers University of Technology
Gothenburg, Sweden

ISBN 978-981-96-1909-2 ISBN 978-981-96-1910-8 (eBook)
https://doi.org/10.1007/978-981-96-1910-8

This Palgrave Macmillan imprint is published by the registered company Springer Nature Singapore Pte Ltd.
The registered company address is: 152 Beach Road, #21-01/04 Gateway East, Singapore 189721, Singapore

Acknowledgements

This book is the product of lengthy cooperations among different academic disciplines, i.e. engineering, social and human sciences and also between very different cultures, i.e. Swedish and Japanese ones. During these days of our cooperation, co-authors owe our study to various researchers and practitioners. Swedish co-authors would like to especially thank Bertil Johansson, Dan Jonsson and Ulf Karlsson. Japanese co-authors also would like to express their gratitude to Tetsurou Nakaoka, Michel Freyssenet, Åke Sandberg and Kaisa Ellegård. We also would like to thank GERPISA network, which has given us the opportunity of communication and discussion.

Contents

1 **Introduction** 1
Hikari Nohara, Lars Medbo, and Tomas Engström

Part I Uddevalla Production System and Its Socio-Technical Conditions

2 **Alternatives to Assembly Line and Assembly Line Work: An Exposé of the Swedish Lines of Development with Focus on the Author's Experience and Insights** 21
Tomas Engström

3 **Application of Certain Principles for Education and Shop-Floor Learning Practices for Assembly Work at the Volvo Uddevalla Plant—A Brief Overview** 125
Lennart Nilsson and Tomas Engström

4 **Characteristics of the New Production Concept in Sweden Based on the Uddevalla Plant: Cognition and Action** 195
Yutaka Tamura

5 **Operators' Work Situation in Volvo Uddevalla** 239
Lars Medbo and Carl Wänström

Part II Toyota Production System and Its Socio-Technical Conditions

6 **Labour Management/Labour Relations and Toyota Production Systems** 269
Masaki Saruta

7 **Toyota Production System and Modern Society** 285
Hikari Nohara

8 **Cell Production System and Its Conditions** 323
Uichi Asao

Part III Analytical Comparison of Renewals of Production Systems Between Sweden and Japan

9 **An Analytical Comparison of Learning Strategy** 343
Lennart Nilsson and Hikari Nohara

10 **Comparative Analysis of the Division of Labour in the Assembly Workshop** 349
Uichi Asao, Yutaka Tamura, and Eishi Fujita

11 **The Analytical Comparison of Work-Focused Improvement** 375
Hikari Nohara

12 **The Summary and Points of Discussion for the Future** 391
Lars Medbo and Hikari Nohara

Index 433

Notes on Contributors

Uichi Asao is former Professor in the Faculty of Business Administration at Aichi Toho University (Japan). He specialises mainly in labour and production systems of the Japanese automobile manufacturing industry and the electrical and precision equipment manufacturing industry. His main co-authored books are *Jidōsha Sangyō to Rōdōsha (Automobile Industry and Its Workers)*, Hōritsu Bunka Sha, 1988, *Shakai Kankyō no Henka to Jidōsha Seisan Shisutemu (Changes in the Social Circumstances and Automobile Production Systems)*, Hōritsu Bunka Sha, 1999 and *Japanese Companies*, Trans Pacific Press, 2004.

Tomas Engström is an Associate Professor Emeritus in internal material handling in the former Department of Transportation and Logistics at Chalmers University of Technology (Sweden). He completed a PhD thesis (a monography) regarding material feeding to autonomous workgroups in 1983 in the Department of Transportation and Logistics at Chalmers University of Technology in Gothenburg. During the past four decades, he has mainly developed alternatives to assembly lines within the Swedish automotive industry and several other research and development initiatives in various branches, i.e. apart from automotive matters. The research development work was of a cross and multi-disciplinary nature, together with senior research competencies within ergonomics, psychology, sociology, vocational learning, etc. After that, he received a specially and personally constructed senior research position dealing with socio-technical design within the automotive industry, financed by three Swedish governmental research foundations.

Eishi Fujita is Emeritus Professor in the Graduate School of Humanities and Social Sciences at Nagoya City University (Japan) and has mainly conducted sociological empirical research on labour relations and workers' lives in the automobile industry. His main co-authored books are *Jidōsha Sangyō to Rōdōsha (Automobile Industry and Its Workers)*, Hōritsu Bunka Sha, 1988, and *Shakai Kankyō no Henka to Jidōsha Seisan Shisutemu (Changes in the Social Circumstances and Automobile Production Systems)*, Hōritsu Bunka Sha, 1999, and his recent article is 'Toyota to Chiiki Shakai no Genzai (Toyota and Local Society Today)', *Annual Review of Tokai Sociological Society*, No 7, 2015, pp.66-82.

Lars Medbo is former Associate Professor in Logistics and Transportation in the Department of Technology Management and Economics at Chalmers University of Technology (Sweden). His research focuses on materials supply systems, materials handling and production flows in production system design. The work has resulted in development of new materials feeding techniques, layouts, materials flow structures and work structuring principles implemented in the Swedish automotive industry. He is supervising PhD students and contributing to the Production Leap national programme as coach and teacher in professional education courses. He participated in the development of Volvo's factory in Uddevalla 1985–1993.

Hikari Nohara is Professor Emeritus of Hiroshima University and Nagano University (Japan). He specialises in production system and workers' lives, and currently in difficulties which Japanese universities have confronted (what Bill Readings calls the University in Ruins). He has also been the member of GERPISA since 1992. His main publication is: *Gendai no Bungyo to Hyojyunka (Division of Labour and Standardisation in Contemporary World)*, Kyoto: Takasuga Syuppan (2006).

Lennart Nilsson is a Professor Emeritus in the Department of Pedagogics at Gothenburg University (Sweden) and a specialist in vocational learning matters, and he has been involved in various research and development initiatives. He started his career at the University of Gothenburg in 1967–2011. He began as a research assistant (PhD degree in 1982) for three years and then project leader for research and development in the vocational line on secondary school and adult education, associate professor in 1982 and professor from 1990. He was a counsellor, researcher and teacher for the Swedish automotive industry. From 1980 to 2010, he was engaged within the Volvo Company, i.e. automobiles, heavy truck chassis manufacturing and the Volvo Uddevalla projection process. As a Senior Professor in Vocational Didactics at Kyambogo University in Kampala, Uganda, he was involved in developing research and renewal of vocational training in Uganda and South Sudan during 2011–2016.

Masaki Saruta is Professor Emeritus of Chukyo University (Japan). He specialises in personnel and labour management and industrial relations, especially of Toyota. His main publication is: *Toyota Shisutemu to Romukanri (The Toyota system and personnel management),* Tokyo: Zeimu Keiri Kyokai 1995, *Toyota uei to jinjikanri,roshikankei (The TOYOTA Way and Human Resources Management and Labor-Management Relations)* Tokyo: Zeimu Keiri Kyokai, 2007, *Nihonteki Roshi Kankei to Fukushi-kokka (Japanese-style Industrial Relations and the 'Welfare State'),* Tokyo: Zeimu Keiri Kyokai, 2013.

Yutaka Tamura is Ph.D., Professor in the Faculty of Business Administration at Aichi Toho University, Japan. His research primarily focuses on production systems, human resources and lean production in the automobile manufacturing industry, with a comparative study between Japan and Sweden. He participated in research programmes on the activities of Japanese suppliers and their systems in Asia, Europe and America from 2003 to 2018. In 2003, he published *Ford kara Uddevalla e (From Ford to Uddevalla)* through Taga Shuppan. Additionally, he published the paper 'Sweden ni okeru sagyou kenkyu to roudou no Sweden Model', 'Time-Motion Study in the Swedish Model on Work: A Tentative Study from Historical Aspects', in the *Meiji Business Review*, Vol. 66, No. 2, in March 2019.

Carl Wänström is an Associate Professor in Technology Management and Economics and he is working at Chalmers University of Technology (Sweden). His main research area is operations management, where he is an expert in lean management. Wänström uses a socio-technical approach to understand how to design, organise and manage production teams as well as continuous improvement infrastructure, to create a learning organisation based on lean practices. The research is empirically driven, interdisciplinary and carried out in close collaboration with manufacturing companies.

Abbreviations

ACP	Autonomous Complete Process
AGVs	Automatically Guided Vehicles
BNP	Basic Psychological Needs
CPS	Cell Production System
FPS	Ford Production System
GL	Group Leader
JDT	Job Design Theory
KDP	Konstruktions Data Volvo Personvagnar
LPS	Lean Production System
NTPS	New Toyota Production System
PKI	Process-/Inspection Instructions
QA	Quality Assurance
SAS	Standardized Assembly Sequence
SDT	Self-Determination Theory
STS	Socio Technical System
SW	Standardised Work
TL	Team Leader
TPS	Toyota Production System
UPS	Uddevalla Production System

List of Figures

Fig. 2.1 Interior of our experimental workshop with several disassembled automobiles (*Source* The author's photo) 40

Fig. 2.2 The Saab Scania body shop in Trollhättan (*Source* The author's drawing) 47

Fig. 2.3 An example from the Volvo 740 automobile model showing three automobile models, i.e. P10. P20 and P70 (the model assembled in Uddevalla), which, metaphorically speaking, was folded along its edges and symmetries. The empty automobile body was regarded as consisting of three boxes featuring different openings (the engine room, with two horizontal openings; the coupé, with two vertical openings; and the luggage compartment, with one horizontal opening) or as illustrated in the folded-out paper box. This schematised automobile leads to hierarchically different views of various areas of the empty automobile body, as was constructed by Mikael Hedin at our experimental workshop (*Source* original illustrations from our experimental workshop) 79

Fig. 2.4	Our unique system of illustrations depicts the hierarchy of the assembly-oriented product structure constructed for the automobile model assembled in Uddevalla. As defined verbally, the assembly-oriented product structure (top) has sometimes been denoted 'Final Assembly Functional Groups' since it is a figure retrieved from an earlier publication. The fragmented PKIs are at the bottom of the illustration (*Source* Engström & Medbo, 1991)	81
Fig. 2.5	The layouts used for the small parallel workgroups at the Volvo Uddevalla plant (*Source* The author's drawing)	87
Fig. 2.6	Layout of one of the most refined versions of Japanese assembly lines (*Source* The author's drawing)	89
Fig. 2.7	Overview of the Volvo Uddevalla plant, which featured six parallel assembly workshops arranged as two three-leaf-clover clusters of buildings grouped around two media-filing and roller-testing facilities and a common material store. These parallel assembly workshops embraced a 'cinnamon-bun-shaped' administration building (*Source* The author's drawing)	91
Fig. 2.8	The rather complex hierarchy of the intra-workgroup pattern of four operators was implemented in the Volvo Uddevalla plant (*Source* The author's drawing)	109
Fig. 2.9	An illustration of how the holistic learning approach was engineered at the Volvo Uddevalla plant, which involved employing an assembly-oriented product structure that gradually permitted access to the two most detailed and extensively long assembly sequences (ultimately to the fragmented PKIs). Some of the concepts or terms in the figure may be inaccurate since the figure was retrieved from an earlier publication (*Source* Engström & Medbo, 1991)	113
Fig. 3.1	An operator performing assembly work on a tilted automobile body in one of the five parallel assembly workshops used for assembly work, i.e. fitting components underneath (*Source* The Volvo Automobile Company, the Volvo Uddevalla plant)	134

Fig. 3.2 The figure illustrates the five phases of the expanded assembly work. The mutual communication between operators in the roles of instructors and newly recruited operators should gain some degree of stability regarding the broadened and enhanced work content through guiding questions, as illustrated in the figure above and also contribute to reinforcing professional skilled identity through such work content in assembly work. Assembly, therefore, involves something more than the act of assembling. Note that this is an old version of this table published earlier, i.e. it could be more coherent with the main text (*Source* Ellegård et al., 1992) 157

Fig. 3.3 Selection of the most prominent components needed for the assembly work of one-fourth of the work in and on the automobile bodies of the Volvo 740 automobile model. That is the fourth and last of the four main groups of the most prominent components comprised in the assembly-oriented product structure for this automobile model, the Interior. The illustration in the figure is an example of the hierarchical system of illustrations constructed in our experimental workshop in Gothenburg (*Source* Ellegård et al., 1992, an original illustration from our experimental workshop) 169

Fig. 3.4 Above is a schematic or symbolic illustration of how the product should be perceived according to various explanations and reasoning presented in this and earlier sub-chapters, namely as one short assembly sequence and as one specific relation (these are in the figure denoted 'sequence' and 'casual connection'). The illustration suggests that these phenomena derive from hierarchies of a partly more taxonomic nature, which define, for example, product architectures and variations of the automobile. Examples of the specific coordinates and symmetries discussed are also illustrated (remember the explicitly and implicitly defined specific coordinates and symmetries) (*Source* Ellegård et al., 1992) 176

Fig. 3.5 This is an example of how the hierarchical illustration systems were constructed in our experimental workshop in Gothenburg. In the case of the figure above, the components were successfully removed from a heavy truck chassis and documented. At the same time, the outline was used to document their original positions and other purposes, as was the case during similar research and development work for the Volvo Truck Company that followed the Volvo Uddevalla plant projection process by approximately two years (*Source* Original illustrations from the experimental workshop) 189
Fig. 4.1 The role of maps: Overview and visualisation of cognitive processes 212
Fig. 4.2 The role of maps: Overview and visualisation of cognitive processes 213
Fig. 5.1 A socio-technical model to understand operators' work environment based on the defined dimensions 243
Fig. 7.1 Toyota Production system as a chain of nexus between ends and means *Note* (1) An arrow means a nexus between ends and means. The beginning of an arrow is a means, and the arrow tip shows an end. (2) SW is regarded as One Best Way for assembly at the moment. Therefore, it is also regarded to contribute to realising high quality. (3) *Jidoka* includes Stop Code System and a machine using an automatic stop device. This concept requests us to pay attention to human and machine. (4) In the case of the principle that defects should be solved when and where they occur, the defects mean defects of products (complete car, component and part) and machines. Therefore, this principle asks us to pay attention to two aspects, namely, a man and machine. The principle also implies three points. That is to say, at first, they have to find not only the defects of products, and components and parts, but also the defects of machines and equipment. Second, they have to find causes of defects and find measures against their recurrence. Third, they have to repair products and equipment 291
Fig. 12.1 Schematisation of results and discussions in the book's chapters and how they relate *Note* The numbers indicate chapters 410

List of Tables

Table 2.1	Examples of less common concepts or terms used in this chapter are listed to clarify their origin and expressed as bold text with underlining. Note that some concepts or terms regarding information, product information, etc., have partly consciously been synchronised with another chapter in this anthology written by the author together with another co-author	27
Table 2.2	Examples of concepts or terms used in this chapter are listed above to clarify their origin, i.e., they are not marked out in specific ways	29
Table 2.3	Acronyms used in this anthology chapter	30
Table 3.1	Selected less-common concepts or terms from this chapter clarify their origin in bold text and underlined. Note that some concepts or terms on central and local information have partly intentionally been synchronised with another chapter of this anthology authored by one of the two authors	131
Table 3.2	A table of selected concepts or terms from this anthology chapter clarifies their origin, which are not marked in specific ways	132
Table 3.3	Acronyms used in this chapter	132
Table 5.1	Coded interviews of the main groups based on the theory section	244
Table 5.2	Socio-technical system impact on operators' autonomy	258
Table 5.3	Socio-technical system impact on operators' competence	260

Table 5.4	Socio-technical system impact on operators' relatedness	262
Table 7.1	Comparison between the TPS and FPS	292
Table 8.1	Comparison of production systems (Company N's assembly plant)	326
Table 8.2	Typology of assembly cells (Company N)	327
Table 8.3	Comparison of production systems (assembly shop for one model of Product X)	328
Table 10.1	The classification of operations involving assembly workgroups in the production preparation and mass production process	351
Table 10.2	Assignment of assembly operations (rear power seat)	358

CHAPTER 1

Introduction

Hikari Nohara, Lars Medbo, and Tomas Engström

This book makes a comparative analysis of the Toyota Production System (hereafter abbreviated as TPS) and the Uddevalla Production System, Volvo (hereafter abbreviated as UPS).[1] From the perspective of the Japanese co-authors, UPS is very humane in terms of abolishing the conveyor belt assembly line which is machine-paced, and of extremely long cycles of work; however, they are sceptical regarding its efficiency.

[1] Generally speaking, as well known, TPS is regarded as the most excellent production system, while UPS is famous as the system without a belt conveyer and with extensively long cycle work.

H. Nohara (✉)
Hiroshima University, Higashihiroshima, Japan
e-mail: nohara@nagano.ac.jp

Nagano University, Ueda, Japan

L. Medbo · T. Engström
Department of Technology Management and Economics, Chalmers University of Technology, Gothenburg, Sweden
e-mail: lars.medbo@chalmers.se

T. Engström
e-mail: tomasengstr@gmail.com

H. Nohara and L. Medbo (eds.), *Technology Convergence and System Divergence*, https://doi.org/10.1007/978-981-96-1910-8_1

In addition, the Japanese co-authors were surprised that Toyota managers criticised the segmented, fragmented and repetitive nature of assembly work on the line in the early 1990s. Therefore, they began finding similarities between Toyota's cautious improvement of assembly work on the line and UPS despite the mainstream opinion that UPS and TPS are opposite to each other. Meanwhile, the Swedish co-authors, who were involved in the development of UPS, wanted to explain and demonstrate UPS' potential compared with that of TPS by understanding the basics of TPS. Could a lack of knowledge about alternatives such as UPS contribute to TPS becoming generally the only prevailing production system/strategy in the world? By comparing these production systems, we believe that this book can provide essential insights in the development of today's production systems.

In this introductory chapter, we will explain several preconditions and the background informing this project. In addition, we will demonstrate how the following chapters are related to each other.

1.1 Why Do We Focus on a Production System?

1.1.1 Historical Background

In the early 1980s, Japanese automakers had just begun competing against the American Big Three, even in the American domestic market. In those days, American auto companies, the US government and most American people assumed that this was the result of protective trade policies and Japanese labour conditions such as low wages and long working hours. Therefore, they insisted that the market competition was unfair. They required that Japanese automakers build their plants in North America for free competition in the same market under equal preconditions. They believed that without the protection of the Japanese government and Japanese-style labour management, Japanese automakers would fall behind in the competition.

In these conditions, Japanese automakers had no other option than transplanting their factories to North America. However, since the mid-1980s, the Japanese auto transplants in North America have shown better performance and product quality compared with American companies, without the protection of the government and Japanese social relations. Worldwide, many observers were shocked by this phenomenon. They had to recognise that this achievement did not the result from low wages or dumping at a low price, but from the excellence of the mechanism, that is, the production system. Businessmen and researchers worldwide were embarrassed because, for a long time, most people in Japan and the West had believed that Japan was far behind the West in a lot of fields.

What was the reason for this marvel? A common question was whether the Japanese production system was really superior to the traditional Ford Production System (hereafter abbreviated as FPS)—also called the post-Fordism debate. One of the answers to such a question came from a rather different technology-oriented context, from an influential popular book, *The Machine that Changed the World* (Womack et al., 1990). In this book, the TPS was considered as a representative of Japanese production systems. It was given a more general name—"lean production". After this book was published, lean production was widely accepted as the most up-to-date and efficient production system, whereas the UPS in the 1980s–1990s was considered the opposite.

Simply speaking, most Western management was only interested in technical mechanisms and not in work contents, which were regarded to be an industrial engineer's domain in terms of a time and motion study. Furthermore, management had no knowledge of an operator's work content because they did not have a manufacturing engineer—a category of engineers common in Japanese plants. Under these conditions, an MIT research project was conducted, leading to the publication of Womack et al. (1990).

Why did this book consider UPS as the opposite to TPS? In this regard, we need to remember the debates and experiments conducted in Europe during the 1970s and 1980s (see Chapter 2 regarding the Swedish case). It should be noted that when the book was being written, UPS was in its design phase and only its pre-production had been completed in the factory. The book saw UPS as a step back to the craftsmanship methods used for vehicle manufacturing before the introduction of the assembly line by Ford. As a matter of fact, the long cycle time caused by alternative learning strategies can connect humanisation with the production system's efficiency improved. However, such relationship was never discussed.

1.1.2 *Overlooked Characteristics: Beyond Lean Production and the Transformation of Production Strategies*

In contrary to the above understanding that UPS and TPS were opposite to each other, since late 1980s through the mid-1990s, Toyota themselves had begun to pursue the next step beyond TPS for production, as mentioned in Chapter 7 in the book. That is to say, the Toyota management made a number of attempts to improve work content as they were facing a labour shortage, even considering UPS.

One day in 1990, the Japanese co-authors were surprised to hear a Toyota manager's statement that the operators' current situation on the assembly line was almost identical to the situation in Charlie Chaplin's 'Modern Times'. We knew about the criticism and debates against Ford's introduction of the conveyor line in assembly work, which arose mostly because of the inhuman nature of the work (see Hounshell, 1984). However, these comments had almost disappeared after several decades since its introduction. Nevertheless, such criticism has recently revived in contemporary Japan, not by the labour union but by Toyota management themselves.

During the 1990s, how did the public perceive UPS? The abolishment of conveyor lines, the unmoving workstation and extensively long work cycles were most distinguishable for outsiders. These were immediately obvious when one visited the plant. However, abolishing the conveyor line was just a tool to reform work contents, resulting in an extensively long work cycle. The core point of UPS had been very hard to understand from an outsiders' viewpoint. In the 1990s, almost no outsider understood the core point. Additionally, the plant closed in 1993. Therefore, in 2000s, almost everyone had become indifferent and forgotten about UPS.

The core points were, based on the knowledge of inherent human cognitive capability, a change in learning strategy from additive to holistic learning, the materialisation of this new strategy on the shop floor and reorganisation of work content and of material handling, as described in Chapters 2, 3 and 4. However, unfortunately, the aforementioned knowledge was not obvious to the public. Therefore, it will be very useful to recall such knowledge for reconsidering the man/machine (computer) interfaces within the automotive industry, especially for assembly work, as described in Chapter 4.

1.1.3 The Production System as a Common Platform

As the last remark on production systems on the shop floor, we refer to one more reason for why we focus on the production system. We want to find a common discussion platform not only among Western and Japanese researchers and practitioners but also among admirers and critics of the Japanese manufacturing system. However, the manufacturing system as a whole is a huge, rather vague entity because it includes not only the production system on the shop floor but also the labour relations, market conditions and social customs that are essential for the functioning of the production system. In this regard, the elements that should be considered would be an endless debate. This especially happens when we attempt to compare the manufacturing system in one country with that of another. On the contrary, the production system on the shop floor is clear and evident, as seen in Sect. 1.4.

In this regard, the production system can be a common platform for an inter-country discussion beyond the differences of real and concrete situations. However, even in the comparison of production systems among different countries, some elements/backgrounds have to be considered. As described in Chapter 7, when transferring a certain production system within countries, two conditions must be focused on. Firstly, the societal and technical constraints faced when transferring. Some modifications are necessary for seamless transfer, such as the transition of FPS to TPS in post-war Japan. Secondly, the conditions enabling the transfer must be considered. In the case of Toyota, peasants from rural areas took the functional role of European immigrants for FPS. Therefore, a production system cannot be transferred without corresponding conditions. However, if it is properly modified or if the necessary functional equivalent is found, the original production system may be transformed with modifications in a different society. Therefore, we will discuss the positives and negatives with regard to the production system itself, considering constraints and functional equivalents.

Based on the above, when we want to transfer a manufacturing system, we must divide the manufacturing system into three elements: the production system itself (the mechanism how to manufacture) on the shop floor; the societal and technical constraints faced by the production system in the country in which it is to be introduced in; and the technical conditions enabling the transformation of the original production system into the modified production system. In this context, the production system could be a common inter-country discussion platform.

1.2 The Three Swedish Authors from Chalmers University of Technology

1.2.1 The Swedish Industry

Sweden was a small wealthy country during the early 1960s and 1970s, protected from the ravages of the Second World War. This created opportunities for industry and society to gradually rectify crucial issues.

For example, decision-makers within the automotive industry were looking for opportunities to improve work conditions in their plants. Large automotive products such as automobiles and heavy truck chassis naturally had assembly lines, and these work practices followed the manufacturing paradigm of scientific management (these plants were in focus due to their particular characteristics, that is, repetitive short work cycles were dominating; such strives were mainly achieved within Volvo and Saab Scania). Nevertheless, the plan of action and methods of organisation became critical questions that remain valid even today. The initiatives and methods have changed over the past decades. However, due to various recent industrial trends, traditional assembly systems are rarely practised today, and a particular Swedish line of development regarding research on non-traditional assembly systems has ended. Among other things, preconditions and labour laws for manufacturing have recently reformed in Sweden substantially (for example, temporary workers and customer-oriented work-time scheduling).

In sum, the previous Swedish preconditions have recently been adapted to international development lines. The manufacturing paradigm of scientific management has reappeared with alterations, having new, rather different characteristics. Simultaneously, lean production has been introduced and developed in the Swedish context.

1.2.2 Initiatives Within the Swedish Industry

In the early period, non-human factors were emphasised heavily. From a social science perspective, the democracy of work and workers' influences in management decisions were considered necessary. Meanwhile, from a technical science perspective, improved ergonomics and improved work environment factors (pollution, sound, vibrations) were proposed. Such initiatives proved possible to attack partly in symbiosis with the universities.

During this period, the affluent Swedish industry not only acknowledged but also celebrated these initiatives. External work environment factors such as leisure centres, vibrant work environments, nearby pause rooms and profit sharing were seen as attractive approaches to enhance the conditions for assembly line work.

After some delay, decision-makers within the Swedish automotive industry became interested in alternatives to assembly lines and assembly line work, that is, non-traditional assembly systems.

1.2.3 The Volvo Uddevalla Plant Projection Process

The message above introduces the research and development work in connection with an industrial projection process discussed in at least two chapters in this anthology. The most exciting and prominent non-traditional assembly system in Sweden was the Volvo Uddevalla plant.

In this case, unlike earlier initiatives, human factors were prioritised to consider workers' interactions with each other and with the manufacturing equipment on the shop floor by means of recapitulating earlier experiences and insights from socio-technical research practices (a research practice that had earlier been prominent in Gothenburg).

In this case, such initiatives meant that technical changes synchronously followed social organisational reformation. In sum, it proved possible to simultaneously attain efficiency and humanisation.

However, such a feat requires established alliances between individuals beyond standard industrial practices, legitimation of the researchers within the company, industrial resources and specific industrial key individuals with excessive insights and knowledge, who are named below. In the case discussed, such feats were conducted as joint-venture cooperation procedures between university researchers and automotive industry practitioners.

In brief, the case provided experiences and insights regarding crucial aspects of humans' inner lives, which had to be grasped differently than before. In the most recent example of the Volvo Uddevalla plant were certain aspects of vocational learning, particular knowledge regarding regulation and work organisation elements and internal material handling were required at crucial periods. Such elements were constructively merged during the projection process of the non-traditional assembly system in Uddevalla's assembly plant (see Chapter 2, where such topics are explored further).

1.2.4 Socio-technical Approach

In 1978, a researcher arrived at Chalmers with a PhD from the Department of Sociology at Gothenburg University; he had worked at the body shop at the Saab Scania automobile plant in Trollhättan during his PhD studies. Namely, Ulf Karlsson (PhD) was an engineer and sociologist who had been working within the Swedish industry from the beginning, even before his studies. His research area was work organisation, focusing on regulating small parallel workgroups. With extensive practical and theoretical experiences from the body shop, the analysis relied on socio-technical research practices.

In short, Karlsson brought a complete research context during his tenure at the Department of Industrial Management. This context included valid industrial research questions, valued industrial contacts and necessary legitimisation within the Swedish automotive industry.[2]

At Chalmers, this meant an immediate creation of direct channels into the Swedish automotive industry (this contrasted other government-financed organisations such as Arbetslivscentrum Stockholm that occasionally struggled to gain constructive industrial access at this period). During the early period, the Gothenburg area was a focal point for industrially oriented socio-technological research practices with specific normative elements (several influential and internationally renowned senior social science researchers worked at Gothenburg University). Chalmers was not involved in this early period. In many aspects, it is a situation created by industry closeness for research on working life (see Chapter 2 for further details).

Consequently, an informal network of specialists within the Swedish automotive industry was created regarding material handling matters for prospecting socio-technical research practices with an overarching aim to develop alternatives to assembly lines and assembly line work.

[2] In 1979, Karlsson became an associate professor in the Department of Industrial Management at Chalmers University of Technology and a professor in 1982.

A long-time friend to the researchers at Chalmers, with essential contributions to this anthology and a crucial individual in this network, was the deceased Bertil Johansson from the Volvo automobile and truck companies. In most aspects, he was the most experienced and knowledgeable individual regarding alternatives to assembly lines and assembly line work in the Swedish automotive industry. For example, he was involved in the long-hidden temporary Volvo Arendal assembly workshop used for complex heavy truck chassis (see Chapter 2 for further details). (Several other Volvo personals and external consultants were naturally also important, but we will not treat thins topic in depth.)

The other author of this anthology who entered the projection process of the Volvo Uddevalla plant was Lars Medbo (M.Sc.). He entered the now-established research area at Chalmers between 1985 and 1986. Before he became a PhD student, Medbo was an already reputable individual with whom Engström had worked together before.

Due to the two authors' participation in extensive research and development projects combined with the already explained industrial legitimation, various cooperation procedures were gradually established with new researchers, leading to the growth of co-authoring initiatives, such as scientific publications (see some of the references given in the authors' contributions in Chapters 2 and 3). The connections due to the two authors' industrial involvements within both social and technical sciences were and still are partly accessible (see Chapter 2 for further details).

1.3 Humanised Work in Relation to Established Theory

As described above, it is commonly assumed that Toyota in the 1990s and 2000s and Uddevalla in the 1980s and 1990s represent opposite poles of efficiency.

However, in our research as co-authors from Japan and Sweden, we have found some interesting facts that contradict this mainstream view. We found that Toyota and Uddevalla faced the same shortage of young workers in the 1980s and 1990s and took the same approach to solve it—work-focused improvement. The tools used in solving this issue were similar or identical work-focused devices and mechanisms, as described later in this book. These facts forced us to reconsider the mainstream understanding and opened the way to study towards an efficient and humane production system.

The overall objective of this project is to make efficiency compatible with work humanisation within the framework of a capitalist firm in the history of production system development. Our focus is on Sweden and Japan, mainly from the mid-1980s to the 2000s. Moreover, we examine whether humanised production is more or less efficient than non-humanised production. If we focus on inherent human cognitive capabilities, we see that efficient work is strongly correlated with humanised work. Therefore, several other features were also important as are treated in some of the Swedish authors′ chapters. The authors refer to those points in their contributions later in the book (see Chapters 2, 3, 4, 7 and 11).

Nevertheless, focusing on the topic of efficiency, TPS' efficiency on the shop floor is embedded in devices and mechanisms, whereas UPS focuses on the efficiency and product quality of individual workers achieved within the small parallel workgroups, which is integrated into the worker's body and mind (see Chapters 7 and 11). Toyota in the 1990s and 2000s and Uddevalla in the 1980s and 1990s appear to employ very different engineering techniques but eventually developed the same or rather similar technical devices and mechanisms, as described later on.

The subjects of our analysis are as follows: for Sweden, we mainly examine Volvo, especially Uddevalla in the 1980s and 1990s; for Japan, we mainly refer to Toyota domestic plants in the 1990s and 2000s. The products examined in this book are automobiles and electrical products in the case of cell production systems. We focused on these two specific eras as they witnessed concentrated work-focused improvements.

Finally, we must mention the reason for our emphasis on assembly workshops. In this regard, empirically speaking, we know that most Swedish authors were deeply involved in building and maintaining the process of UPS, whereas Japanese co-authors had so much impressed by the ongoing assembly work reform at Toyota during the 1990s. However, why are such work reforms meaningful from a social reform viewpoint? In this context, one of the editors, Nohara will demonstrate one perspective, although it may not be necessarily common even among the co-authors. In modern society, why do we not liberate the class divisions that have existed between the ruling and ruled classes for such a long time? What results in this distinction between the rulers and the ruled? The answer is the vertical division of labour—the division of conception and execution. Simply put, people monopolising conception work constitute the ruling class, whereas the people excluded from conception and forced to execute those concepts constitute the ruled class (see Weil, 1934).

In this regard, machine-paced assembly operations on the belt conveyor line in the automobile industry have been a stark representation of the vertical division of labour in modern society. Nevertheless, most people think such division of labour is inevitable for maintaining the society in which most people enjoy material wealth—for instance, for the construction of a low-priced car. However, we have witnessed the work reforms at Uddevalla and the Toyota assembly line. Based on these observations, we believe it necessary to clarify why and how these division of labour occurs in automobile factories.

1.4 What Is a Production System?

The production system on a shop floor works only when its operators have the necessary competence. Additionally, the system can be introduced to a certain society only when society has the necessary societal and technical conditions to utilise the system. Therefore, we should focus on three aspects of a production system when we think from the viewpoint of combining efficiency and work humanisation: the technical system itself, the required competence to function—in other words, the learning strategy for operators to acquire the necessary capabilities—and societal and technical conditions to activate the system.

In this regard, when we take the concept of the production system without referring to its societal conditions and labour relations, it is kind of abstraction and different from reality. Therefore, when discussing the transferability of one specific production system, we should focus on whether it can be realised without the societal and labour conditions.

This is the reason why, although other contributions other than Chapter 5 focused on the production system, this book involves Saruta's contribution (Chapter 6). In his contribution, Saruta focuses on labour relations, which are preconditions to the operations in Toyota plants. He warns people who focus solely on the production system, clarifying that the actual labour relations in Toyota plants are complex and beyond the imagination of people focused only on the production system.

Regarding the definition of a production system, in this book, a 'production system' on the shop floor comprises the manufacturing of a product with labour subjects (workers), labour means (tools and machines) and labour objects (materials). In other words, a production system is the organisation of activities, processes and resources used to transform input into output.

In any production system, manufacturing is done collectively. Therefore, any production system implies some work-related social relations (operator–operator relations, operator–engineer relations and operators' relations with the supervisor or manager, in other words, the vertical and horizontal division of labour on the shop floor). However, we do not directly focus on what kind of difference in wages and in promotion systems, what kind of difference of treatment between regular employees and temporary workers and what kind of change of power balance between labour-management relations influence work-related social conditions, when discussing a production system.

1.5 Research Background and Difficulty of Cooperative Research

Of the Swedish co-authors, Nilsson, Engström and Medbo were deeply involved in the process establishing UPS, but only Wänström belongs to the next generation to three other Swedish co-authors. All Japanese co-authors—Nohara, Saruta, Asao, Tamura and Fujita—have jointly researched management, production system, labour relations and workers' lives. The research groups met each other around 1993 and have since worked together until now. The GERPISA network began in 1992, through which we communicated with many automobile-related researchers internationally.

The Swedish and Japanese co-authors have conducted prolonged cooperative research in Japan and Sweden. Our work aims to achieve more than interdisciplinary research—its goal is the integration of engineering and social science and although they are very different, an amalgamation of Swedish and Japanese ways of thinking.

However, the above cooperative works were hard to practice in real life. Let us take two examples. Firstly, for social science, the clear characterisation of what happens is very important; therefore, simplification or exaggeration of the events sometimes happen. Here, the key term is 'understanding'. On the contrary, for engineering science, reproducibility appears to be key. A social scientist is satisfied when they understand the characteristics of UPS, but an engineering scientist wants to obtain a detailed description that makes it possible to reproduce UPS after reading about UPS. Hence, the research audience seems to be different.

Therefore, some misunderstanding may have happened between different research disciplines.

Secondly, TPS strongly influences workers from the viewpoint of Japanese social scientists. Therefore, they are very sceptical of the good reputation of lean production in the west, because it generally represents TPS. However, Westernised lean production differs from the original Japanese lean production. In this case, the term 'lean production' has different connotations between Swedish and Japanese researchers.

As described above, it takes time to build mutual understanding and communication among people with different backgrounds.

1.6 Chapter Structure

1.6.1 Relationship Among Respective Chapters

This book project consists of three parts. Although all of them consider implicit or explicit comparisons between TPS and UPS, the first and second parts mainly focus on UPS and TPS, respectively, and their trajectory. The third part compares TPS and UPS. The authors of Chapters 2 and 3 are deeply involved in establishing UPS. Nilsson proposed a change of learning strategy from additive to holistic learning for the purpose of getting the breakthrough beyond Kalmar experience. Without this proposal, UPS would not have happened. However, the proposal was not enough to make a production system. Materialising the proposal was a complex endeavour. Of course, Nilsson, Engström and Medbo worked together on site. Chapter 3 is the first experience of the joint writing as the result of the collaborations among the person who proposes a holistic learning strategy and persons who actualise the strategy on the shop floor.

Chapter 5 refers to what really happened on the shop floor at Uddevalla plant during these days, based on interviews with workers. This is the first and may be the last appearance of the voice of workers who were really working inside UPS.

Now we will refer to the relationship between Chapters 2 and 3, and Chapter 4. With regard to UPS, the authors of Chapter 2 and 3 are not only researchers but also the heavily involved practitioners, while the author of Chapter 4 (Tamura) is a Japanese outsider. His field is social science and not engineering science, the field of Chapter 2's author. Furthermore, one of Tamura's main fields of research is Swedish management. The relationship between the authors of these chapters reminds us of the relationship between actors and very enthusiastic audience while the play being performed. In this case, an actor is in the theatre with the audience appreciating the theatre from the outside. Because of these differences in viewpoint, the landscape between the actors and the audience will be different. Both views, in their respective landscapes, will make up for each other to access the reality of UPS. To the best of Nohara's knowledge, Tamura is the first external researcher to deeply grasp the inner structure and logic of UPS. In this regard, his challenge on understanding UPS is very exciting.

Part 2 consists of three chapters. Chapter 6 emphasises the real situation and backgrounds on which Toyota factories operate. Chapter 7 demonstrates TPS and its trajectory. Finally, Chapter 8 delves into cell production. Although this chapter refers to the case of the electronic industry, in terms of abolishing the conveyor line, cell production seems to have similarities with UPS. Therefore, we need to investigate the content of cell production by comparing with UPS. Finally, Part 3 (Chapter 9-12) provides the analytical comparison between TPS and UPS, mainly focusing on conceptual work.

1.6.2 Projects in Each Chapter

Now, let us introduce each chapter.

Chapter 2, authored by Engström, explores alternatives to assembly line and assembly line work, providing a comprehensive overview of the Swedish line of development. The author and his colleagues share their insights and experiences accumulated over several decades of research and development work carried. While the author primarily focuses on the Volvo Uddevalla assembly plant's case, he also mention other assembly plants and systems. For instance, the reason why the extensively long work cycle times proved feasible is explained. Material feeding technique, group work and layouts were essential areas. The industrial value of the collaboration between the author and his colleagues with Volvo personnel is also discussed.

Chapter 3, co-authored by Nilsson and Engström, treats education and shop floor learning principles from historical and scientific perspectives. Theories and practices were amalgamated during several decades of research and development work within the Swedish automotive industry before the Volve Uddevalla projection process. This process intensified further during the Volvo Uddevalla projection process, which was mainly accomplished in a temporary workshop in Uddevalla (though some parts of the prehistory are also cowered). This workforce was later moved to an entirely new assembly plant with material feeding by materials kits and small parallel workgroups, who single-handedly completed the construction of automobiles with long work cycle times—the advanced and unorthodox Volvo Uddevalla assembly plant. This projection process later reached a full-scale manufacturing pace and required both authors' involvement in various aspects.

Chapter 4, authored by Tamura, raises the following question: How do operators build a complete car in a stationary situation without a line system? After the 1970s, Sweden developed a production change based on the socio-technical systems (STS) perspective. The world's attention was drawn to the Uddevalla plants for their no-conveyor system: the UPS. This chapter focuses on the cognitive aspects of UPS and compares it with TPS. Tamura shows that UPS is an epoch-defining production model that utilises cognitive capability as an indispensable factor in achieving high flexibility.

Chapter 5, co-authored by Medbo and Wänström, discusses the factory operators' social and technical work situation. As the Uddevalla factory was seen as the future of sustainable manufacturing in the automotive industry, understanding the operators' work situations and contents is crucial for defining the prerequisites for successful production. The discussion is based on interviews with operators in assembly workshops and reflects their perceptions and views. In the analysis, a cross-disciplinary framework is used, based on the self-determination theory, job design theory and socio-technical systems theory.

The design with parallel product flows and extensive work cycles and contents was promising for high performance and improved working conditions. Further, the organisation of the production teams created good opportunities for operators to fulfil their basic psychological needs. However, this potential was not fully utilised, especially with respect to productivity. A leadership that focused on the design and needs of the production system would have significantly improved the workers' well-being and performance.

Chapter 6, authored by Saruta, focuses on the human resource management and labour in the TPS, and reveals the reality of the 'Hitozukuri' (human resource development) underpinning Toyota's 'Monozukuri' (manufacturing) process.

Chapter 7, authored by Nohara, consists of two parts, namely the study of TPS itself and its influences over society. The former consists of three subparts, documenting the transformation of FPS into TPS, Toyota's response to workers' aversion to short cycle line work and Toyota's approach towards mitigating differences in conception and execution. The second part provides an overview of the influences of characteristics of TPS over society.

The author's primary concern is that the excellence of TPS probably has substantial influence over the human way of thinking. In other words, as far as the production system itself is concerned, TPS or lean production is the crystallisation of modern rationality. However, modernity includes ambivalence. That is to say, if we focus on the production system itself, TPS is one of the best examples of modern rationality. But once we widen our perspective and go beyond the production system towards society, we can find the realisation of the danger embedded in modern rationality.

Chapter 8, authored by Asao, discusses assembly systems without conveyor equipment (the cell production systems), which have been introduced in Japanese electrical and precision equipment assembly workshops since the early 1990s, and identifies two main points. Firstly, the cell production system (CPS) has four basic characteristics compared with traditional production systems. Secondly, five social and technological conditions necessitated and enabled the conversion from traditional production systems to CPS. Where these five conditions or similar conditions exist, it is expected that conversion from traditional production systems to CPS will keep progressing.

Chapter 9, co-authored by Nilsson and Nohara, discusses a comparison of learning strategies among FPS, TPS and UPS.

Chapter 10, co-authored by Asao, Tamura and Fujita, analyses the division of labour structure of individual operators or workgroups in three different assembly systems: parallel production at the Volvo Cars Uddevalla plant in Sweden, the Autonomous Complete Process (ACP) at the Toyota Motor Corporation in Japan and the CPS at the Japanese electrical and precision equipment manufacturing company N. Two main findings are revealed. Firstly, the standard operations tended towards recovering functional cohesion or wholeness as the operators moved from traditional to ACP-implemented assembly lines and CPS or Uddevalla's parallel production. Secondly, in systems where standard operations are executed over long cycles and one or a few operators assemble the whole product, the possibility of delegating non-standard operations to individual operators or workgroups has been expanded. To realise this possibility, this chapter points out the importance of rethinking the belief that a thorough division of labour is the only way to increase efficiency.

Chapter 11, authored by Nohara, characterises the comparison between TPS and UPS as technical convergence and system divergence. According to the author, Toyota and Volvo faced the same problem—labour shortage. In terms of solving it, both organisations had the same work-focused orientation. Therefore, their technical resolution methods were similar or identical and their experimentations appear to technically converge.

However, the organisations had different targets that were realised in the respective production systems as a whole. Toyota's target was to realise more comprehensive work, meaning that work recovered the contextually meaningful connection among series of operations. Meanwhile, Volvo's target was to realise integrated work—indicating a system divergence.

In the final chapter, Chapter 12, a set of analytical results of the above chapters and future discussions are presented, along with a conclusion of the study's findings.

References

Hounshell, D. (1984). *From the American system to mass production, 1800–1932*. The Johns Hopkins University Press.

Weil, S. (1934). 'Réflexions sur les causes de la liberté et de ľ oppression sociale', in S. Weil, *Oppression et liberté*. Gallimard (1995).

Womack, J. P., Jones, D. T., & Roos, D. (1990). *The machine that changed the world*. Palgrave Macmillan.

PART I

Uddevalla Production System and Its Socio-Technical Conditions

CHAPTER 2

Alternatives to Assembly Line and Assembly Line Work: An Exposé of the Swedish Lines of Development with Focus on the Author's Experience and Insights

Tomas Engström

2.1 Overview

This chapter explains how a socio-technical research tradition successfully developed alternatives to assembly lines and assembly line work. These initiatives substantially improved efficiency and flexibility, including advanced work organisation with true group work in many different respects that are unique in a global context.

In the Volvo Uddevalla case discussed, assembly time per automobile and product quality outperformed the assembly line. However, this required several already-established features in combination with various non-traditional ones. In addition, specific learning and training principles

T. Engström (✉)
Department of Technology Management and Economics, Chalmers University of Technology, Gothenburg, Sweden
e-mail: tomasengstr@gmail.com

H. Nohara and L. Medbo (eds.), *Technology Convergence and System Divergence*, https://doi.org/10.1007/978-981-96-1910-8_2

needed to be developed. Such feats were accomplished simultaneously with substantially upgraded work and work conditions, which has been an area of research interest for the author for the last four decades.

This research approach established the foundation for the author and some of his closest colleagues' research and development work, which began for the author in the late 1970s and ended with the involved researchers' retirement. During this period, things accelerated as we increasingly engaged in some of the Swedish automotive industries' projection processes; the case discussed was not the sole industrial case worth mentioning.

*

The most prominent and protracted projection process involved the now defunct Volvo Uddevalla plant, which started in 1985 and closed down in the spring of 1993. This assembly plant was gradually transformed into one of the two most advanced and unorthodox assembly systems worldwide. These featured parallel product flow construction and small parallel workgroups with extensively long work cycle times (the other was at the assembly docks at the Volvo Tuve heavy truck chassis plant). The assembly plant discussed also utilised an advanced material feeding technique with material kitting fixtures containing the components required for each assembled automobile.

This chapter reports the history of the evolution of the construction of this assembly plant, including malfunctions and merits. Furthermore, this chapter details how we and others could interpret the holistic learning approach explained engineering-wise, i.e., the common denominator between the social and technical science disciplines treated here.

This feat was achieved even though most of the prevailing industrial contexts within the Volvo Automobile Company were constructed for assembly lines and assembly line operations. In various ways, existing central operations, information systems, practices, etc., needed to be more suitable for the advanced and unorthodox assembly systems practised in Uddevalla. These parameters have been organised and refined to fit an overarching logic based on assembly lines and assembly line work throughout the past decades.

The distinctions between central and local operations were decisive for this and other reasons. However, because it was a new plant with special governmental funding located separately from the main Volvo Torslanda automobile plant in Gothenburg, several new non-traditional features could be developed and implemented in Uddevalla. This assembly plant operated at full scale during the end of its short existence; practising such non-traditional manufacturing principles in an existing assembly plant would likely be impossible.

*

This chapter explains the author and his closest research colleagues' research and development work, which was partly conducted at our experimental workshop, with support from the projection process organisation and partly at a temporary learning and training workshop in Uddevalla, which started before the assembly plant was completed (the operators were transferred to new building facilities). Fortunately, we gained access to this experimental workshop from 1986. We also secured access to company internal expertise and senior research competencies from several academic fields.

A supply of several new automobiles, combined with what became an extensive amount of company information, was also crucial to our research and development efforts. The automobiles were disassembled according to gradually established disassembly procedures, mirrored against all available information, mostly company internal information, but also, necessarily, against general automotive literature from external sources.

For example, numerous tractor paper printouts of the central product information contained in the central information system, the KDP (central product information, in Swedish 'Konstruktions-Data Volvo Personvagnar') and numerous paper copies of the central assembly process information, the PKIs (central assembly process information, in Swedish 'Process/Kontroll-Instruktioner') were required. Note that product information is the information in focus here, while other types of more or less general information are also discussed, usually in more general senses.

This product information and other documents were prepared for systematically organised disassembly procedures. They were identified into manageable portions and reorganised on large, elbow-high, easily movable tables.

Such procedures required the researchers to understand and sometimes rectify prevailing administrative idiosyncrasies. These peculiarities would have otherwise been undetectable (such procedures had not been performed before inside the Volvo Automobile Company; to be frank, neither had such methods been required previously).

*

This chapter comprises six sub-chapters. Sub-chapter 2.2 is an introduction that begins with an overview of the author's research context and how the research questions have evolved over the years. Sub-chapter 2.3 contains a report on the early Saab Scania body shop in Trollhättan, an intriguing example of an advanced work organisation in the Swedish automotive industry.

Sub-chapter 2.4 discusses a detailed summary of the administrative prerequisites within the Volvo Automobile Company . The reader must comprehend these prerequisites to fully realise how they were transformed to fit the distinct demands created by Uddevalla's new local assembly plant.

Sub-chapter 2.5 provides a brief theoretical background on how the holistic learning approach was treated from an engineering perspective. This fact proves that insights from social science disciplines can be amalgamated with technical science disciplines. Such efforts were essential for the author and his closest research colleagues.

Sub-chapter 2.6 is a brief description of the Volvo Uddevalla plant. This description covers both the plant's merits and malfunctions. The effects gained were either expected or discovered. However, some practitioners' expectations differed from those of the engaged researchers. The complex construction was hampered by an automatically guided vehicle system and an awkward manufacturing scheduling system interacting somewhat non-transparently (partly due to internal conflicts and partly for other reasons). Unfortunately, these interactions were never rectified during the assembly plant's short lifespan.

Finally, Sub-chapter 2.7 provides some conclusions, including a concise excursion towards competitive assembly line systems in Japan.

*

To summarise, from a slightly different perspective, the content in Section 2.4.3 explains precisely why the practitioners within Volvo assumed that the automobile and its product variants were far too complex, numerous and different. Unfortunately, the explanations in this section are both tedious and intricate, although these are required to convey the message of this chapter.

The reason for this dilemma was that the content and organisation of the central information systems had distorted the perception of the product, and no possible stable and suitable wholes were available to find or construct anywhere, as was required for an engineer-wise interpretation of the holistic learning approach. Our systematically organised disassembly procedures, i.e. meticulous reverse engineering process, was the only way to transform the content in the central information system, the KDP and central assembly process information, the PKIs. The researcher manually performed various detailed excursions into the secrets of the central information systems and retrieved other relevant automotive documents, leaving almost no stone unturned in our quest for understanding.

That is, identifying the removed components in combination with the different excursions and grasping the unique characteristics of each component was critical. Processes that ultimately led to the construction of stable and suitable wholes were revealed when, for example, the removed components were spread out in various ways on the shop floor (see Section 2.5.2). The criteria for arranging the components into main and sub-groups of the assembly-oriented product structure was based on hierarchically and spatially discriminating component characteristics, thus creating the grouping criteria.

A suggested approach that the shop floor, personnel, or manufacturing engineers should have sorted out the required wholes is partly Utopian. In the case discussed, such feats required several years of work in our experimental workshop, for which the author and his closest research colleagues are naturally thankful for the help and resources provided.

2.1.1 Some Selected Concepts or Terms

Due to the nature of this chapter, the author has paid extra attention to the vocabulary; this is required since this anthology, in many respects, is a result of cross- and multi-disciplinary research and development work based on intricate and long-term collaboration between researchers in social and technical science disciplines.

*

A table containing a selection of (1) potentially complex concepts or terms from this chapter (see Table 2.1) has been compiled. These concepts or terms are boldfaced and underlined upon their first appearance in the main text. However, in some instances, the context may sufficiently convey the meaning, and the principle of marking the first occurrence of these concepts or terms was not strictly observed.

The author has determined that such extra attention is required for the reader since the concept's origins or terms may or may not need clarification. The aim was to separate those concepts or terms regarded as more common knowledge from those that are inventions of the author and perhaps also by his research colleagues, all presented subsequently and denoted as established or not established. Moreover, company-specific nomenclature is sometimes a nuisance since the author can only sometimes assume that the reader knows these.

In addition, (2) some less cryptic concepts or terms are presented in another table to inform the reader of their origin since these may appear strange for various reasons. However, these concepts or terms are judged not to be specifically emphasised (see Table 2.2). Finally, (3) a third table presents essential acronyms (see Table 2.3).

Such editorial approaches are sometimes required to crystallise the primary concepts or terms from what any author of most scientific publications—consciously or not—regards as outside of the everyday language mastered by most readers. As is the case here, these editorial approaches are unavoidable for cross- and multi-disciplinary efforts such as this chapter.

2.2 Introduction

This introductory chapter briefly discusses the background and beginning of the Volvo Uddevalla plant projection process and touches on the author's background.

2.2.1 Background

The following two sections of this introduction briefly sketch the author's scientific background from a longer-term perspective. Some research colleagues from other scientific disciplines involved in the assembly plant are also mentioned. As is partly evident in some of the references and several other publications by the author and his closest research colleagues, different scientific disciplines should have also been mentioned.

Table 2.1 Examples of less common concepts or terms used in this chapter are listed to clarify their origin and expressed as bold text with underlining. Note that some concepts or terms regarding information, product information, etc., have partly consciously been synchronised with another chapter in this anthology written by the author together with another co-author

Concept or Term	*Comment*
Administrative autonomy	Established
Advanced assembly work instructions	Not established, i.e. the author's choice
Advanced material feeding technique	Not established, i.e. the author's choice
Assembly-oriented product structure	Established (in the Volvo Uddevalla plant case product structure was used locally to describe the product architecture and product variation from an assembly work perspective
Autonomous workgroups	Established
Autonomy	Established
Balance losses	Established
Central assembly process information (the PKIs), i.e. 'Process/Kontroll-Instruktioner'	Established, in the Volvo Automobile Company case the assembly work was defined centrally
Central product information, i.e. the design-oriented product structure (KDP) and the central assembly process information (the PKIs)	Established in the Volvo Automobile Company case
Design-oriented product structure (KDP), i.e. 'Konstruktions Data Volvo personvagnar'	Established (in the Volvo Automobile Company case the product architecture and product variation was defined centrally by such a product structure.
Experimental workshop	Not established, i.e. the author's choice
Extensively long assembly sequence unique for each particular product variant	Not established, i.e. the author's choice
Extremely short assembly sequences in standardised forms	Not established, i.e. the author's choice
Extremely short assembly sequences in the PKIs.	Not established, i.e. the author's choice

(continued)

Table 2.1 (continued)

Concept or Term	*Comment*
Function Group Register	Established; a general register of an obsolete product, i.e. a compulsory standard within the Swedish automotive industry on which the design-oriented product structure used within e.g., the Volvo Automobile Company was based
Handling losses	Established
Holistic learning approach	Established
Inherent flexibility	Not established, i.e. the author's choice
Intra-workgroup pattern	Not established, i.e. the author's choice
Machine-paced product flow	Established
Mechanistic learning approach	Established
Most advanced and unorthodox assembly systems	Not established, i.e. the author's choice
Most refined versions of Japanese assembly lines	Not established, i.e. the author's choice
Non-machine-paced product flow	Established
Operator concentration	Established
Operators' inevitable stochastic variation	Established
Product architecture and product variation	Not established, i.e. the author's choice
Product change orders	Established
Product information and information	Established but what usually is not observed, product information is information concerning the product supplied from central operation to local assembly plants, while just plain information Is usually means a more general concept or term
Product variant codes	Established

(continued)

Table 2.1 (continued)

Concept or Term	*Comment*
Progressively organised material and information processing	Not established, i.e. the author's choice
Smallest work elements in the PKIs	Not established, i.e. the author's choice
Surplus competencies	Not established, i.e. the author's choice
System losses.	Established
Technical autonomy	Established
Tilting devices	Established
Unique system of illustrations	Not established, i.e. the author's choice
Variant losses	Established
Work module	Not established, i.e. the author's choice
Workgroup cohesion	Established
Working-up	Established

Table 2.2 Examples of concepts or terms used in this chapter are listed above to clarify their origin, i.e., they are not marked out in specific ways

Concept or Term	*Comment*
Common scheme for codifying all product variants	Not established, i.e. the author's choice
Edges and symmetries	Not established, i.e. the author's choice
Learning and training principles	Established
Socio-technical research tradition.	Established
Structure elements	Established
Structure links	Established

A local tradition within the Gothenburg area in research on working life.

Close cooperation between industry practitioners and researchers has always been a distinctive trademark for various research initiatives, especially Scandinavian research on working life. Such approaches have often created opportunities for researchers to discuss issues of common interest with experts directly. These collaborative approaches have also enabled researchers to gain access to accurate shop floor information and other types of data that would otherwise be difficult to obtain.

Table 2.3 Acronyms used in this anthology chapter

Acronym	*Comment*
KDP	Acronym for 'Konstruktions Data Volvo Personvagnar', which within the Swedish automototive industry is organised according to the Function Group Register; a Swedish standard for the automotive industry according to Volvo terminology, i.e. a central information system comprising the design-oriented product structure, which in turn is based on each automotive company's interpretation
PKI	Acronym for 'Process/Kontroll-Instruktioner', according to Volvo terminology, i.e. central process engineering documents from the central Volvo Manufacturing Process Engineering Department

In many respects, the Gothenburg area was and remains, to some limited extent, a focal point for an industrially oriented socio-technical research tradition with specific normative elements. In many ways, this unique environment was created by closeness to the industry. Furthermore, many of the large industrial companies in Sweden were and still are located within the Gothenburg area. However, this environment is due to the local research tradition from Gothenburg University.[1]

*

Applied disciplines within social science disciplines, such as sociology and psychology, were relatively new in the mid-1950s and early 1960s. Thus, they had to assert themselves against other scientific disciplines already established within natural science. The predecessors felt the need to overemphasise the development and use of scientific methods to be accepted; thus, house-training applied science researchers in natural science perspectives was deemed essential during this period.

[1] Professor Edmund Dahlström (deceased) and Professor Bengt Rundblad (deceased) initiated this local research tradition in working life. These two pioneers were responsible for sociology, and Professor Sigvard Rubenowitz (deceased) was in charge of psychology.

The pioneers were instrumental in developing these applied scientific disciplines from the early 1960s to the mid-1980s. They were followed by researchers within sociology, such as Hart and von Otter (1972), Karlsson (1979) and Hörte (1983), as well as within psychology, such as Johansson (1996) and Nonås (2005) at Gothenburg University. Researchers within materials handling and logistics at Chalmers University of Technology also worked in this tradition. Engström (1983), Johansson (1989), Christmansson (1997) and Medbo (1999) exemplify this. (The works cited are doctoral theses.)

However, during the early 1970s and mid-1980s, this local research tradition focused on organisational matters from a social science dimension, emphasising psychological and sociological perspectives. Furthermore, this trend also characterised the procedures for collaboration between industry experts and researchers. Consequently, sociology and psychology became the centre of attention in matters like democracy, group work, meaningful work, participation and work organisation (Dahlström et al., 1970).

Early Gothenburg researchers were influenced by Norwegian socio-technical research traditions (Thorsrud & Emery, 1964, 1969; Emery & Thorsrud, 1976). However, in the late 1970s, the technical science dimension embedded in this and similar research approaches were successively recognised as critical for dealing with working life; this was often the case when group work, i.e. autonomous workgroups, was introduced at the shop floor within the automotive industry.

*

In the following sub-section, the author will expand on what has been briefly explained in the context introduced above. This background developed into an engineering-wise approach regarding working life. It was particularly true for a socio-technical research tradition with a local base within the Gothenburg area.

Integration between the social and technical science dimensions: A trademark of the author and his closest research colleagues' research and development work.

This sub-section explains how the local research tradition in the Gothenburg area was and still is tied to the author's research and development approach, as has been the case for both near and more or less peripheral research colleagues and industry experts.

As stated, this local research tradition entailed collaboration between researchers from applied scientific disciplines within social and technical science disciplines. Such collaborations have led to the integration of social and technical science dimensions.

Professors Emeriti Jan-Åke Granath and Kajsa Ellegård approached the projection process discussed through their own scientific disciplines' perspectives, i.e. design theory and time geography, respectively, in the doctoral thesis of Granath (1991) and the publication of Elllegård (1989). Several other examples of cross- and multi-disciplinary collaboration procedures are also worth mentioning, especially those involved within or on the border of several scientific disciplines that deal with ergonomics in working life. Such integration was and is still performed simultaneously, as the purer social science and technical science dimensions were naturally developed further (as hinted, several other researchers should be cited for reference).[2]

Another very early key figure and close friend is Professor Emeritus Dan Jonsson from the Department of Sociology at Gothenburg University, with whom the author and his closest research colleagues have achieved joint venture publishing. This cooperation gained momentum in the early 1990s.

In discussing the Volvo Uddevalla plant, reference must be made to the research and development work on vocational learning matters that have been and still are partly conducted at the Department of Education and Educational Research at Gothenburg University. A proponent and long-time friend, Professor Emeritus Lennart Nilsson at the mentioned department (Nilsson, 1981), has partly been involved in our various Volvo

[2] Jan-Åke Granath (today Professor Emeritus at the School of Architecture, Industrial Architecture and Planning at the Chalmers University of Technology) actively participated in the Uddevalla projection process. Among other things, he organised buildings of scaled-down models of different constructed assembly systems that temporarily halted the then ongoing projection process at a crucial early period (e.g., Asterhall & Åkesson, 1986).

Furthermore, Granath, the author, and Medbo were involved in a collaborative initiative for some time. These particular efforts later allowed us to participate in suggesting revisions to the assembly system at the main Volvo Torslanda Plant. This effort (together with a cadre of Volvo engineers in Gothenburg) improved the Uddevalla parallel product flow construction for a substantially larger scale manufacturing volume.

After a period, the ex-top manager of the Volvo Uddevalla plant (Sven Date), led the following projection process, as he later took charge of 'Ny Montering Torslanda' (NMT), i.e. the Torslanda projection process, which from the beginning was initiated by an influential union representative (Björn Mattsson, Swedish Metalworkers' union).

For several different reasons not treated here, the top managers of Volvo Automobile Company ended the NMT projection process. There, unfortunately, Granath had, so to say, been left alone with his closest research colleagues and some Volvo personnel to fight the traditionalists at the main plant.

Granath's PhD thesis describes this projection process from a design-theoretical perspective while replicating a socio-technical research tradition (Granath, 1991).

After a while, the authors and his closest colleagues were recommended to concentrate their research and development work at the Volvo Uddevalla plant. Granath's and others´ efforts to launch an improved parallel product flow construction to the main Volvo Torslanda plant, which proved impossible, are treated in Granath's PhD thesis (Granath, 1991).

research and development projects, including the assembly plant discussed in this chapter (Nilsson is also a co-author of this anthology).

2.2.2 *Approaching the Projection Process of the Volvo Uddevalla Plant*

The following three sections discuss how the author's research questions have evolved from the original ones to those that became critical during engagements in the Volvo Uddevalla projection process.

The author's original research questions from the late 1970s

The author's original research questions in the late 1970s addressed the material feeding technique used in parallel product flow assembly systems for manufacturing automotive products. An articulated industrial requirement was derived from the experience at Saab Scania's body shop in their automobile plant in Trollhättan.

In this case, a parallel product flow manufacturing system (grinding and welding work) used advanced work organisation, a genuine socio-technical initiative that clarified crucial aspects. A then-doctoral candidate from the Department of Sociology at Gothenburg University conducted the research and development work, and the insights gained were conveyed mainly in academic circles (Karlsson, 1979).

Regarding the author's early research and development work, a similar construction with small parallel workgroups and an advanced work organisation may be implemented for assembly work, which was assumed to be characterised by overly complex products, extensive material volumes and a far too extensive number of different product variant.

During this period, a general assumption within the Swedish automotive industry was that the material volumes were too large to manage. This assumption was falsified in hands-on methods by the author's inventories of all the components needed for only one product, as was accomplished along two assembly lines in Gothenburg and Trollhättan, i.e. one inventory for an automobile and one for a heavy truck chassis. To the practitioners' surprise, the material volumes for one (single) automotive product were far lower than expected, given that the product information in the central information systems reported excessive material volumes, etc.

Such research and development work was first accomplished and later academically presented as a doctoral thesis at the Chalmers University of Technology's Department of Transportation (Engström, 1983).

Elaboration of the author's original research questions from the late 1970s and how these questions changed because of engagement in the Volvo Uddevalla plant projection process

The author's original research questions were tackled under the supervision and tutoring of then Associate Professor Ulf Karlsson (now professor emeritus) at the Department of Industrial Management at the Chalmers University of Technology. These two researchers' practical and theoretical results (Engström & Karlsson, 1982) were financed by several years of extensive funding from various governmental foundations.

This is just one exciting example to which the reader will return, which will be explained later in the chapter. This research and development work proved that shop floor realities within the Swedish automotive companies investigated were dissimilar to the computerised world in some critical respects. While the material volumes proved manageable, questioning the prevailing information systems of the time was impossible for the researchers, i.e. a far too extensive research area outside our capabilities and scope.

Moreover, at this time, the practitioners within the automotive industry proclaimed an all-embracing assumption of an overly complex product encumbered by numerous product variants. Some parts of this assumption were disproved early, whereas others needed to be sufficiently clarified; no constructive method was available at this early stage.

The author and his closest research colleagues discovered explanations hidden in the central information systems, which were not easily influenced then. However, these information systems were later appropriately rectified during the projection of what gradually became one of the two most advanced and unorthodox assembly systems. This assembly plant is the focus of this chapter. Among other things, the researchers had to contend with the same questionable assumptions explained previously, which remained relevant during the projection process of the Volvo Uddevalla plant.

Note that in this chapter, the author distinguishes between two different types of information. 'Product information' refers to information about the product, supplied by central operations to local assembly plants, but 'information' is also used in a more general sense.

As noted, questioning and revising the prevailing computer technology during the early period was impossible, even though both researchers had initially approached the author's original research questions in various ways through the existing information technology. Physical inventories along assembly lines focusing on one product provided the only way to obtain the correct material volumes.

Overall, this meant that the practitioners within the Swedish automotive industry were still maintaining erroneous interpretations of the research areas outlined here. Projecting any parallel product flow assembly system was definitively considered Utopian in the early period of the projection process in focus. Choosing work cycle times that exceeded a certain length was especially controversial.

New and far more comprehensive research questions that were valid during the engagement in the Volvo Uddevalla projection process

Compared with the author's original research questions regarding material volumes, where new and far more comprehensive ones gradually appeared, they later became valid from other perspectives with our deepening involvement in the projection process of the Volvo Uddevalla plant. Examples of questions identified that proved challenging for the research and development work conducted in our experimental workshop include the following:

1. Can the product architecture and product variation be analysed and operationalised? The author refers to understanding the physical product as it appears on the shop floor and how it is shaped and expressed in central information systems (sometimes this concept or term is synonymous with 'product design').
2. How may such insights benefit practitioners involved in constructing a new assembly plant featuring a new assembly system?

Expressed in another way, the new and far more comprehensive research questions may also be summarised as follows, partly by referring to the terminology used by the author to explain the automotive company's administrative world in this chapter, while ignoring many other issues also addressed during the projection process:

3. Specifically, the author and his closest research colleagues had to understand how to edit and transform the central product information. In the automotive company's vocabulary, this refers to the central product information, the KDP ('Konstruktions Data Volvo Personvagnar'), and the central assembly process information, the PKIs ('Process/Kontroll-Instruktioner instructions'), to conform to the material on the workstations on the shop floor work.

   In other words, the advanced material feeding technique mirroring the product architecture and product variation formed a starting point for assembly work. Note that product information is just one type of information that is in focus here, while other types of more general information are also discussed, but in a more general sense.

An explanation that implies that such approaches improved learning and training of what finally ended up being the extensively long work cycle times implemented for automobile assembly work conducted on an industrial scale worldwide.

2.2.3 *The Early Period of the Volvo Uddevalla Projection Process*

The earlier research and development work was continued and expanded due to unique opportunities gained when a new assembly plant with a new assembly system was envisaged. The following four sub-sections provide a quick overview of how these opportunities were identified and partly why this was the case.

Entering the Volvo Uddevalla Projection Process

After initiating the Volvo Uddevalla projection process in 1985, the author got opportunities to resolve the paradox briefly described previously, i.e., how and exactly why the central information systems were misleading, successively became clear.

The author became increasingly involved in this projection process due to contacts, mostly with experts within this automotive company, the most prominent of whom was deceased engineer Bertil Johansson, a key individual in Volvo automobile and truck companies. He and the author had been close friends for a long time.

Johansson was the most knowledgeable individual in Volvo regarding work organisation, including group work. He had hands-on experience in an early, successful alternative to assembly lines and assembly line work at the Volvo Truck Company's temporary Arendal workshop (Blackler & Brown, 1978). The exceptional results achieved at this temporary workshop were kept secret within the Volvo sphere since no plausible explanation was offered at this period for the surprisingly short assembly times gained and substantially improved product quality. However, nowadays, this dilemma is resolved if time losses are recognised as a theoretical frame of reference (see Section 2.3.1)

*

The author's continuous involvement in the projection process started with Lars Medbo, a period master of science who later became associate professor emeritus at Chalmers University of Technology. Karlsson and the author previously collaborated on research and development initiatives (e.g., the assembly of sailing yachts, a project led by Karlsson and funded by the National Swedish Board for Technical Development).

Launching our experimental workshop was an essential key initiative discussed at the start of this projection process. As the author saw it, this initiation was unavoidable since some practical matters urgently required attention (the projection process organisation was occupied with traditional industrial matters in temporary office premises, without any mechanical resources immediately available).

The engaged practitioners had never been involved with parallel product flow assembly systems, which finally became the result. From the author's point of view, the organisation of the projection process and most other operations within Volvo was saturated with several bothersome assumptions (assumptions concerning the product, choice of learning and training matters, length of the work cycle times, etc.).

Initial manning and brief exposé of the research and development work conducted in our experimental workshop.

Apart from engaging Medbo for work at our experimental workshop, the author also engaged architect and doctoral student Mikael Hedin from the Department of Design Methodology at Chalmers University of Technology. During our engagement through the Chalmers University of Technology, these research colleagues were initially hired by the hour. Incidentally, Medbo's first work in our experimental workshop was to paint the shop floor in a somewhat rundown building provided by Volvo. At the time, he was not yet a postgraduate student.

*

Hedin was recruited because, for various reasons, the author and his closest research colleagues realised that some new and dissimilar illustrations were urgently required. A critical and notable initiative was to obtain several completed automobiles for disassembly, i.e., subjected to systematically organised disassembly procedures while removing components that, e.g. contributed to Hedin's work with our unique illustrations. In addition, these components served other crucial purposes, as will later become evident to the reader.

The partially incomprehensible and unsystematically constructed illustrations of assembly work available in the PKIs constructed by the central Manufacturing Process Engineering Department were unsuitable for immediate use for our specific purposes (as will be fully clarified in the following).

Furthermore, Hedin's involvement was a rather fortunate initiative that created our unique system of illustrations. Constructing these unique illustrations proved to be a demanding but fundamental task for us working in our experimental workshop. This initiative was required since the product information in the KDP and PKIs provided by the central operation needed to be more cohesive and more suitable for assembly operations in Uddevalla, even though they all contained essential and valid information.

Our unique system of illustrations, created mainly by Hedin but also in collaboration with the other research colleagues, was critical for several reasons, which will be detailed further elsewhere.

*

Crucial information for our purposes was frequently unavailable. For example, coherent assembly work instructions for longer work cycles or files of work tasks performed in workstations on the assembly lines at the Volvo Torslanda automobile plant were not available anywhere.

The same dilemma was later encountered for assembly time consumptions for various product variants (as was required by us later on in the summer of 1987 at the period of the final choice of layouts in the semi-completed Volvo Uddevalla plant, as is treated elsewhere). In all, this was a somewhat unfortunate situation despite the accelerated assistance received by the author and his closest research colleagues from the various individuals involved in the projection process and various Volvo resources outside the projection process.

The reason for this situation took time to determine. However, after a few years, the reason became apparent, which will also successively become evident to the reader as they assimilate the message conveyed in this chapter.

The prototypes of material kitting fixtures were constructed by the author and his closest research colleagues in our experimental workshop, and they were used almost from the start of the projection process discussed. Each pair of these fixtures has a unique configuration stemming from component differences and demands from the assembly work. Subsequently, an advanced material feeding technique was developed.

These material kitting fixtures remained unchanged for two of the proposed assembly systems. Three very dissimilar assembly plants were projected, and two assembly plant proposals were rejected by the overarching steering committee, even though both required considerable work. Ellegård (1989) explains the projection process discussed from the point of departure of her scientific discipline, time geography.

The temporary office premises of the projection process organisation before its relocation to Uddevalla

Participants in the Volvo Uddevalla plant projection process, composed of Volvo employees and consultants, convened in temporary office premises at a closed-down shipyard in Gothenburg some years before the organisation finally relocated to Uddevalla, where another closed-down shipyard was converted into a facility for the assembly plant.

The construction of the new assembly plant on the planned site in Uddevalla was yet to be readily available. This was because the closed-down shipyard business left the target site in a state that required preparation before any construction could begin. It was a vast industrial property with several building facilities, some of which had become dilapidated.

However, the temporary office premises initially used by the projection process organisation in Gothenburg were situated just above several dilapidated workshops previously used for various shipbuilding matters. The

Fig. 2.1 Interior of our experimental workshop with several disassembled automobiles (*Source* The author's photo)

author and his closest research colleagues' experimental workshop were one of these. This location was sufficiently close to the projection process organisation but isolated enough not to be unnecessarily disturbed (see Fig. 2.1).

*

The activities in our experimental workshop gradually increased as the projection process progressed. Interest in the research and development work conducted in this workshop accelerated progressively. Therefore, we slowly obtained several other research and development assignments from companies within the Volvo sphere, including those for the central Manufacturing Process Engineering Department, Volvo Torslanda automobile plant and Volvo Truck Company.

The temporary learning and training workshop in Uddevalla

Answers to various conceivable and urgent questions that were essential to us and others during the early stage of the projection process became attainable. This situation was especially true since insights were continuously gained in the temporary learning and training workshop in Uddevalla, which had the human resources and several newly recruited operators who required prompt support.

For example, the material kitting fixtures were prototypes that needed further refinement and had to be filled with correct components and used in practice. Furthermore, as the duration of the work cycle times increased incrementally, a general lack of sufficient knowledge created a challenging situation in Uddevalla.[3]

The temporary learning and training workshop personnel needed an appropriate information system from the start, as did assembly lines and assembly line work, but this was initially unachievable. Note that the development of the Uddevalla plant's unique, fully integrated (local) information system began in the early stage of the projection process and was finalised when the construction of the building facilities was almost completed.

However, the situation in Uddevalla provided critical opportunities for the author and his closest research colleagues to handle several urgent matters that sometimes became necessary. For example, appropriate assembly work instructions were unavailable, and most learning and training principles still needed to be determined. Certain aspects of material feeding issues also called for urgent action.

In particular, feeding small-sized components was a decisive matter tackled by Bertil Johansson. The unique manufacturing equipment required had to be developed. Prototyped in our workshop was the first emergent manufacturing equipment, constructed out of Johansson's daughter's scrapped vacuum cleaner and a copper funnel found by the author in a garbage can at the Chalmers University of Technology. This initiative was passed over to others, i.e. consultants and Volvo employees.

*

The temporary learning and training workshop was established in Uddevalla in the spring of 1986 before the building facilities for the advanced and unorthodox assembly system were completed. The workshop started approximately one year before the incremental start-up of the first parallel assembly workshops.

However, the material kitting fixtures were delivered to this workshop from a large remnant building at the closed-down shipyard premises

[3] In the early years of our experimental workshop, the author and his closest research colleagues were occasionally required to take quick action. After our experimental workshop was established, components were delivered to Uddevalla from the central Volvo department, responsible for providing spare parts for the KSO (Knocked Down Operation at Torslanda). For example, the initial operation of the common material store in Uddevalla was guided by photos from disassembled automobiles in our experimental workshop. KSO was, for several reasons, unable to provide any appropriate product information whatsoever.

that became the common material store ('Kärraverken'). At the start of the projection, the building was initially left idle. However, it was later substantially rebuilt as the common material store throughout the assembly plant's short life.

The temporary learning and training workshop required authentic components in the material kitting fixtures to begin the operator's learning and training. To facilitate this, the first 50 automobiles studied were identical, i.e. a white Volvo 740 aimed at the Scandinavian market that was not fitted with any option. An identical product variant was subsequently systematically disassembled in our experimental workshop.

The use of the six parallel workshops featuring parallel product flow construction was still being decided at this early stage. The length of the work cycle was a contentious issue within the projection process organisation. Because of this, the layout selection in the six parallel assembly workshops was postponed for as long as possible.

The common material store functioned reasonably well while the temporary learning and training workshop took time to function properly. (To be fair, the common material store was readily available first, and the assignments differed, while the learning and training principles were something totally new).

The final configuration of these material kitting fixtures was just one of several important tasks for employees in the temporary learning and training workshop.[4] Thus, aside from the more apparent primary task of learning and training the newly recruited operators, this was an assignment for Professor Emeritus Lennart Nilsson, who had already become involved.

[4] Another example of a vital assignment was distinguishing which sub-assembly tasks should be included in the workgroup's assembly operations from the ones the common material store should address.

One criterion for this categorisation of tasks was that the sub-assemblies would save time extensively if several components were assembled simultaneously. Such sub-assemblies were organised in an area in the common material store, i.e. at the centralised sub-assembly area, like fitting locking knobs on the rod for the door locking mechanism, which was facilitated by using a fixture for several rods where knobs were attached to them consecutively. The long and slinky brake line was another example of components addressed to this area.

2.3 The Early Saab Scania Body Shop in Trollhättan

Professor Emeritus Ulf Karlsson contributed to the author's original and crucial research and development work on alternatives to assembly lines and assembly line work, the development of advanced work organisation, at an early stage, specifically before the author started his career as a researcher. At this time, autonomous workgroups were a somewhat widespread concept or term within the Swedish automotive industry that needed to be fully understood. Essential early research and development work was conducted at the body shop at the Saab Scania automobile plant in Trollhättan. Sometimes, the then-new manufacturing system focused on in this sub-chapter was referred to as the 'Line-out' (Karlsson, 1979).

The Saab Scania body shop insights influenced the author and his closest research colleagues' research and development work, such as the Volvo Uddevalla plant projection process and several research initiatives from other researchers across scientific disciplines. Therefore, these early insights were applied in the assembly plant discussed in this chapter, especially the parallel product flow construction and the relevant insights regarding advanced work organisations gained earlier.

The author assimilated several of the topics discussed in the following during his research and development work with his long-time friend Karlsson.

2.3.1 A Brief Explanation of Time Losses

Karlsson's doctoral thesis was based on practical industrial insights that were also academically accepted. The thesis explored a manufacturing system for grinding and welding on empty automobile bodies by proving the merits of parallel product flow construction, which simultaneously achieved efficiency and humanisation. Among other things, the findings from the Saab Scania body shop explicitly on the shop floor and implicitly in the doctoral thesis revealed the importance of the theoretical frame of reference of time losses (Wild, 1975). However, the importance of regulating small workgroups was the essential message conveyed.

The same theoretical frame of reference explains the efficiencies reported from the Volvo Uddevalla plant, as do several other earlier academic publications (Blackler & Brown, 1978; Eckerström & Södahl,

1981). Nonetheless, this theoretical frame of reference is not acknowledged in the explicit manners reported from the Saab Scania body shop, as it was by Karlsson (1979), i.e. the theory of time losses was unknown within the Swedish automotive industry and elsewhere in this early period.

Whether the Swedish automotive industry has accepted such explanations and facts now in the 2020s remains to be determined (unfortunately, the theory of time losses is somewhat difficult to comprehend and explain). For example, such recognition was not included in the internal or public debate that followed the closing down of the Volvo assembly plant. However, the awkward manufacturing scheduling system meant that the available efficiency potential was never fully exploited, and the merits were never accepted internally (see Engström, Jonsson & Medbo 1996a for further details).

While the author urges that the reader consult the original references, some of the author and his research colleagues' earlier publications discuss these time losses. See the original graphs used to illustrate the time losses as published by Wild (1975), as these are enlightening.

Time losses are mainly classified into three different types with one sub-category: (1) balance losses, (2) handling losses and (3) system losses. Balance losses include the sub-category variant losses.

The three types of time losses

Time losses are influenced not only by the amount of variation in required work time but also by the shape of the frequency distribution. An operator performing non-machine-paced and repetitive work cycle assembly work while standing alone, e.g., on a sub-assembly workstation, has a skewed statistical distribution of work times (Dudley, 1968); this means that the work varies within the frames defined by the work cycle time, i.e. 66% of the work assignments are shorter than the mean value.

This reasoning means that most work assignments are performed faster. In this case, such work follows a skewed frequency or Weibull distribution. Thus, the **operators' inevitable stochastic variation** is significant.

Despite common assumptions, such work is not a matter of normal distribution, which is symmetrical with similar numbers of short and long work assignments. Less learned and trained operators and a higher degree of machine-pacing generate a more symmetrical distribution.

For these reasons, combining several operators along an assembly line's product flow with no buffer positions between workstations unavoidably creates waiting times, i.e., **system losses.**

For example, just three operators in a series without buffers intersected between workstations may generate 14%–17% system losses. All three losses are expressed as percentages based on net assembly time consumption. The net assembly time is an ideal hypothetical situation wherein no time losses occur. The problem is analogous to an operator completing a product 'as god gives them the correct components at the exact moment required'. (Remember, the reader is recommended to study the original enlightening graphs).

System losses manifest themselves as (1) waiting times or (2) a need for additional human resources (depending on the type of assembly line, as will be explained later). Otherwise, (3) a higher work pace may be required if the assembly system was unfortunately constructed based on the assumption of normal distributed manual work. Thus, the operators are forced to overwork, as was the situation for the diesel engine assembly at Saab Scania in the mid-1970s (Note, within the Swedish automotive industry, an upper limit for the work pace has been established due to labour legislation).

*

Difficulties in evenly distributing the product's assembly time consumption along all available workstations on an assembly line will result in another type of time loss, i.e. **balance losses** with the sub-category **variant losses**.

It is impossible to achieve 100 per cent utilisation of all the operators since the product cannot be divided or balanced among the workstations. Some fitted components must be kept intact. Achieving an even and smooth product flow requires mathematical procedures for this and other reasons. Furthermore, the most time-consuming product variants will cause extra waiting times since they pace the product flow, resulting in variant losses.

*

The last type of time losses are the **handling losses**. These time losses occur because a proportion of the operators' work assignments are always spent handling components, tools, and other manufacturing equipment.

Shorter work cycle times mean that such tasks require a greater proportion of the work cycle time. For example, these time losses even out after some minutes, i.e., approximately 4%–6% on the graphs that illustrate them.

2.3.2 Detailing the Saab Scania case

The Saab Scania case was an evaluation of a rebuilt body shop that, in several respects, raised various concerns on the rationality of assembly lines and assembly line work. Specifically, the manufacturing work studied involved grinding and welding on the automobile bodies before these were transported to the paint shop. The original transfer line, characterised by a high degree of machine-pacing and short work cycle time manufacturing work, was replaced with a parallel product flow construction (a transfer line is characterised by serial product flow with a mix of automated and manual workstations).

The findings from comparing these two dissimilar manufacturing systems demonstrated that substantially improved efficiency could be achieved by implementing a parallel product flow and advanced work organisation through buffer mechanisms (the efficiency gained was due to reduced time losses). Because autonomy was taken into account, this reformation simultaneously improved ergonomic and psychosocial work conditions. That is, the often assumed contradiction between efficiency and humanisation proved possible to overcome.

Although the traditional work organisation in the old transfer line required extra attention, what was needed was an open question at the beginning of this early period. Moreover, this was the reason for engaging a doctoral student from the Department of Sociology at Gothenburg University (Karlsson lived and worked for four years at the automobile plant in Trollhättan during his doctoral studies).

The original concern was a broken work organisation that needed explicit and specific agreements and rules. However, it is essential to note that discussing this particular situation in detail is outside the scope of this chapter.

Figure 2.2 illustrates the schematised layout of the Saab Scania body shop layout of 18 small parallel workgroups. The work cycle times implemented were approximately 45 minutes, compared with 3–4 minutes for the old transfer line. The sophisticated overhead power-and-free conveyor system is illustrated in the bottom-right corner of this figure. Each small parallel workgroup contained two workstations in series, i.e. an extra workstation that served as a buffer mechanism (a decreased operator concentration suggested by the crossed-out automobiles in Figure 2.2).

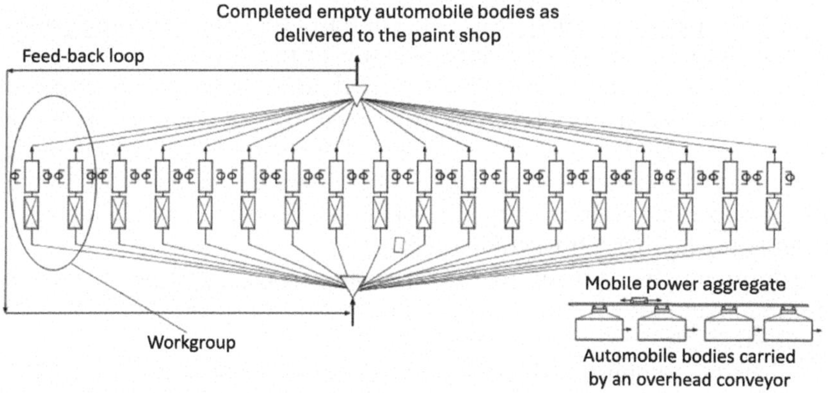

Fig. 2.2 The Saab Scania body shop in Trollhättan (*Source* The author's drawing)

Briefly explained, Karlsson's doctoral thesis evaluated this parallel flow manufacturing system, which, among other things, relied on academically established insights into time losses.

Construction of assembly and manufacturing systems based on considerations for autonomy

Karlsson's research and development work entailed considering autonomy. The various types of autonomy thus gained benefited the small parallel workgroups in the body shop, a direct consequence of the parallel product flow construction, implementation of buffer mechanisms, and advanced work organisation. The sophisticated overhead power-and-free conveyor system's construction contributed to the necessary regulation.

This manufacturing system was constructed with engineering considerations and social implications in mind. This evaluation, among other things, helped refine the concept of autonomy. In short, understanding how the small parallel workgroups were autonomous was essential.

*

As articulated more meticulously, **technical autonomy** could be attained in carefully constructed assembly or manufacturing systems. However, when discussing assembly lines and assembly line work, a notable difference exists between whether or not the operators were working at a **machine-paced product flow**. In contrast, the **non-machine-paced product flow** is distinctly different in terms of the effects.[5]

In machine-pacing, an operator may work more or less machine-paced using tools or other manufacturing equipment during the work cycle. An example is when the operator has to wait for a tool or manufacturing equipment to complete the progressive operation or when a more significant proportion of the operators' work assignments are spent, allowing a multiple torque wrench to complete its work.

Administrative autonomy involves an operator's dependency on various organisational routines, specifically, freedom from being steered administratively.

Achieving administrative autonomy requires technical autonomy. For example, doing more machine-paced work throughout a regular working day eliminates any opportunity for self-administration and, thus, administrative autonomy.

As previously stated, the commonly assumed incompatibility between improved efficiency and improved work and work conditions, e.g. ergonomic and psychosocial work conditions, can thus be clarified. As Karlsson (1979) demonstrated, the technical autonomy gained was an incentive for accepting increased responsibilities, which were closely linked with the possibility of learning and training in supplementary work tasks. Such considerations are fundamental to advanced work organisations for all types of group work, especially those considered autonomous.

Some assembly or manufacturing systems can be engineered to create and maintain autonomy for these and other reasons; however, such complex systems often require intriguing constructions. In conclusion, creating true group work for assembly lines and assembly line work is almost impossible.

Critically, technical autonomy must be placed on a higher workgroup level than what can be obtained individually. An important principle is to create **workgroup cohesion** suitable for not-too-large workgroups with a low labour turnover; large workgroups with a medium or high turnover must be avoided. Such engineering applications provide critical incentives

[5] The effects of time losses on an assembly line differ between machine-paced assembly lines and non-machine-paced ones. In the former cases, extra control and rectification (adjustment) personnel are required; however, more operators are needed for the latter to compensate for the time losses.

essential for advanced work organisations. They are, thus, another cornerstone that should be considered at the start of any projection process for assembly and manufacturing systems.[6]

*

From a slightly different perspective, for the Saab Scania case discussed in this sub-chapter, the effects described previously were achieved by combining organisational changes with engineering aspects, such as by introducing autonomous workgroups with engineering in the form of a reformed product flow combined with buffer mechanisms. As stated, this was accomplished in intriguing ways, indicating that something very unorthodox is required.

The Saab Scania case required reforming grinding and welding work on completed but empty automobile bodies, not assembly work per se (for this, appropriate material feeding techniques needed to be included, but they were not invented during the early period discussed now). Nevertheless, the insights obtained remained applicable and valid, such as in any reformed assembly line work.

The reorganisation of the body shop introduced **inherent flexibility** with technical and social dimensions, a concept or term that will be discussed and exemplified later. Consequently, this reformation was underway while the traditional work organisation was reformed into genuine autonomous workgroups; this meant that these small parallel workgroups were given enhanced and defined responsibilities to fulfil manufacturing goals in the form of scheduled manufacturing volumes at the specified product quality. External and internal agreements and rules were also established and maintained (exchanging freedom for responsibility).[7]

[6] As discussed in the main text, placing technical autonomy on a higher workgroup level creates workgroup cohesion if the workgroup size is manageable and the internal turnover is controlled (too large workgroups must be avoided). Furthermore, this is a critical prerequisite for gaining administrative autonomy.

[7] Agreements and rules are required between (1) the local management and union representatives on one side of the negotiation table and, on the other, (2) each workgroup's representatives. This was required to penetrate matters regarding competence, ergonomic and physical duties concerning other workgroup members, job rotations, performance and product quality, among others (responsibilities also include preventing strong, dominating men to take the best work, leaving women to work on the jobs left over).

Measures that moved away from a hierarchical and traditional type of group work were implemented, e.g. supervisors leading the work while allowing the product flow to dictate the pace of each operator's short cycle time.

Such necessary regulation is inconvenient and difficult to introduce and maintain. However, this presupposes that (1) representatives of the management and the union are negotiating from 'one side of the negotiating table' and that (2) each workgroup representative is 'situated on the other side' (note that the Swedish union has an exceptional formal position according to labour legislation, which is seldom observed in other countries). Such negotiation procedures must be regularly accomplished to ensure the required control, meaning the established external and internal regulations must be periodically checked.

2.3.3 Engineering Characteristics of Autonomous Workgroups in the Saab Scania Case

Autonomous workgroups are associated or synonymous with increased operator involvement. In analogy, grinding and welding work on empty automobile bodies were also complemented by supplementary work tasks of different ranges and qualities. These work tasks were integrated into the regular workload.

Thus, this type of group work entailed extra tasks complementing the regular task of grinding and welding. Examples of these tasks were cleaning the work site, dealing with product quality issues, maintaining tools and other manufacturing equipment and performing product quality correction work.

Some of these extra work tasks served as a buffer mechanism, which is only briefly referenced elsewhere in this chapter. Such cases are specific supplementary work tasks, i.e., the ones free to carry out independently of the product flow (e.g., usually some cleaning and maintenance).

*

Earlier in the 1970s, management and the union agreed on the demand for new assembly and manufacturing systems (e.g. Saab Scania in Trollhättan). These agreements generated a list itemising the inputs of the responsible manufacturing engineers (agreeing on the goals is not the same as constructing a new manufacturing system; Engström and Karlsson, 1982). This initiative established a more constructive environment.

In the Saab Scania body shop, recirculating the empty automobile bodies created workgroup cohesion, particularly for those with product quality issues. Such returns were performed using the sophisticated overhead power-and-free conveyor system.

In some cases, the specified product quality was not achieved, as determined by an external operation's sampling of the empty automobile bodies. If an external review confirmed the existence of product quality issues, such automobile bodies were returned.

Quite intriguingly, the two operators responsible for the product quality issues were not responsible for receiving the rejected automobile body. Instead, such automobile bodies were recirculated to any workstation operated by what was, in this case, a small parallel workgroup (see Fig. 2.2). Thus, a crucial aspect of individual and collective responsibilities was regulated mechanically. As stated, most other aspects of the necessary regulation were performed manually, with regular meetings between managers and union representatives.

Product quality issues were a question of prompt hands-on feedback in 'raw form' since the empty automobile bodies were rejected and returned. Codifying and writing down any information regarding product quality issues was not required. Such routines have particular merits, as discussed elsewhere and exemplified in the Volvo Uddevalla plant case.

These feedback routines were made possible by the sophisticated overhead power-and-free conveyor system, which features separate mobile power aggregates. These aggregates oscillated back and forth along the conveyor rails in the ceiling, allowing them to address and transport each empty automobile body to any workstation in accordance with the sophisticated system's programming.

In addition, buffer mechanisms were established in the form of products located before and after each small parallel workgroup, which the sophisticated overhead power-and-free conveyor system handled. These products also functioned as buffer mechanisms. Consequently, the small parallel workgroups were allowed to vary work pace, such as working faster to recover lost manufacturing volumes and creating opportunities to be ahead of the planned manufacturing volumes.

Therefore, they could accumulate and release specifically defined amounts of accumulated **working-up**, which meant that technical autonomy was used as one of several incentives (in addition to, e.g., the construction of the salary system that incorporated collective and individual stages promoting various matters). In conclusion, technical autonomy was constructively used as an integrative feature for the operators. Such mechanisms were also applied subsequently in construction, e.g., the layouts for the small parallel workgroups at the Volvo Uddevalla plant.

Because of these buffer mechanisms and other features, some operators were temporarily allowed to leave the workgroups in the Saab Scania body shop. Naturally, the remaining co-operators needed to increase their work pace temporarily. Therefore, such procedures had to be agreed upon externally and within each small parallel workgroup. In contrast, all operators must always be available to manufacture any product on an assembly line.

2.3.4 Some Conclusions Regarding the Saab Scania Case

In conclusion, the Saab Scania body shop integrated several intriguing features (some requiring further explanation for the uninitiated visitor). These features were needed to avoid having specific supervisors lead the workstation and to let the product flow dictate the work pace of each operator's short work cycle.

These features are briefly explained here and elsewhere. Some were engineering-related, while others involved organisational matters.

*

Alternatively, returning to the specific local research tradition within the Gothenburg area described in the first sub-chapter, this was partly achieved by reforming the technical science dimension of an assembly or a manufacturing system (remember the technical science dimension was later proved possible to integrate with the social science dimension).

Empty automobile bodies were recirculated in this body shop until they met specifications, pushing each small parallel workgroup to gradually become sufficiently knowledgeable and proficient in delivering empty automobile bodies according to the planned manufacturing volumes at the specified product quality standards.

As also mentioned, such hands-on feedback in 'raw form' is usually more suitable and precise for performing work involving various complex judgments, such as regarding product quality and eventual product quality issues. Such decisions are often difficult to communicate effectively and rationally (as pointed out before, it was discussed in the body shop case, not the requirement to codify and type any information regarding potential product quality issues). The benefits of hands-on feedback in raw form are, to some delimited extent, also applicable to assembly line work.

2.4 Prerequisites for the Volvo Uddevalla Plant Regarding Administration of Information

This sub-chapter explains several crucial administrative facts regarding assembly lines and assembly line work compared to the most advanced and unorthodox assembly systems.

At assembly lines and assembly line work, the movements of the products along a product flow and their variation define what is going on, as discussed elsewhere, precisely for everyone on the shop floor.

In contrast, operators' movement patterns are crucial for advanced and unorthodox assembly systems. The intra-workgroup pattern is a rather complex hierarchy that defines the division of labour within the small parallel workgroups at the Volvo Uddevalla plant. The operators' movements between sub-assembly workstations and work positions along the automobiles expressed and determined what is going on.

The latter phenomenon is not easily grasped by an uninitiated visitor since hardly anything important appears to happen. Managing these assembly systems is more complex than its opposite pole, i.e., assembly lines and assembly line work. Besides, as explained in the previous sub-chapter, regulating workgroups is rare knowledge within the Swedish automotive industry today. Such facts are elucidated in another chapter.

2.4.1 Product Structures Within the Automotive Industry

We will now discuss so-called product structures used in all automotive industries to define product architecture and product variation.

Every automotive company must formalise the work the central Product Design Department performs digitally through the company's overarching product structures. These hierarchies extend from completed products to the most minor components and vice versa. In particular, a product structure interlinks structure elements through structure links, pointing out more or less aggregated components.

However, product architecture and product variation involve external and internal company operations, as do internal and shared intra-company product development initiatives and cooperation protocols with suppliers. Assembly work also needs to be defined similarly, although these administrative mechanisms are addressed to local assembly plants in a generalised form.

*

Within the Volvo Automobile Company, these administrative mechanisms and the relevant product information communicated are the responsibility of the central Design and Manufacturing Process Engineering departments, which both define the product and product variants in the most overarching or general sense.

The central Product Design Department was responsible for (1) the information system KDP regarding product architecture and product variation, communicated through the central information system. In contrast, the central Manufacturing Process Engineering Department generated the fragmented PKIs.

The researchers working in our experimental workshop were required to handle such facts in practical terms since we were initially responsible for constructing appropriate assembly work instructions during the projection process. This research and development work created the **advanced assembly work instructions**.

This accomplishment was considered a challenging assignment, especially since the task was unknown, especially considering that it involved learning and training **extensively long assembly sequences** unique to each product variant assembled. It would have been impossible if this assignment had been interpreted as requiring humans to remember such sequences composed of component numbers and mutilated component denotations according to the automotive company's practices. Other approaches to addressing this situation were required (by the way, the capacity of the human memory was not the valid reason for the limitation of the ability to extend work cycle times over the passing decades; it was something else, as will be clarified in the following).

The KDP and PKIs were created by the two central departments within the Volvo Automobile Company.

As partially explained previously, the central information system, the KDP, dealt with product architecture and product variation and was the base for the PKIs. During the Volvo Uddevalla plant's short existence, this central information system communicated the computerised stream from the KDP which the central Product Design Department provided. Volvo's internal courier service delivered the PKIs somewhat erratically as paper printouts (two very different streams of central product information reached each local assembly plant).

Each of the two streams of product information was characterised by its own partly distinctive logic and moderately dissimilar validity for shop floor work. Therefore, these two streams were not automatically and digitally synchronised and were valid within dissimilar time perspectives for this and other reasons.

The KDP and PKIs, were supplied and delivered to each local assembly plant operating at a full-scale manufacturing pace. The local manufacturing engineers processed these two information streams by identifying valid product information for the locally assembled product variants. If necessary, product information was compared to eliminate suspected annoyances and anomalies.

Thus, excessive central product information arrived at every local assembly plant, i.e., more than required internally to be used for the local shop floor work. The computerised KDP was manageable, whereas the paper printouts in the form of the PKIs were delivered erratically by the internal couriers and were inconvenient to administer and process.

Because each local assembly plant assembled only a subset of the total number of different product variants, the central administrative mechanisms must also be able to codify product variants; i.e., a common scheme for codifying all product variants must exist. For example, this demand meant that within the Volvo Automobile Company, each component must be authoritatively designated by a series of incomprehensible product variant codes consisting of multiple and single **product variant codes**.

*

The administrative approaches to managing product architecture and product variation to construct local assembly work instructions involved combining the product information communicated by the information system, the KDP, which was expressed by a **design-oriented product structure** with the PKIs' describing in detail how to carry out the assembly work. This proved to be complicated and confusing for any advanced and unorthodox assembly system, as well as also for the work to administrate assembly line and assembly line work. These problems occurred because these administrative protocols, used to administrate assembly line work, were constructed as a fragmented, one-way communication to local assembly plants (an excessive and vast amount of fragmented product information was received, which unavoidably had to be selected and processed further).

The two information streams were synchronised administratively solely by using four-digit plus four-digit numbers (the PKI-number), the first four digits of which referred to the numbering of the design-oriented product structure. However, the KDP comprised far more nuanced structure-oriented component denotations, facilitating navigation within the product structure if the user was sufficiently knowledgeable (the PKIs were more disorganised and challenging to deal with).

On the other hand, the PKIs did not specify the structure elements of the design-oriented product structure and only identified the components by component numbers and a series of incomprehensible product variant codes (remember, the PKIs were communicated through paper printouts). The extensively mutilated component denotations and other intricate matters meant that the PKIs were rather useless for direct use on the shop floor of any assembly plant, no matter how it was constructed.

The PKIs provided access to fragmented specifications of assembly work. However, the process was complicated, which created several difficulties for the local manufacturing engineers, including those in the temporary learning and training workshop and our experimental workshop. Each PKI featured a specific codification (with the PKI numbers substantiated by the PKI names). Thus, the two streams of product information needed to be sufficiently administratively synchronised locally, even though they partly dealt with similar matters of the same automobile model, and these two streams were valid within different time frames.

As mentioned, the incoming KDPs and PKIs cannot be used directly on the shop floor, and this was also the case in the main Volvo Torslanda automobile plant, which was also assigned a selection of product variants.

As previously stated, the two streams of product information discussed needed to be processed, synchronised and reformed locally to provide the appropriate product information for constructing appropriate local assembly work instructions. These also needed to suit the specific assembly system in place.

Delivery of the KDP and PKIs to different local assembly plants during the Volvo Uddevalla plant projection process

If differently constructed assembly systems were used at each local assembly plant, the matters discussed previously would become more complicated and confusing. Closely similar assembly lines and assembly line work would mitigate this intricate dilemma. Conversely, having only one advanced and unorthodox assembly system among several traditional ones complicates administrative processes for any automotive company; this was especially the case for the Volvo Uddevalla plant, which had to deal with this dilemma.

The traditional assembly line plants within Volvo were adapted to the situation described. These plants were, in practice, constructed and organised for a fragmented and one-way supply of the KDP and PKIs.

During the said period, the material used for small-sized components was singled out in the design-oriented product structure and addressed to specific workstations at the Volvo Torslanda automobile plant, which made things troublesome. For example, identical screws were located in different positions of the automobile, an unnoticeable situation because some important components were impossible to define unambiguously in the two streams of product information and material delivered and supplied to any Volvo assembly line plant. Information denoted as so-called material control codes had to be revised in ways not required for assembly line work (the author avoids to treat this topic further for reason of space).

Engström et al. (2000) describe how this dilemma was tackled constructively by far more refined methods than those practised during the projection process now discussed, i.e. during the projection process of the successor of the Volvo Uddevalla plant, the Autonova plant in Uddevalla.

These now-addressed issues were exacerbated further because, for several reasons, the KDP and PKIs were constantly changing due to customers' demands and varying markets, mistakes, and feedback from the shop floors.

*

Simultaneously, managing assembly lines, assembly line work, and advanced and unorthodox assembly systems at an industrial scale was challenging. As exemplified by the Volvo Automobile Company situation during the period discussed in this chapter, such conditions are not ideal for any automotive company in aspects addressed, such as supplying product information. Furthermore, this was Volvo's first new assembly plant in several decades.

*

In conclusion, as the reader must now understand, the administrative methods for managing the delivery of the KDP and PKIs were constructed for the existing assembly line plants. Therefore, the Volvo Uddevalla plant's projection process needed to manage the KDP and PKIs appropriately and locally, which was challenging.

At the time, the design-oriented product structure and aged computer technology communicated a fragmented product architecture and product variation suited for equally fragmented assembly work. This unfortunate situation impacted the entire operation. As described here and elsewhere, tunnel visions of crucial matters marred the company organisation.

2.4.2 Two Different Product Structures were Implemented Internally for the Volvo Uddevalla Plant and Centrally by the Volvo Automobile Company

This section discusses two different product structures. Unfortunately, during their implementation in the automotive company, one of these structures did not provide any possibilities to construct the desired stable and suitable wholes, as was required for researchers' efforts to develop and implement the holistic learning approach for the learning and training; no comprehensive understanding was achievable for the advanced and unorthodox assembly plant. This fact is elaborated upon subsequently. The opposite learning approach is mechanistic, characterised by, e.g. short or relatively short work cycle times (this approach uses the other product structure).

The design-oriented product structure has been used centrally within the Volvo Automobile Company for a long time.

The product structure implemented in Volvo has always been design-oriented. In this case, this product structure is based on the even more widely accepted Function Group Register. This register is characterised by relying on seven main groups of components, numbered from 2 000 to 8 000, and these seven main groups partially disregard existing valid interconnections. For example, numerous product systems are composed of components from more than one of the seven main groups of components.

This register is a Swedish standard within all automotive companies. However, each Swedish automotive company has its tradition for interpreting this particular register. Such a product structure was used for the Volvo 740, a four- and five-door automobile with rear-wheel drive assembled in Uddevalla.

Unfortunately, this design-oriented product structure is inhomogeneous (not coherent) for reasons detailed elsewhere because it is based on or emanates from an obsolete automotive product.

Such product structures are characterised by unforeseen and varying holes. 'Implicit holes' is a concept or term coined by the author to refer to the holes, missing structure elements and missing structure links that appear here and there in these Swedish interpretations of the Function Group Register (i.e. the design-oriented product structure). These implicit holes have been propagating to the PKIs because of the way they were codified by four-digit plus four-digit numbers. The absence of some components for particular products or product variants causes these missing structure elements and structure links.

For example, a heavy truck chassis requires a separate frame, which fills the design-oriented product structure (or Function Group Register) more evenly than modern automobiles. In contrast, modern automobiles feature welded-together bodies composed of stamped sheet metal. Furthermore, designing modern automobiles within Volvo, the last main group of the design-oriented product structure, works as a garbage bin for components deemed unsuitable for inclusion in the other six main groups.

The assembly-oriented product structure specifically implemented within the Volvo Uddevalla plant

Fortunately, the Volvo Uddevalla Plant was later implemented, before reaching full scale pace, with a fully integrated local information system unique to Uddevalla, which meant complete control of all information provided inside this assembly plant. The assembly-oriented product structure is the one that the author and his research colleagues were required to construct for the Volvo Uddevalla plant using systematically organised disassembly procedures. This product structure consists of four main groups of components of more or less interconnected components for assembly in and on the automobile body, which includes the sub-assembly work. Conversely, the sub-assemblies are viewed separately (for the automobile model assembled in Uddevalla, these include the doors, engine, instrument panel and sunroof).

Such a product structure was required to achieve extensively long work cycle times. These product structures may serve as the basis for the division of labour and configuration of material kitting fixtures; these are based on the product architecture and product variation defined from an assembly work perspective, as discussed in more detail elsewhere. For reasons explained, this feat was unattainable with the design-oriented product structures implemented within the Swedish automotive industry.

*

As a consequence of the facts and explanations presented in this chapter, the operators in the Volvo Uddevalla plant generatively, intuitively and progressively learned and trained to master the extensively long work cycle times under an advanced work organisation of small parallel workgroups.

Consequently, the operators working on the shop floor were more or less inevitably trained by acquiring several distinctive human skills, such as communication, manual, motor, and perception skills; this was accomplished in ways that allowed both differentiation and integration, combined with prospection of two other mental mechanisms, as will be discussed in a forthcoming sub-chapter. Other merits of projection process organisation efforts, both foreseen and unforeseen, became feasible, such as cheaper administration and lower annual cost of model changes, as discussed elsewhere.

Marton (1970) proposed that learning and training were developed to demonstrate how humans improved and gradually mastered complex knowledge in the case discussed; this accomplishment will be addressed in depth elsewhere. Note the apparent similarities to the concept or term natural work, which the co-author and the author discuss in another chapter of this anthology.

*

These achievements contradict the traditions of assembly line work because the multi-skilled learning and training principles discussed provide opportunities for expanded learning and training. However, learning and training for assembly lines and assembly line work focus mainly on restricted manual or motor skills (including manual dexterity).

The short work cycle times also severely limit some distinctive human skills that can be mastered through another type of learning and training.

In addition, various other mechanisms regulate the assembly lines and assembly line work. These are prospecting several other distinctive human skills. However, these restricted human skills inhibit the extensive expansion of the operator's knowledge and skills beyond certain limits, at least not in the ways that characterise the case discussed.

From the assembly line operators' perspective, the work involves acknowledging the state of product flow and the positions of your nearest co-operators while remaining under control. Various displays, lights, sound signals and information boards all influence human action.

The possibilities for expanding knowledge and skills within a correctly constructed advanced and unorthodox assembly system are almost limitless, simultaneously allowing differentiation and integration while enabling the prospection of other mental mechanisms applied in the Volvo Uddevalla plant case. However, these are impossible to achieve in traditionally constructed assembly systems, as discussed in more detail elsewhere in this anthology.

*

In conclusion, a thorough grasp of the design-oriented product structures that were mediated by the information system KDP, which, in turn, defines the product architecture and product variation centrally, is crucial. Such insights were obtained during the Volvo automobile case, promoting an understanding of the administrative mechanisms that regulated the PKIs, in particular, how the products and the subsequent assembly work axiomatically are defined in the company's central information systems and finally formally are established by various administrative practises and routines.

The author was responsible for constructing a product structure that would facilitate advanced work organisation with extensively long work cycle times. Small parallel workgroups at the Volvo Uddevalla plant were to prove its practical value on the shop floor. Similar experiences at the Volvo Truck Company corroborated similar results (e.g., the assembly docks at the Tuve heavy truck chassis plant outside Gothenburg).

However, this presupposes that an advanced material feeding technique is implemented and that the small parallel workgroups adhere to the standards established for autonomous workgroups, as discussed elsewhere.

2.4.3 The PKIs from the Central Manufacturing Process Engineering Department

This sub-chapter clarifies some crucial and tedious details concerning the supply and delivery of the KDP and PKIs during the early period of the Volvo Uddevalla plant projection process. During this time, for the reasons stated thus far, the PKIs provided the only accessible information regarding assembly work. These were available through Volvo's internal courier service, which delivered the paper printouts of the PKIs. No appropriate information system was established between central operations, the temporary learning and training workshop, or our experimental workshop.

We had to manage using paper printouts of the design-oriented product structure on tractor paper. These printouts were collected in 14 binders representing the KDP frozen in time, valid for a specific period for a particular automobile model.

The temporary learning and training workshop may have adopted a similar approach, but this was not considered. In addition to this workshop, the components were obtained from the KSO (Knocked Down Operation at Torslanda), an operation that supplied components for assembly abroad, but this was not possible to use appropriately for immediate assembly because the components were organised according to component numbers in various packages, requiring for export to small foreign assembly plants for reasons to avoid import taxes was thus impossible to couple them to individual automobiles.

Thus, suitable information was not available anywhere inside Volvo. Identifying the correct assembly work for the forthcoming product variants scheduled for assembly in the temporary learning and training workshop was impossible (the reader may have started to suspect why).

Examples of inconsistencies in the PKIs

The PKIs from the central Manufacturing Process Engineering Department caused various inconsistencies during the early period of the Volvo Uddevalla plant projection. PKIs with inconsistencies persisted throughout the Volvo Uddevalla plant's short life.

The PKIs were fragmented. These non-digital, central documents consisting of paper printouts were (1) codified by four plus four-digit PKI-numbers and named according to the assembly work specified by each PKI (PKI-name).

As said, the first four digits expressed the position in the design-oriented product structure, while the last four digits after the hyphen were department-specific. The PKIs were ideally organised in binders following the first four digits, i.e., according to the seven main groups of components in the design-oriented product structure (or interpretation of the Function Group Register).

The PKI numbers and names were provided with specific index pages. These were delivered together with the PKIs to be used for keeping track of every PKI inserted into a set of binders (at the end were numerous sets of binders of the PKIs available in the gradually completed assembly plant in Uddevalla, with at least one set in each parallel assembly workshop but numerous photocopies of individual PKIs were scattered throughout the shop floor).

However, (2) the PKIs were continuously changed following the variable life of the assembly-oriented product structure. Furthermore, (3) the names of the PKIs needed to be more consistent. In some cases, specific product variants were mentioned; in others, this was not the case. Several other discrepancies were also observed regarding the choice of the PKI-names.

(4) The common scheme for codifying all product variants was expressed as a series of incomprehensible product variant codes introduced in particular sections of the standardised forms of the PKIs.

Consequently, keeping track of and constantly reorganising the content in a complete set of PKI binders, including managing the overflow, took time. All of the PKIS were only valid once. At the same time, a continuous and complementary stream of product change orders arrived in the form of colour-coded PKIs (the paper printouts themselves were printed on papers with different colours); these were provided as warnings, such as for future product changes and, consequently, influenced the future assembly work. Volvo's internal courier service delivered envelopes containing 1–20 PKIs to any local assembly plant each regular business day. Similar massive, somewhat erratic deliveries were also made to the temporary learning and training workshop during its existence and to our experimental workshop.

As said, the PKIs dealing with the product change orders were colour-coded, indicating different time points; this was required to keep the local assembly plants meticulously informed of present and future products. Some of these product change orders needed urgent action, while others had to be queued up for future processing or sent back to the central Manufacturing Process Engineering Department due to, e.g., findings by the local manufacturing engineers or on the shop floor.

*

Each PKI consisted of two different types of pages. The first type of page contained somewhat incomprehensible and unsystematically constructed illustrations (a maximum of two pages but usually one), and the other type contained standardised forms of several sections of product information (extremely short assembly sequences, tools and other manufacturing equipment, tolerances, torque, etc.). These standardised forms sometimes required an extensive number of pages.

However, the most extensive standardised forms consisted of numerous pages, not because the required assembly work was excessive (this was never the case) but because of how product variants were codified. These extensive standardised forms, containing a series of incomprehensible product variant codes, had exploded due to the choice of computer technology implemented earlier in Torslanda. However, the product information collected in other sections was clear for any competent manufacturing engineer (remember that the assembly work specified was fragmented).

For example, as discussed previously, these particular sections of the PKIs contained up to 40 pages of standardised forms with an excessive number of incomprehensible product variant codes, such as the work to fit the external rear-view mirrors (while the author can explain precisely why, this particular explanation was excluded for reasons of space).

Administrative effects of the common scheme for codifying all product variants expressed in particular sections of the PKIs

The illustrations in the PKIs were rather rough drawings that needed to be standardised. These illustrations were usually drawn from the operators' point of view, as the operators were assumed to view the assembly work standing on the shop floor. As said, the standardised forms consisted of sections that detailed assembly work.

Each PKI described approximately half-a-minute or less of assembly work through **extremely short assembly sequences** in standardised forms for a defined number of selected components coupled to the **smallest work elements** in the PKIs, each assigned a series of incomprehensible product variant codes.

The smallest work element in the extremely short assembly sequence relies exclusively on component numbers combined with the appropriate series of incomprehensible product variant codes. Product variant codes consist of multiple and single codes assigned to each component with a component number.

These extremely short assembly sequences in the PKIs featured mutilated component denotations because the space in these specific sections of the standardised form was restricted, often resulting in a confusing language (by reading several different PKIs simultaneously, the author and his closest research colleagues surmised that several different central process engineers were involved in constructing these fragmented central assembly work instructions, i.e. the PKIs).

Prospecting the PKIs for the systemically organised disassembly procedures performed in our experiment workshop

The facts presented in the previous sections were distressing for others and us. However, creating appropriate clusters of components required and thereby identifying the corresponding assembly work expressed using the PKIs as a basis was intriguing.

The mutilated component denotations hampered our experimental workshop's research and development work on identifying components that were removed or supposed to be removed from completed automobiles; this was the reason for the systematically organised disassembly procedures, which included certain initiatives and specific tools, e.g., general automotive literature, paper printouts of the PKIs, small cards with shrunken illustrations from the PKIs, unique literature from Volvo and large, elbow-high, easily movable tables).[8]

Again, because of the ways the design-oriented product structure was formalised by the KDP and PKIs, the research and development work in our experimental workshop was a frustrating endeavour (remember, the 'implicit holes' that created inhomogeneity (non-coherency) in the design-oriented product structure were propagated across into the PKIs because of the first four digits in the four-digit plus four-digit numbers assigned to the PKIs).

[8] The small cards mentioned in the main text were crucial for the research and development work. However, they were time-consuming to fabricate. Each such PKI illustration had to be processed by a photocopying machine. This machine had to shrink the illustrations in steps, which after that had to be cut out and glued to small sheets of cardboard to which also corresponding PKI numbers and names were transferred (approximately 800–1000 ones if the author still remembers this correctly.)

Numerous PKIs (1) needed to be collected to identify components and perform more extensive work tasks in and on the automobile. Several of the smallest work elements in the PKIs acquired from the extremely short assembly sequences defined by each PKI were needed (remember that, often, only a subset of the smallest work elements was valid for a particular product variant; the product variant codes split up the extremely short assembly sequences).

Furthermore, to make matters worse for the research and development work in our experimental workshop, (2) these PKIs were often collected from more than one of the seven main groups of components in the design-oriented product structure (remember, the 'implicit holes' that caused the inhomogeneity (non-coherency) of the design-oriented product structure).

As emphasised before, (3) these sections of the standardised form in the PKIs had exploded due to the choice of computer technology. As we discovered in our experimental workshop, these administrative mechanisms had never been explored in the necessary ways. However, for the practitioners within Volvo, the series of incomprehensible product variant codes was an accepted reality, i.e., it contributed to supporting the prevailing assumption that the number of product variants was enormous and incomprehensible.

*

Thus, to summarise, the (1) design-oriented product structure practised within Volvo has been used to specify the information content in the KDP, which defines the product architecture and variation, resulting in crucial administrative inconsistencies.

Such crucial administrative inconsistencies were aggravated by (2) how product variant codes defined the product variants. These administrative processes were interlinked by the four-digit plus four-digit PKI numbers and the series of (3) product variant codes to the PKIs.

In the case discussed, the manufacturing engineers at each local assembly plant had to identify the product information appropriate for their particular product variants.

*

Overall, the one-way process successively and inevitably fragmented the KDP and PKIs supplied to the local assembly lines. At such plants, the manufacturing engineers were also required to keep track of dissimilar time perspectives simultaneously.

Discussing the construction of appropriate assembly work instructions

Creating extensively long, unique assembly sequences for each product variant proved possible at the Volvo Uddevalla plant. As mentioned, the author and his closest research colleagues were initially appointed to construct appropriate assembly work instructions. Such assembly sequences were interesting, in contrast to the short ones that were valid for assembly lines and assembly line work.

The short assembly sequences used for assembly line work were unstable and not applicable to us (1) because of the frequent rebalancing required to achieve an even and smooth product flow in an assembly line plant. Such sequences were also unstable because of (2) constant changes to the KDP and PKIs. (Keep in mind that the dissimilar time perspectives of the two streams of product information and the continuous and complementary stream of product change orders created extremely short assembly sequences because the incomprehensible product variant codes limited the validity of some of the smallest work elements in the PKIs assigned to one particular product variant).

Instead, extensively long and unique assembly sequences were implemented on the shop floor of the Volvo Uddevalla plant, with a unique sequence for each product variant; this meant that the product variants that took more time to assemble, e.g. because its many different components caused varying assembly sequences and assembly time consumption. These unique assembly sequences were feasible to learn and train for several other reasons. One reason for this feat was the configuration of the material kitting fixtures that embraced the automobiles' components in specific ways. Furthermore, the automobiles were allowed to remain within each small parallel workgroup while being completed in all possible respects.

Even if the advanced material feeding technique supported the extensively long work cycle times and the type of group work performed, appropriate assembly work instructions were still needed. The automobiles assembled are required to meet the company's centrally determined specifications. The local manufacturing engineers needed to create constructive prerequisites to assemble each automobile according to a particular valid time window by prospecting precisely correct product information for themselves and the shop floor operators.

Thus, in Uddevalla, they had 'to keep several structure weeks in the air simultaneously' in the unique, fully integrated (local) information system. The appropriate product information for a particular product variant had to be synchronised with a specific time window and also coupled with material and authorized assembly work instructions, which unfortunately changed constantly.

To complicate things further, the product variants on the shop floor were not always assembled in the manufacturing sequence planned for each small parallel workgroup for the reason treated in this chapter and several other reasons (something always happened, and the awkward manufacturing scheduling system made things worse). In contrast, this situation is far more manageable for assembly line work (the author challenges the reader to discover why).

*

In conclusion, the author and his closest research colleagues found constructing the advanced assembly work instructions challenging, especially regarding the probably far too detailed topics explained in this sub-chapter. To summarise:

- What can be done to address the constant changes in the two streams of product information to support the assembly work on the shop floor?
- Furthermore, how can the extensively long work cycle times be addressed, especially those with extensively long assembly sequences unique to each product variant?

2.5 Theoretical Background on Engineering the Holistic Learning Approach

Three theoretical frames of reference were essential for the engineering-oriented interpretation of the holistic learning approach.

For context on the reasoning described in this sub-chapter, the author invites the reader to imagine the inside of our experimental workshop. It was filled with partly disassembled and completed automobiles and numerous removed components on the shop floor. In addition, an abundance of general automotive literature, paper printouts of the

PKIs, unique literature from Volvo, binders, books, drawings, pamphlets, photos, etc., were also found throughout this workshop.

Furthermore, several elbow-high, easily movable tables were also used in the workshop. Some were empty or partly empty, while others were cluttered with stacks of papers dealing with ongoing and interrupted activities. The research and development work were a difficult and tedious search process, and numerous methods had to be tested (from cutting and pasting tractor papers to customised computer printouts for hands-on support for the systematically organised disassembly procedures of automobiles).

The inside of the workshop would easily resemble a gigantic puzzle of automobiles at different stages of completion, with removed and labelled components and handwritten, typed, or photocopied papers, including an extensive number of PKIs and their copies in various constellations used, e.g., to mimic preliminary assembly work instruction or to facilitate our disassembly work.

2.5.1 Brief Explanations of Three Theoretical Frames of Reference Applied

Here, we briefly summarise three crucial theoretical frames of reference. However, other frames of reference were also essential, as reported in another chapter. That chapter mainly discusses vocational learning matters, as advocated and explained by Professor Emeritus Lennart Nilsson, a senior research competency from Gothenburg University who was also heavily involved in the projection process. Moreover, the construction of the assembly-oriented product structure is presented at the end of this sub-chapter.

The human capability of mastering complex knowledge through differentiation and integration introduced crucial insights into possibilities to use extensively long work cycle times

The author and his closest research colleagues began theorising by progressively interpreting the holistic learning approach on the assumption that humans gradually learned and trained to master complex knowledge in specific ways.

Briefly explained, humans gradually advance towards increased understanding by simultaneously moving in two diametrically different directions (Marton, 1970). However, Marton also explains that such learning begins with an undifferentiated, poorly integrated understanding of a whole.

The following slightly paraphrases Marton's observations:

- One direction is integration, which begins with a vague understanding of certain types of wholes. The second direction is differentiation, which leads to a greater understanding of details, as explained further in this particular theoretical frame of reference.

*

Consequently, in the author and his closest research colleagues' understanding, the assembly-oriented product structure was a crucial instrument for human success, as explored in this chapter. The design-oriented product structure, which manages the KDP and PKIs within Volvo, was not an option for reasons already explained.

Like all product structures, the assembly-oriented product structure is hierarchical. Integration is achieved by moving upwards in this hierarchy towards the more aggregated components or elements. Moving towards the bottom creates differentiation since smaller and smaller elements or components or elements are identified.

The possibility for operators and their co-operators to develop an adequate and enhanced alignment between details and wholes was facilitated on the shop floor by initially empty automobile bodies. These bodies were gradually assembled into complete automobiles while on stationary workstations for extensive periods (they moved twice or thrice between workstations).

The operators did this as part of group work. Consequently, they were inevitably caught between several completed automobiles and empty automobile bodies, with all utilised components harboured in step-wise flows of material kitting fixtures. Three pairs of material kitting fixtures were provided for each automobile.

Therefore, the assembly work was, in most cases, conducted in the operator's direct vicinity. Note that the layouts for the small parallel work-groups were constructed to enable personnel to scan each other's faces, i.e. distances were kept below 11 metres.[9]

[9] The operators' communication and overview were enhanced by carefully constructing the work environment in the six parallel assembly workshops, e.g., using a particular colour

Each material kitting fixture contained different components required for each product variant. These fixtures were constructed to handle the extensively long work cycles characterising the group work discussed. For example, maximising the content of plastic containers (overfilling) was not acceptable since the content had to be identified at first glance.

The human capability of mastering grammar generated crucial insights into extensively long work cycle times based on deep and surface-level structures

The structure and learning of extensively long work cycle assembly work can be analyzed in terms borrowed from Chomsky (1957, 1975), the father of 'generative transformational grammar'. Chomsky claims that such grammar is innate to all humans.

*

This insight was applied to creating the assembly work instructions as follows. The author and his closest research colleagues successively became aware of the importance of a professional language while working on, e.g. the advanced assembly work instructions at our experimental workshop. We avoided initiating any disassembly procedure without being sufficiently prepared, i.e. wrecking expensive, new and shiny automobiles without any suitable information was not allowed. Furthermore, unidentified components had to be avoided.

A textbook on Swedish grammar was an essential and thought-provoking resource for our research and development efforts at our experimental workshop. Our task was (1) to connect extensively long assembly sequences unique for each particular product variant with (2) the product architecture and product variation defined from an assembly work

scheme. The essential elements on the shop floor, such as the automobiles and operators, were purposely multi-coloured.

In contrast, the background was painted more discreetly in white and shades of gray, depending on how mobile the artifacts there were. Thus, various shades of white and gray marked the background against which the more colourful automobiles and the operators appeared. Furthermore, dangerous manufacturing equipment or parts thereof were painted orange.

This approach was required since, perception-wise, humans are more adept (or attentive) at recognising and remembering patterns than what is possible to create using intensive colours (actually, the author consulted with an architect who was a specialist in colour schemes at Chalmers University of Technology who visited the Volvo Uddevalla plant during the projection process).

perspective. The critical insight discovered was that assembly sequences and sentences shared certain similarities.

This theoretical frame of reference implies that other matters underlie any defined assembly sequence (i.e. sufficiently long assembly sequences can consequently be assumed to involve even the product architecture and product variation). The differences between what Chomsky denotes as 'deep' and 'surface structures' became a guiding principle for developing appropriate assembly work instructions.

The methodology for creating the advanced assembly work instructions enabled (1) more detailed information to be obtained. At the same time, it was feasible (2) to obtain more aggregated information, which in turn created possibilities to connect to the division of labour of each small parallel workgroup, which in turn (3) gave access to the assembly-oriented product structure, as discussed elsewhere.

*

Therefore, discovering generative transformational grammar enabled differentiation and integration, allowing information gaps to be addressed or grasped. However, such a feat presupposes certain preconditions, such as the unavoidable need for correct and meaningful component denotations.

The human capability of gaining 'dual perceptions capabilities' creates crucial insights regarding extensively long work cycle times

According to Gestalt psychology, humans can attain a dual perception capability. This mental mechanism allows humans to instantly shift between figures formed by details and wholes. Berelson & Steiner (1964) described this mental mechanism, which was later interpreted to explain the assembly work discussed in this anthology (Engström et al., 2001).

Based on the author's understanding, Volvo's operators attained this capability as it gradually became accessible during their learning and training. The fully learned and trained operators can instantly switch between figures and recognise the small-sized components, whereas the newly recruited operators initially perceive the large-sized components. This dual perception capability enables immediate and simultaneous perception of several hierarchies of differently sized components by the fully learned and trained operators.[10]

*

In conclusion, dual perception capability can be attained because of the product architecture and product variation since the components in automobiles are variously interconnected. Small-sized components need large-sized ones that belong to specific product systems and feature particular combinations of colours, finishes, textures, etc.

Conclusions regarding the three theoretical frames of reference

To summarise, according to the three theoretical frames of reference differentiation and integration processes create connections between the deep and surface structures of the assembly sequences. Furthermore, such connections are linked to active human perception processes. All of this is because of the more or less interconnected automobile components.

Product architecture and product variation is essential; however, in all cases within the Swedish automotive industry, they are not defined from an assembly work perspective (to repeat, the design-oriented product structure was not an option because it did not provide any stable and suitable wholes due to inhomogeneity and disruptive non-coherence).

Components of different sizes were positioned into the material kitting fixtures. During the group work, all components were gradually used,

[10] Note that all automotive products, such as automobiles, buses and heavy truck chassis, feature some common characteristics. For example, such products have a centre line running from the rear to the front of any vehicle. Some components appear in pairs around this centre line and product systems or sub-systems thereof are located on or around this centre line (some product systems are wound around this centre line). Therefore, automotive products present several 'generic characteristics' that offer insights into the relationships between components of various sizes and semi-completed or completed products.

The products exemplified above are bilaterally symmetric, i.e., structured like a human body, with a centre line along which limbs and organs are arranged in pairs. However, some organs or organ systems are divergent.

More generally, automotive products exhibit eight 'generic relationships ', which are conceptually related to either single or multiple products (see Engström et al. 2001 for further details). These 'generic relationships' are bilateral intra-product symmetries, bilateral inter-product symmetries, contingent inclusion relationships, genealogical links, intra-system synergies, part-whole relationships, proximity of components and similarity of components.

which made sense for the operators as automobiles were assembled one after the other.

As mentioned, the relationship between details and wholes was repeated but from a different point of departure. Identical or nearly identical assembly work was executed but with a particular variation of a 'theme' in ways consistent with a piece of music. Such 'themes' were developed by repeatedly handling clusters of components and performing the corresponding assembly work of more or less interconnected components.

These components were initially included in the material kitting fixtures before accurately fitting in and on the sequentially assembled automobiles. The varying clusters of components and assembly work between these two opposite poles (empty automobile bodies and completed automobiles) were subjected to a series of activities (preparatory work before initiating a session of assembly work, fitting and possibly re-fitting in and on the automobile bodies, inspection and perhaps correction at the discretion of the nearest co-operator and after that by other workgroup members).

However, the correct and meaningful component denotations and communication with nearby co-operators become essential. Notably, the construction of the assembly system promoted such activities (remember the earlier discussions regarding technical autonomy, which was crucial for creating workgroup cohesion). Furthermore, note the implication of and similarities with the concepts or terms 'expanded assembly work' and 'inner monologue'.

*

The author hopes the reader understands the engineering of the holistic learning approach, namely, by providing several different hierarchies, which in turn allows differentiation and integration (for both blue-collar and white-collar employees). As said repetitively, Volvo did not offer any stable and suitable wholes; they did not have access to any (remember, the consequences of the design-oriented product structure in turn based on the Function Group Register).

Whether some of the responsible practitioners in leading positions sufficiently grasped what happened practically and theoretically during the projection process and at the full-scale manufacturing phase at the end of the assembly plant's short life remains uncertain (to admit, it was a complex, intriguing and sometimes strained projection process that was not entirely completed until the closing-down decision), a statement the author will return to at the end of Section 2.7.1 after having substantiated this statement further.

Learning and training of the small parallel workgroups at Volvo Uddevalla

The operators were inextricably linked to the successively consumed material kitting fixtures and gradually completed automobiles. They could not avoid becoming more generatively, intuitively and progressively knowledgeable and trained with time. Such unavoidable situations were amplified by the need to coordinate with their co-operators (fetching components, handling tools and other manufacturing equipment, initiating internal transportation of empty automobile bodies and completed automobiles, reorienting automobile bodies during the assembly work, etc.).[11]

As hinted before, when required, opportunities to communicate with co-operators were always available (in fact, this was necessary). Furthermore, several automobiles were always available nearby in different states of completion. These automobiles served as complementary learning and training aids, allowing operators to observe the preceding and succeeding assembly work in ways inaccessible to assembly lines and assembly line work.

Together, all these shop floor phenomena were enhanced further through professional language substantiated by correct and meaningful component denotations, nicknames and synonyms, as were provided by the advanced assembly work instruction and other measures.

[11] In our experimental workshop, an enlightening experiment was performed with a 16-year-old trainee, which was vital since it validated our previous findings (not least about holistic learning). After two weeks of learning and training, the trainee managed to assemble one fourth of an automobile at 150% of the estimated (ideal) net assembly time (although without quality assurance) compared to initially approximately 1,200% in the temporary learning and training workshop (see Engström et al. 1994 for further details).

The trainee was more or less forced to explore several auxiliary materials and tools available at our experimental workshop (completed and semi-completed automobiles, illustrations from the automotive company and our unique system of illustrations, pre-structured binders of PKIs, small cards with shrunken illustrations from the PKIs, unique literature from Volvo, etc.).

Aside from the time-proven performance, the trainee assimilated knowledge from a comprehensive and widespread mix of sources. The most crucial sources were derived from our unique illustrations and the assembly-oriented product structure. This situation later proved possible to replicate on the shop floor in Uddevalla.

The struggle to sustain such learning and training was boosted further on the shop floor in intriguing ways, as was the case due to the construction of the layouts. When the operators improved their skills, their ability to accumulate working hours increased, i.e., working up. As discussed elsewhere, gaining technical autonomy is a necessary precondition for achieving administrative autonomy; freedom from being machine-paced is thus an essential condition for administering your and your colleagues' work (Karlsson, 1979).

*

In conclusion, the learning and training performed in the temporary learning and training workshop and then in the gradually completed assembly plant were inevitable for several reasons.

However, in the earlier case, such initiatives were hampered by a lack of access to appropriate assembly work instructions since the Uddevalla plant's unique, fully integrated (local) information system only became operational after the workforce moved to the first parallel assembly workshop completed. Only the PKIs were available during the temporary learning and training workshop's life.

This lack was vital since it meant that the operators within this learning and training workshop acquired knowledge and skills from the PKIs. The insight they brought with them as they were transferred to the newly completed assembly plant helped them handle administrative matters on the shop floor. They were schooled in two worlds: their shop floor world and the automotive company's intriguing and prevailing administrative world.

2.5.2 *Construction of the Assembly-Oriented Product Structure Through Systematically Organised Disassembly Procedures and Its Realisation Through Our Unique Illustration System*

As explained previously, the design-oriented product structure needed to be reformed in one-way or another to suit the assembly work in Uddevalla. Therefore, the author and his closest research colleagues recognised that another product structure was required, a product structure, which proved possible to achieve trough editing and transforming the KDP and PKIs. This new product structure must be based on how the automobile was perceived by members of the small parallel workgroups of operators working together on the shop floor and should mirror the product architecture and product variation of the automobile model in question from an assembly work perspective.

The author and his closest research colleagues thus had to construct an assembly-oriented product structure that metaphorically folded the automobiles along their edges and symmetries to fit within each material kitting fixture (see Fig. 2.3). In the case discussed the Volvo 740 automobile model became the point of departure. Implementing these material kitting fixtures was an axiomatic decision for Volvo, which was carried over to the chosen assembly system from a previously rejected proposal. To repeat, for several reasons, these kitting fixtures were nevertheless considered to be used also for the subsequent proposal?[12]

*

The desired product structure had to be balanced. Approximately half of the automobile's assembly time consumption needed to be represented on either side of the middle of the total assembly work in and on the automobile body. Thus, each of the four main groups of components must correspond to one-fourth of an automobile.

This particular criterion for the construction was derived from earlier facts during the projection process, namely, the two rejected and projected proposals for assembly plant construction. Initially, an assumption pervaded the projection process organisation that the maximum work cycle time was 20 minutes and that three operators were supposed to work together and alter between two automobiles, which by practitioners sometimes is denoted as double-docks (a questionable assumption, according to the author), this pervading assumption was slowly modified based on the experiences and insight gained during the temporary learning and training workshop.

*

[12] A folded paper can be grasped in its entirety if one understands that the paper is complete (repesents a whole) before being folded. The lines of edges and symmetries for folding can sometimes also be understood or at least anticipated. This analogy corresponds to the fact that all components for one specific product variant were collected in the three pairs of material kitting fixtures that had to be fitted and used (ultimately, empty kitting fixtures verify the assembly work).

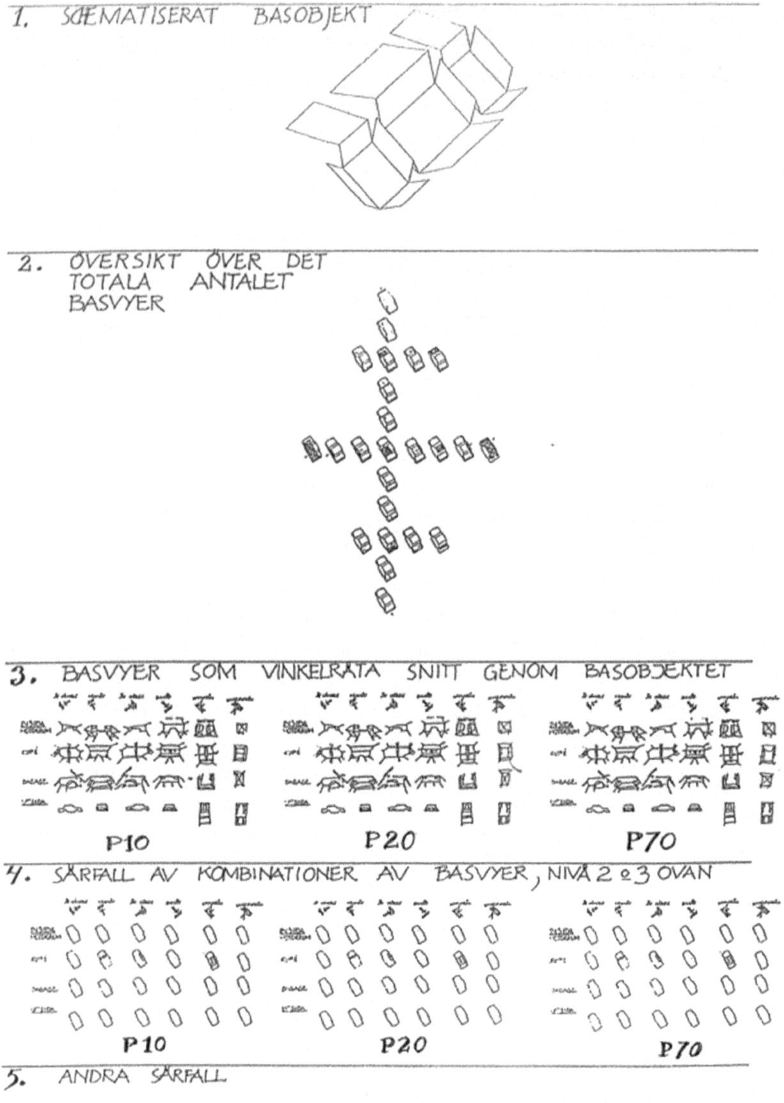
1. SCHEMATISERAT BASOBJEKT
2. ÖVERSIKT ÖVER DET TOTALA ANTALET BASVYER
3. BASVYER SOM VINKELRÄTA SNITT GENOM BASOBJEKTET
P10
P20
P70
4. SÄRFALL AV KOMBINATIONER AV BASVYER, NIVÅ 2 o 3 OVAN
P10
P20
P70
5. ANDRA SÄRFALL

◂**Fig. 2.3** An example from the Volvo 740 automobile model showing three automobile models, i.e. P10. P20 and P70 (the model assembled in Uddevalla), which, metaphorically speaking, was folded along its edges and symmetries. The empty automobile body was regarded as consisting of three boxes featuring different openings (the engine room, with two horizontal openings; the coupé, with two vertical openings; and the luggage compartment, with one horizontal opening) or as illustrated in the folded-out paper box. This schematised automobile leads to hierarchically different views of various areas of the empty automobile body, as was constructed by Mikael Hedin at our experimental workshop (*Source* original illustrations from our experimental workshop)

The assembly-oriented product structure was based on systematically organised disassembly procedures. Automobiles were disassembled, and the removed components were labelled and positioned on the shop floor, where they were moved around until we were satisfied with the results.

Four main groups of components were identified: (1) Leads for electricity, air and water; (2) Driveline; (3) Décor; and (4) Interior. Several sub-groups of components were identified under these main groups (five to seven sub-groups for the assembly work in and on the automobile body and four for the sub-assembly work). The large sub-assemblies are not discussed here since work on the automobile bodies was prioritised.

These main groups and sub-groups were each assigned a colour and a suitable denotation. Orange was used for the sub-assembly of the door, while yellow, blue, red and green were chosen for the four main groups for the work in and on the automobile. Principally speaking, the overarching assembly sequence assumed that one operator should be able to assemble one main group of components.

Furthermore, the assembly-oriented product structure was demonstrated using our unique illustration system, printed on coloured papers according to the particular colours assigned for each main group of components. For this reason, the removed and labelled components were arranged in the appropriate configurations (depicting appropriately organised components often required specific arrangements, e.g. the internal locking mechanisms and other components inside the doors needed to be hung by strings from above).

Criteria for the construction of the assembly-oriented product structure

The overall criteria for grouping the components on the shop floor of our experimental workshop were to construct a product structure that was hierarchical but also allowed discrimination between constructed clusters of components. The components contained in each main group of components should be as self-explanatory as possible. While trying to make sense of the gigantic jigsaw puzzle on the shop floor of our experimental workshop, which included dealing with an overwhelming number of PKIs, hierarchically and spatially discriminating component characteristics were used as criteria for arranging the components into the main groups and sub-groups.

Both the operators on the shop floor in the temporary learning and training workshop and ourselves in our experimental workshop focused on analysing large- and medium-sized components that were perceived almost immediately, intuitively resulting in a hierarchical and spatial perception. The assembly-oriented product structure may, to some limited extent, be of a taxonomic rather than an absolute character, i.e. in practice, a limited number of components were necessary to regard but impossible to assign to each main group unambiguously (in practice on the shop floor was due to division of labour certain minor inevitable divergence necessary).

In brief, long and slinky components were assigned to the '1 Leads for electricity, air and water' group, while the more significant components utilised for power generation and transmission were assigned to '2 Driveline'. Components more or less intuitively associated with bling and decoration were categorised under '3 Décor'. Finally, colour-codified components with specific textures that cultivated comfort and pleasure inside the coupé and luggage compartment were placed under '4 Interior' (see Fig. 2.4). The construction was somewhat more complex, and different component characteristics were regarded simultaneously but these varied between the main groups and sub-groups.[13]

[13] Product variation was grasped by employing the concept or term generativity, a way for the author to understand the 'products' inner logic ', i.e., how apparent the product architecture and product variation also became apparent for the shop floor operators.

This approach meant categorising product variation into several levels based on how it related to two extremes (1) what was constituted and evident both before and after the assembly work by observing several empty automobile bodies and completed automobiles and based on information derived from some crucial components included in the material kitting fixtures, or (2) product variation that cannot be understood by observing the above-mentioned circumstances.

For example, the colour of the automobile body and number of doors (four or five doors) represented the former extreme, but the type of spare wheel, windshield wiper, etc. represented the latter extreme.

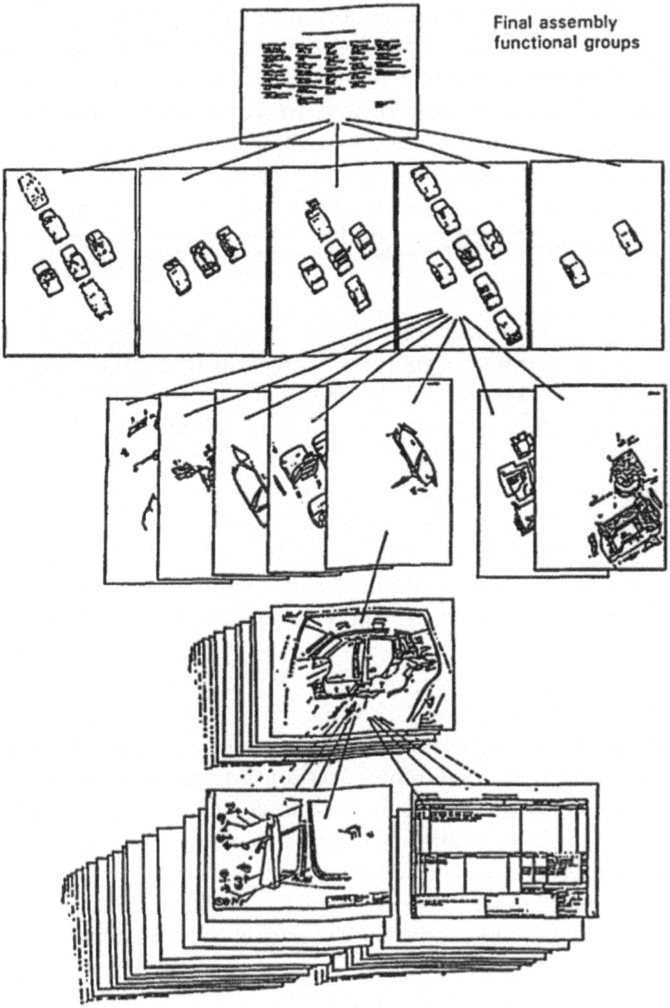

Fig. 2.4 Our unique system of illustrations depicts the hierarchy of the assembly-oriented product structure constructed for the automobile model assembled in Uddevalla. As defined verbally, the assembly-oriented product structure (top) has sometimes been denoted 'Final Assembly Functional Groups' since it is a figure retrieved from an earlier publication. The fragmented PKIs are at the bottom of the illustration (*Source* Engström & Medbo, 1991)

Note that the construction of an assembly-oriented product structure and the necessary research and development work have primarily nothing to do with functions or product function, but the product variants had to be re-codified into levels according to generativity, which to some extent refers to product function, as treated elsewhere (see Engström et al. 1995b, which explains these matters in depth). (Note, it is impossible to divide the automotive products discussed evenly on the shop floor according to product functions; besides, this criterion or approach must be sufficiently stable for the construction of an assembly-oriented product structure.)

During the author and his closest research colleagues' professional lives, such product structures were constructed during several diverse, long-term industrial collaborations within the Swedish automotive industry involving different automobile models and heavy truck chassis at Volvo, as well as large-sized diesel engines at Scania (Engström et al., 2000 for further details). As before, the latter case involved disassembling the large diesel engine and spreading out the removed components on a shop floor while, simultaneously, all relevant automotive information was gathered and used as guidelines for the systematically organised disassembly procedures and construction of stable and suitable wholes.

Our unique illustration system

As discussed previously, components in an automobile can be arranged in spatially discriminative hierarchies according to their placement in and on the automobile body. This approach allows sufficiently knowledgeable and trained operators to distinguish the particular main groups and subgroups from other conceivable components, i.e. to navigate within the assembly-oriented product structure depicted (see Fig. 2.4).

Each illustration in our unique illustration system depicted groups of the most prominent components in a standardised way, specifically as these are located exploded outside several partly invisible automobile bodies (i.e. an outline and a specific perspective were used). Our

Engström et al. (1995b) explain how the product variants must be re-codified into levels between these extremes according to generativity, i.e. how the product variants variants were more or less relevant in terms of characteristics determined from an assembly work perspective. Such re-codification would have resolved the awkward manufacturing scheduling system, as suggested by Engström & Medbo (1991).

unique illustrations formed a hierarchy that successively became more schematised towards the top.

These depicted components and the partly simulated automobile bodies were described this way. The automobile bodies were simultaneously observed by several imaginary operators working together in small parallel workgroups on a shop floor.

The outline standardising these illustrations deviated from an exact reproduction, i.e. as prescribed by correctly applying the chosen perspective (a method that allowed the components to be moved around without disturbing the perspective).

For example, the sides of the automobile body were purposely tilted a few degrees, exposing more of the right side of the illustrated automobile to depict the components clearly (as an assumed driver was entering the automobile or as considered to be overviewed by a fictional small workgroup on the shop floor). The paper format was thereby also better utilised. The illustrator (architect and doctoral student Mikael Hedin) would thus sometimes consciously distort an illustration to allow it to achieve its purpose (see Engström et al., 1992 for details and further explanations).

*

The research and development work involving constructing our unique illustration system, achieved with Hedin, was exciting but time-consuming. This initiative emphasised how a general lack of such illustrations previously had hampered the understanding of the automobile within Volvo.

One example of our efforts was the very time-consuming task of sewing each small-sized component together with our unique illustrations and each sub-group of components in the assembly-oriented product structure on several large-human-sized paper sheets.

Furthermore, the positions of each small component were related to their positions in and on the automobile body (this was indeed an intriguing task for us since corrected positions were critical due to the various winding areas exposed in and on the naked automobile body).

Moreover, a selection of shrunken illustrations from the PKIs was also utilised on the large paper sheets, which made things difficult. Unfortunately, this initiative was not worth the extra labour on our part since we had to collect these large paper sheets from the garbage cans outside the temporary learning and training workshop after some weeks (the second-class pupils were discarding the first-class pupils' learning media since these were considered worthless after being used).

However, this initiative also made us appreciate the mental stimulation of being surrounded by completed and disassembled automobiles in an environment filled with appropriate automotive documents (obviously, such masses of partly interrelated learning and training aids dealing with automobiles were explicitly and implicitly successively making sense as very different tasks were carried out in this environment. Some of these learning and training aids were consciously inter-connected by our colour-coded unique illustrations, strategically positioned here and there in our experimental workshop).

In addition, we were sometimes inspired by interested visiting individuals asking thought-provoking questions, mainly members from the projection process organisation and occasionally the highest-ranking managers. Overall, we had numerous diverse opportunities to contemplate and discuss the product architecture and product variation of the Volvo 740 automobile model and ponder on our achievements.

Further development of the research and development work to construct assembly-oriented product structures beyond the Volvo Uddevalla plant

The research and development work to construct assembly-oriented product structures was proven possible, even if the search process could have been more streamlined (the meticulous disassembly procedures, combined with the depicting, were time-consuming). Similar approaches performed by the author and his closest research colleagues not only dealt with automobiles and heavy truck chassis (in a gearbox case for Saab, we needed to revise the codification of product variants to resolve some product quality issues). The systematically organised disassembly procedures produced deep insights into administrative mechanisms (including an understanding the intriguing and prevailing administrative idiosyncrasies).

These methods were developed further during the projection of the Autonova automobile plant, the successor of the Volvo Uddevalla plant. The process became far smoother and faster because of a computerised analytical database, assistance from the employees of this automotive company, and the use of small cards with shrunken illustrations from the PKIs, as practised at our experimental workshop.

In this case, we were assisted in disassembling an old product variant supported by numerous specially constructed paper printouts representing three different product variants. An analytical database was developed by considering the components delivered from the KSO (Knocked Down Operation at Torslanda) and was then used for the subsequent automobiles manufactured at full-scale manufacturing pace. This time, the necessary product information was due to these procedures already available in Uddevalla long before the components (see Engström et al., 2000 for an in-depth elucidation).

2.6 A Brief Description of the Volvo Uddevalla Plant, One of the Two Most Advanced and Unorthodox Swedish Assembly Systems

This sub-chapter briefly appraises the Volvo Uddevalla plant by describing the layouts for the small parallel workgroups. As a contrast, the most refined versions of Japanese assembly lines are subsequently reviewed. The Volvo Uddevalla assembly system involved work cycle times that were far more extensive than what was adopted for assembly line work in other countries and were, in fact, much longer than anywhere else in the international industrial scale automotive manufacturing industry, i.e. apart from the assembly docks at the Volvo Tuve heavy truck chassis plant (see Engström, Blomquist & Holmström 2004 for further details).

An exhaustive explanation of several other aspects of the assembly plant in Uddevalla follows after the brief discussion on Japanese assembly plants, such as how the assembly work was conducted and how the division of labour within the small parallel workgroup was organised. However, several other matters are also touched upon.[14]

[14] The parallel product flow at the Volvo Uddevalla plant was an 'organic product flow 'with more automated manufacturing equipment in high transport frequency locations. This product flow thus features a gradually decreasing transport frequency as the product flow becomes more and more parallel until it finally reaches the periphery (like the branches of a tree). The extent of manual work gradually increases until, finally, at the edge, the manual work is taken on by small parallel workgroups. If required, these workgroups can organise themselves differently (note that they are thus unrestricted by more automated tools and manufacturing equipment).

Furthermore, this principle is essential for selecting tools and other manufacturing equipment (which should give the reader something to reflect on while trying to think 'outside the box' about the construction of more or less automated devices in relation to the choice of product flow design).

*

From the author's and his closest research colleagues' perspectives, the projection process was an opportunity for us to influence the construction of a completely new assembly plant. The industrially oriented socio-technical research tradition with particular normative elements laid the foundation of this initiative. As already discussed, the projection process engaged several individuals from the local research tradition, which included collaboration among researchers from applied social and technical science disciplines (several other researchers should be mentioned with names; see references).

Specific circumstances and personal ties within the Volvo sphere were also critical. Our subsequent access to our experimental workshop was also crucial. These opportunities allowed us to convert earlier and recently gained insights and experiences into something practical at an industrial scale in exciting ways.

2.6.1 The Small Parallel Workgroups at the Volvo Uddevalla Plant

This assembly plant consisted of building facilities comprising six parallel assembly workshops ('product workshops' in Volvo terminology) and one common material store. Figure 2.5 shows the physical and schematised layouts of the small parallel workgroups at the Volvo Uddevalla plant.

The first parallel assembly workshops utilised for full-scale manufacturing were nos. 1–3. Automobile bodies were transferred between workstations after approximately half of the total assembly while assembly work in and on the automobile bodies was completed. The automobiles were moved from one of two tilting devices to be positioned on one of two available lifting tables, i.e. from the first pairs of workstations to the second pairs. The tilting devices enabled the re-orientation of the automobile bodies in different ways during the assembly work, i.e. through height adjustments and rotations.

Conversely, when parallel assembly workshop nos. 4 and 5 were activated later, the automobile bodies were instead positioned in one of three or four tilting devices, i.e., up to four workstations were used. The

Implementing the holistic learning approach for assembly line work at high-frequency sections of the product flow is almost impossible due to the high transport frequency of the products combined with the short time window allowed by the work cycle times; machine-paced work at an assembly line does not provide sufficient freedom.

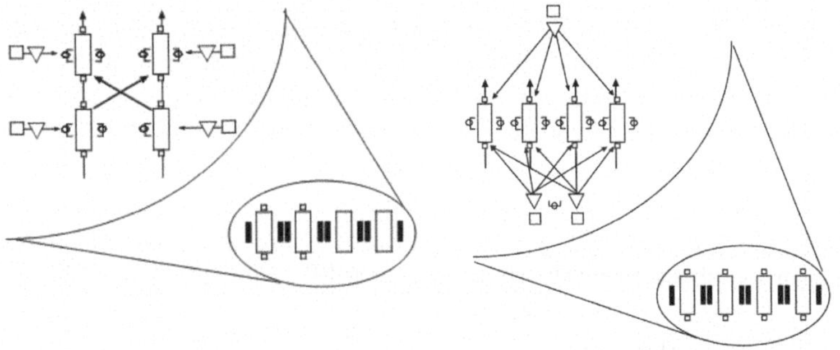

The schematic and physical layouts of the small parallel workgroups within the assembly workshops, workshop nos. 1–3, involved seven operators

The schematic and physical layouts of the small parallel workgroups within the assembly workshops, workshop nos. 4–5, involved nine operators

Fig. 2.5 The layouts used for the small parallel workgroups at the Volvo Uddevalla plant (*Source* The author's drawing)

number of tilting devices depended on the stage of this assembly plant's life (the continuous evolution of advanced work organisation was one of several other reasons).

In addition, the locations of the integrated sub-assembly workstations differed between the first and last assembly workshops activated. However, these sub-assemblies were still an integrated part of each small parallel workgroup's regular work in both cases. The sixth parallel assembly workshop (assembly workshop no. 6) was never used for full-scale manufacturing and was thus never fully manned. Instead, it was used for learning and training purposes.

The work cycle time implemented was at least 80 minutes for assembly workshop nos. 4–5 and 100 minutes for assembly workshop nos. 1–3. However, substantially longer work cycle times were also practised. At the end of the Volvo Uddevalla plant's short life, approximately 25 operators could single-handedly assemble completed automobiles according to a formal test, to which operators could apply voluntarily (one of these was female).

*

Accordingly, based on this brief introductory explanation, an uninitiated visitor would still be unable to follow what was going on on the shop floor. Thus, these assembly systems need to be carefully explained. For example, the shop floor work may have appeared less efficient, especially compared with assembly lines and assembly line work, but as will be described, it was far more efficient. These considerations apply to any uninitiated visitor who needs to increase familiarity with or to be fully informed about several less obvious measures being practised on the shop floor.

For example, employing intermediate buffers in an assembly line is self-explanatory, especially if the automobiles are automatically hoisted up in the ceiling, making them inaccessible for any assembly work. However, several other buffer mechanisms were engineered into the assembly systems discussed.

Another example of less observable measures is how the KDP and PKIs were locally complemented, edited, and transformed into suitable information. This, in turn, made the necessary advanced assembly work instructions possible for local manufacturing engineers to construct, who were helped out by comparable, more skilled operators.

2.6.2 Small Workgroups in the Most Refined Versions of Japanese Assembly Lines

To compare what is explained previously about the most advanced and unorthodox alternatives to assembly line work, some matters regarding the **most refined versions of Japanese assembly lines** are described first in this section, as proved possible due to a joint venture research cooperation.[15]

*

During their study visits, the author and his closest research colleagues observed that the most refined versions of the Japanese assembly lines featured manufacturing sections that were extensively longer than those implemented in Swedish assembly lines.

[15] 'The End of Mass Production and the Emergence of New Production System: From Serial Flow System to Parallel Stationary System' was a four-year research project managed by our Japanese research colleagues. This research project included several research tours and brief study visits to Japan. In turn, our Japanese research colleagues visited Sweden (financed by 'The Scientific Research Foundation' in Japan and referred to as 'Basic Study B1').

The project leader was Professor Emeritus Hikari Nohara, who was first employed at the Social Institutions Graduate School for International Development and Cooperation at Hiroshima University and later worked at Nagano University.

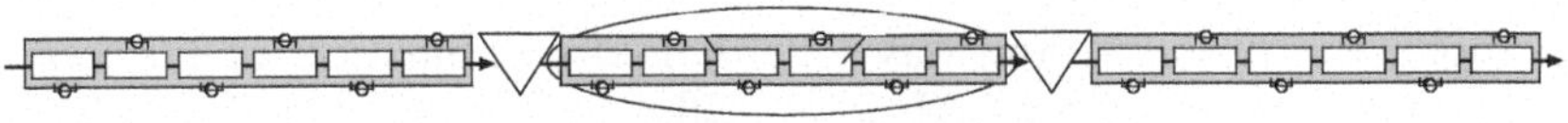

Fig. 2.6 Layout of one of the most refined versions of Japanese assembly lines (*Source* The author's drawing)

The extensive length was due to (1) the zigzagged placement of operators within the relatively small workgroups positioned one after the other along product flows. These workgroups were also (2) interspersed by inaccessible intermediate buffers, which were impossible to work in because the automobile bodies were hoisted into the ceiling. (3) The operators' work was defined by 11 marks inscribed on the shop floor and (4) an 'Andon system' consisting of lights that changed colour and several other intriguing measures. The operators initiated each short work cycle at the first mark and completed it at some of the last marks.[16]

These and other measures aimed to reduce system losses (remember, the operators' inevitable stochastic variation). Scope for administrative and technical autonomy that proved relevant in some Swedish cases were not considered (see Fig. 2.6).

*

[16] The inaccessible intermediate buffers along the most refined version of Japanese assembly lines were inserted for technical reasons. These buffers do not allow the accumulation and release of working-up to permit operators of workgroups to vary the work pace in the ways explained elsewhere in this chapter (remember, the reasoning concerning technical autonomy as a prerequisite for administrative autonomy).

Another essential aspect to recognise in this context is that the work pace is not pre-limited for Japanese assembly lines. In contrast, according to negotiated labour laws, Sweden has an upper limit for work speed set ('MTM-avtalet', in the Swedish language).

In Sweden, assembly work is sometimes conducted in (accessible) intermediate buffers. Moreover, sometimes, the operator is responsible for using the time represented by the intermediate buffer volumes, which are employed for various reasons (both social and technical reasons).

For example, a specific percentage of this time, represented by the intermediate buffer volumes, is allocated to smoothing out the effects of product variants. The rest of the time may be used for other purposes.

However, this presupposes that appropriate information on assembly time requirements, product variants and manufacturing scheduling principles, among others, is provided beforehand and that the operators trust this information. A better understanding of buffer mechanisms beyond what is explained in this chapter is needed (see Engström et al, 1996b, which is an attempt to discuss several Swedish assembly systems relevant to what is touched upon here in this footnote and elsewhere in this chapter).

In short, the product flow was a hybridised combination of machine-paced and non-machine-paced product flows. The product flow within each manufacturing section had only two defined states: (1) the automobiles were moved along continuously at a steady and defined speed while maintaining exact distances between each of the automobiles, or (2) the product flow within a manufacturing section was stopped. Furthermore, (3) all the workstations were simultaneously influenced (as said, no work was performed in the inaccessible intermediate buffers).

However, within a manufacturing section, the product flow was sometimes stopped for a short predefined period, which was implemented by allowing any operator to request attention by pulling a cord. This action was performed while the product flow proceeded continuously at a constant and defined speed. Nonetheless, the automobile bodies were always automatically stopped at the exact location, i.e. at the last inscribed mark on the shop floor.

During our short study visits, we observed the rapid and variable product flow within each manufacturing section, which featured several notable measures. The goal was to create and maintain a defined and specific mode of operation.

An assembly system should be kept in balance with striving towards a 'defined steady-state condition'. These assembly systems were constructed to reach and preserve a mode of operation grasped in full by all employees on the shop floor (with clear and distinctive borders between blue-collar and white-collar employees' jobs).

The mode of operation was characterised by harmonically oscillating operators within each manufacturing section, a very steady and predictable product flow, and several inaccessible intermediate buffers with volumes varying solely for technical reasons.[17]

[17] The operators working along the Japanese assembly lines discussed were monitored using logic similar to how the sophisticated overhead power-and-free conveyor system was programmed for the Saab Scania body shop, allowing for some inherent flexibility. However, as explained previously, this particular mode of operation was instead accompanied by the prospect of gaining administrative and technical autonomy.

2.6.3 A Detailed Description of the Volvo Uddevalla Plant

At the Volvo Uddevalla plant, four pairs of material kitting fixtures were initially used (one for each main group of components in the assembly-oriented product structure), each containing only the components needed for one automobile. However, these pairs were reduced to three after some refinements were carried out at the temporary learning and training workshop. However, specific small-sized components prone to damage were available within each workgroup.

These material kitting fixtures were successively delivered in pairs, and the automatically guided vehicle system managed the internal back-and-forth transportation. This vehicle system also delivered empty automobile bodies and retrieved completed automobiles to and from the small parallel workgroups. Finally, the completed automobiles were filled with media and roller-tested at two facilities in the middle of the two three-leaf-clover-formed buildings (see Fig. 2.7).

The automatically guided vehicle system

As said, the automatically guided vehicle system (1) coupled the six parallel assembly workshops with a common material store to transport both loaded and empty material kitting fixtures. This vehicle system (2) also transported empty and completed automobile bodies back and forth within the assembly plant (two different vehicles were used, i.e. for automobiles and material kitting fixtures).

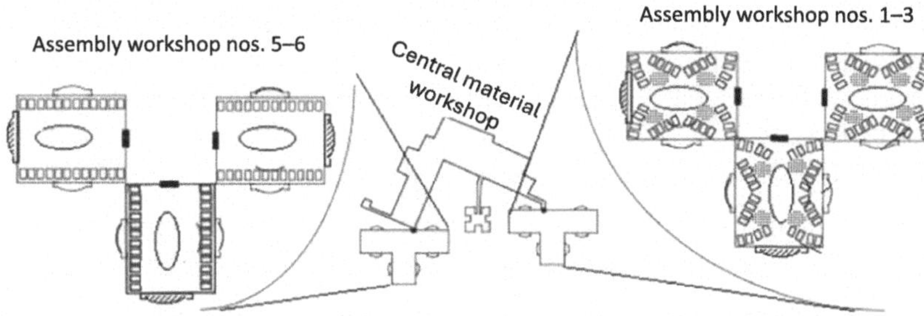

Fig. 2.7 Overview of the Volvo Uddevalla plant, which featured six parallel assembly workshops arranged as two three-leaf-clover clusters of buildings grouped around two media-filing and roller-testing facilities and a common material store. These parallel assembly workshops embraced a 'cinnamon-bun-shaped' administration building (*Source* The author's drawing)

The automatically guided vehicle system delivered each pair of material kitting fixtures at identical roundabouts in the middle of each parallel assembly workshop. Subsequently, the operators manually transferred each pair of material kitting fixtures to be divided or separated within the small parallel workgroups. Each material kitting fixture was then positioned on each side of the automobile bodies on the parallel workstations.

The automatically guided vehicle system projected for the Volvo Uddevalla did not feature any all-embracing intelligence and created several inconveniences that need to be explained more deeply. The system worked like a chessboard where the automatically guided vehicles received restricted transportation assignments from one chessboard square to the next; this was achieved without knowing precisely what the two different vehicles were transporting. As this chapter roughly described, the system did not have a high and stable delivery precision because of the type of computer technology described. This system was projected by and purchased from Auto Carrier System, a subsidiary of Volvo, and it was decided early during the projection process that no alternatives would be considered.

Volvo Automobile Company had already been renowned for automatically guided vehicle systems from the opening of the Volvo Kalmar automobile assembly plant in 1974. However, the older Kalmar plant system was more intelligent than the system implemented in the Volvo Uddevalla plant since each automobile transported was identified individually during the total assembly process, i.e., every automatically guided vehicle recognised each product variant. However, the most modern assembly plant could not dispense with such fashionable modernity, which was somewhat curiously associated with modernity and group work. Thus, this vehicle system was regarded as an axiomatic feature from the start of the projection process.

Buffer mechanisms in the form of decreased operator concentration and integrated sub-assembly workstations

As mentioned previously, the sub-assembly workstations were integrated within the small parallel workgroups, i.e., the operators within each small parallel workgroup were also performing this particular assembly work.

Moreover, the operator concentration was decreased to a value much lower than the usual Swedish practice to create a crucial buffer mechanism. For example, the number of automobiles available for assembly work increased, creating extra free work positions along the automobile bodies. Naturally, all automobiles were retained inside the small parallel workgroups until completed in all possible respects.

This statement implies, roughly estimated and explained, that the assembly time corresponding to four automobile bodies was directly available for work in assembly workshops nos. 1–3. At the same time, only three were needed to meet the stipulated manufacturing volume. The assembly time consumption corresponding to one buffered automobile was smeared out invisibly, and a crucial buffer mechanism was thereby created (yet a feature almost impossible to grasp by an uninitiated visitor) (remember the system losses dealt with elsewhere in this chapter).

Conversely, the integrated sub-assembly workstations were purposely not fully utilised, another crucial buffer mechanism. During appropriate periods, the operators of these workstations, often by necessity, helped other operators working in and on the automobile bodies, and left the sub-assembly workstations temporarily.

As discussed previously, this was possible since the sub-assembly workstations were deliberately not fully utilised (not 100% occupied); thus, the maximum operator concentration on the automobile bodies varied during the total assembly process.

Therefore, a variable operator concentration was constructively exploited, and the operator involved in sub-assembly work regularly helped the other operators working in and on the automobile bodies.[18]

[18] As reported in the main text and elsewhere in our publications, the learning and training period was surprisingly short, a few months, in contrast to sometimes published data, which refers to a period of approximately one year, i.e. this was the period of governmental funding received before the full-scale manufacturing phase.

For example, from the shop floor, newly recruited operators began by learning and training in the integrated sub-assembly work of one automobile door to learn and train in working on the other three automobile doors with similar assembly work. Because of such procedures, the small parallel workgroups could reach full work pace without mastering all tasks needed immediately (regarding the learning and training times achieved, see Engström et al., 1996a for further details).

This publication presents data obtained in a closing-down questionnaire survey (68 operators involved in assembly work and 91 in total) on personal data, i.e. age, education, learning and training times, number of work modules mastered, sex, work perception, etc.

The relationships observed between quantitative work content and self-estimated intellectual and manual competence requirements were statistically analysed. Moreover, a conclusion that confirms the benefits of the holistic learning approach over those of the mechanistic learning approach is presented.

Such activities were feasible because the maximum operator concentration in the last half of the automobile assembly (Décor and Interior) was three operators.

Usually, two operators performed this assembly work. Extra operators were thus welcome since they could work in and on the automobile bodies without disturbing the two regular operators.

In contrast, on the first half of the automobile (Leads for electricity, air and water and Driveline), a maximum of two operators could work together on the automobile bodies; a third operator was not of any help whatsoever. In general, the operators only needed to avoid blocking each other (remember the system losses discussed elsewhere in this chapter).

Fig. 2.7 shows the free space between the two lifting tables, two tilting devices and identical roundabouts for the automatically guided vehicles. As already mentioned, identical roundabouts were also found in the middle of the other parallel assembly workshops.

These interfaces were identical. Thus, the configuration of the material kitting fixtures was always similar, requiring similar fixtures for the same components for each particular product variant.

As noted, at the edge of the identical roundabouts were empty automobile bodies and completed automobiles were collected and manually transported back and forth to the workstations. Furthermore, the automobile bodies were then required to be buffered on the free space available within each pair of the small parallel workgroups.

The interface between the small parallel workgroup and the, automatically guided vehicle system consisted of two lifting devices (or docking stations) located at the edge of the identical roundabouts. Both these devices allowed for the buffering of empty and completed automobile bodies. Such buffering was achievable regardless of whether it was a matter of delivering or fetching completed automobiles or empty automobile bodies (see Figure 1.1 in Medbo 1999 for an overview and further details regarding the layouts and the group work).

This available free space and the two lifting devices (or docking stations) created yet another crucial buffer mechanism. They both contributed to a critical separation between each small parallel workgroup from the, automatically guided vehicle system and prevented them from being paced by this vehicle system.

Figure 2.7 illustrates the Volvo Uddevalla plant's building facilities and the physical layouts of the six parallel assembly workshops on either side of a reduced outline of all the building facilities. Moreover, the two lifting devices (or docking stations) interfaced the identical roundabout for the automatically guided vehicles.

2.6.4 *Elaborating Further on the Configuration of Material Kitting Fixtures*

To elaborate and recapitulate what has been said so far, the material kitting fixtures constructed for the Volvo Uddevalla plant fulfilled several purposes. For example, the material kitting fixtures not only facilitated the learning and training of the extensively long work cycle times required but also reduced the material handling work of the operators. Such fixtures also eliminated the need for extra space on the shop floor required for fetching components when more traditional material feeding techniques are practised, e.g. line stocking instead (as described earlier in this chapter, while discussing the author's original research questions during the late 1970s).

*

As mentioned, the material kitting fixtures comprised the components required for one specific product variant. Furthermore, from the operators' point of view, these components were organised logically in each material kitting fixture, i.e., as a pre-structured jigsaw puzzle.

One crucial consideration was to avoid forcing as many components as possible into any of these material kitting fixtures to attempt to reduce their size. Some plastic containers in the material kitting fixtures contained relatively few components to enrich the operator's understanding of the corresponding assembly work, i.e., the content should be accounted for immediately (which may be perceived as an unnecessary waste by an uninitiated observer).

As discussed earlier, the primary purpose was to ensure that the pairs of material kitting fixtures were filled with the correct components, each with a designated position. As explained further, this method of organisation was complemented by correct and meaningful component denotations following what was pointed out elsewhere in this chapter.

*

Along similar lines of reasoning but from a slightly different angle of approach, taking out just one plastic container from a material kitting fixture meant that the components inside should immediately specify the assembly work required. This way to present the components had to be spontaneously understood by the operators at a glance.

As underlined before, metaphorically speaking, the automobiles were folded along their edges and symmetries. The required components were categorised in several ways to be squeezed into four pairs of material kitting fixtures, which were later reduced to three pairs. This meant that the configuration of each material kitting fixture was unique, and the content was evident for the operators.

A more nuanced and expanded perception of the product architecture and product variation than is possible for any assembly line work was created and developed further.

*

Correct and meaningful component denotations became possible late in the projection process upon computerising the advanced assembly work instructions. Such denotations were crucial for assembling and selecting components in the common material store. This essential vocabulary was derived from computer files retrieved from the central Product Design Department (remember, the fragmented PKIs featured mutilated component denotations used in each extremely short assembly sequence in combination with a far too cryptic language partly due to lack of space on certain sections of the standardised forms and aged computer technology).

The component denotations in the advanced assembly work instructions used in the completed assembly plant were, among other things, complemented and edited locally by varying demarcation lines, fonts, nicknames, and synonyms.

*

In conclusion, the components, in combination with several other features, generated a semantic network, which is essential for the operators' gradual mastering of complex knowledge and provides a three-dimensional perception of assembly work and products.

2.6.5 *The Small-Sized Components, Processing Procedures for Verification Purposes, Awkward Manufacturing Scheduling System, Performance and Administration Effects, and Other Partly Unforeseen Results*

This section discusses five issues that should be addressed more thoroughly when examining advanced and unorthodox assembly systems. One of these were (1) handling the smaller-sized components. A particular advantage of Volvo Uddevalla's unique, fully integrated (local) information system that must be acknowledged was (2) the progressively organised material and information processing procedures.

Furthermore, the reader should observe one specific unfortunate matter: (3) the awkward manufacturing scheduling system. This topic, in turn, leads to a discussion of the (4) performance and administrative effects gained at the end of the assembly plant's short life. Finally, (5) are the space requirements of tools and other manufacturing equipment.

Small-sized components required special attention

Picking the correct small-sized component has always been a dilemma for assembly line work at other local Volvo assembly plants. Numerous identical small-sized components are constantly generously exposed together in their designated material containers on the workstations along a product flow. These components must be fitted to several different product variants, i.e., they should be fitted to particular product variants in unpredictable ways as delivered according to a planned manufacturing sequence. Therefore, all of the small-sized components must be simultaneously available in differently denoted material containers at each workstation. Picking the correct components from each labelled material container during assembly line work has thus proved critical for small-sized components.

These components are easily confused for something else or sometimes deliberately changed by the operator. This dilemma is aggravated by difficulties ensuring that the correct and valid information on small-sized components is synchronised to the correct product variant. Each material container is labelled with such information, but these containers comprise numerous identical small-sized components. Keeping track of these components at the parallel product flow assembly plant discussed became impossible, requiring extraordinary measures to be established (the Uddevalla unique automated manufacturing equipment for composing the small transparent plastic bags). Furthermore, the effects of the frequent product change orders should also be considered (see Johansson & Johansson (1990) for further information regarding how this dilemma was handled).

Rather annoyingly, dropping small components on the shop floor during assembly line work almost immediately makes them worthless since they instantly become unidentifiable.

Consequently, they are usually swept up and thrown out. Identification after droppings is time-consuming and troublesome since, for example, some of the frequent product change orders are sometimes either administrative (the same component with another name) or real (different components), and generously exposed small-sized components kept in their plastic containers' in assembly line fashion create extensive inconveniences for assembly plants like Uddevalla since numerous such component together may or may not designate particular product variants (precisely which one is impossible to know for operators).

The general lack of correct and meaningful component denotations makes things worse since at the Swedish assembly lines, at best, mutilated component denotations and other specific incomprehensible product variant codes complementing the all-pervading component numbers are marked on several labelled plastic containers.

The length of the work cycle is also essential. Shorter work cycles create more fragmented perceptions, thus reducing several possible guiding contexts. Consequently, assembling unidentifiable components becomes almost unavoidable on an assembly line, while use of material containers for small-sized components en masse are, for the reasons considered, impossible for parallel flow systems.

Progressively organised material and information processing procedures for verification purposes

The interaction between flows of components and the Uddevalla plant's unique, fully integrated (local) information system can be explained using the concept or term **progressively organised material and information processing procedures**.

Such procedures always meant a step-wise verification on the shop floor, in contrast to the fragmented, one-way information supply practised at the Volvo Torslanda automobile plant's assembly lines. The Torslanda plant suffered from aged computer technology during the discussed period (remember, the Uddevalla unique, fully integrated (local) information system was not used at the Volvo Torslanda automobile plant).

*

Several verification procedures were always performed in the Uddevalla plant, which was possible due to the information system discussed. This unique, fully integrated (local) information system verified and sanitised all inbound computerised information through steps that involved the component flows that entered this assembly plant.

In brief, the first phase of these verification procedures began with (1) ordering components for various reasons, while the second phase involved (2) verifying the information against the inbound components at the gate of the assembly plant.

Subsequently, (3) the components were verified during storage in the common material store. The subsequent verification (4) required the components to be selected inside this material store and loaded into the material kitting fixtures. Finally, (5) the operators performed the final verification during the assembly work.

Therefore, the component flows were locally step-wise cross-referenced with the previous and current information in the fully integrated (local) information system (no questionable component or product information was allowed to enter the assembly plant).

Any assembly plant always conducts such processing procedures as described, as well as several other administrative processes, some with their local information systems. Nevertheless, these processing procedures were meticulously conducted in Uddevalla, partly through the fully integrated (local) information system. However, the construction of the assembly system and choice of work organisation were most certainly contributing factors.

For these and other reasons, the introduction of new models and product variants (as well as the frequent product change orders) was more efficient and less costly than what was the case for the Volvo Torslanda automobile plant (see Engström et al., 1996a, 1996b for further details).

The comparable, more skilled operators and the relatively closer relationship between blue-collar and white-collar employees, who quickly observed and understood completed and semi-completed automobiles, was crucial. Furthermore, there was a far better understanding of product variants and the common scheme for codifying all product variants as well as expanded experiences and insights into almost everything else.

The awkward manufacturing scheduling system

During the early stage of the Volvo Uddevalla plant's short life, the small parallel workgroups used the regular working days differently. Some workgroups worked continuously throughout regular working days, while others initially increased their output to meet the planned manufacturing volume and the specified product quality. Others more or less waited until the last minute to recoup the predetermined manufacturing volume, depending partly on age and personal preferences.

However, this situation changed as the impact of the awkward manufacturing scheduling system successively became felt on the shop floor due to increased manufacturing volume. (See another chapter of this anthology describing the structured interviews conducted directly after our video recordings of nine different product variants assembled. There, it becomes evident that regulation of small workgroups, as treated in Subchapter 2.2, is somewhat of a lost art within the Swedish automotive industry.)

*

This awkwardness was due to the fact that the assembly time saved for one specific product variant could not be transferred to the subsequent one (the working-up was impossible to accumulate), and this was because the product variants were almost impossible to predict in the manner required. Thus, the hours saved were wasted and the purpose of the parallel product flow construction needed to be thoroughly reviewed.[19]

[19] A similar dilemma was handled differently in the Volvo Kalmar plant. The components available within this assembly plant assigned for one specific product variant were administratively coupled with each automobile body at the last possible moment at the assembly plant gate. Thus, the automobile bodies being transported to this assembly plant partially functioned as an external buffer mechanism since identical automobile bodies would then replace one another.

To some extent, the construction of the manufacturing scheduling system allowed changes in the scheduled manufacturing sequence because of the zigzagged, partly accessible intermediate buffers that allowed specific product variants to be retained, which were overridden by subsequent product variants. In contrast, in the Volvo Uddevalla plant, empty and painted automobile bodies needed to be brought together to the (internal) manufacturing sequence created already at the body shop at the Volvo Torslanda automobile plant, a troublesome and unnecessary step.

This particular difference was caused by the awkward manufacturing scheduling system and the unintelligent, automatically guided vehicle system. As underlined repetitively, Uddevalla's unintelligent, automatically guided vehicle system could not identify what was transported (remember, transportation assignments were received from one square to the next, i.e., like a chessboard).

This unfortunate situation occurred because the product variants were scheduled according to assembly line practices that did not account for differences in assembly work. The assembly line's cryptic common scheme for codifying all product variants created an inexplicable, extensively long series of more than 30 digits that the operators were required to master to schedule their product variant sequences, making it in many respects impossible to schedule the product variants appropriately.

Consequently, the observed and actual assembly time consumption differed from the logged information saved in the Uddevalla unique, fully integrated (local) information system, as became evident from video recordings of nine product variants which we made during the closing-down phase (interviews in connection with the video recording confirmed this).

The authors and his closest research colleagues suggested ways to correct this dilemma. Still, the recommendations were never realised during the assembly plant's short life, partly due to personal conflicts within the projection process organisation and partly for other reasons (like other more urgent problems that required attention). However, the remedy was communicated and published internationally beforehand (Engström & Medo, 1990)

For space reasons, a full explanation is impossible to include in this chapter. Suffice it to say that to remedy this challenging situation the product variants had to be re-codified according to generativity, i.e. according to the relevance of characteristics from an assembly work perspective (see Engström et al. 1995b, which explains these matters in depth, this matter is briefly touched upon in a footnote)

Insights gained on performance and administration at the end of the Volvo Uddevalla plant's short life

The advanced and unorthodox assembly system discussed resulted in both foreseeable and unforeseeable effects compared with assembly lines and assembly line work. The practitioners involved predicted some of these, while other effects proved to be surprises.

The actual and observed assembly times were never presented publicly. These were publicly proclaimed equal to assembly line work during the closing-down phase. This information was partly false; the situation was more complex, as explained in this chapter.

Today, the depth of understanding by the upper management of the situation remains an open question. However, direct observations from our video recordings of nine different product variants showed that assembly time consumption was 2–4 hours shorter than required to assemble the same Volvo 740 automobile model at the Volvo Torslanda automobile plant. The product quality was nevertheless the best for this automobile model within the Volvo Automobile Company, according to the company's investigations.

Reduced time losses contributed to shorter assembly times than in assembly line work. Furthermore, the operators controlled the total tolerance chains between components, leading to improved product quality. They were able to pre-fit several components loosely and subsequently adjust them to the required tolerances or functions. The components were then torqued down, an impossibility on a machine-paced assembly line.

A definite surprise was the low administration and annual costs of model changes, which were evaluated as part of the demands arising from the decision to close down. The yearly cost of model changes per automobile for three years was approximately 25% of those in the Volvo Torslanda automobile plant. This estimation included hard and soft costs (investments, tools and other manufacturing equipment, engineering, training, and try-outs) (see Engström et al., 1996a for further details).

*

We conducted video recordings following each operator's assembly work for nine different product variants during regular working days (see Engström et al., 1996a, 1996b for further details). Since processing all the VHS tapes collected was a massive task, the author and his closest research colleagues had to construct unique analysis equipment at that time (Engström & Medbo, 1997).

Notably, the Volvo 740 automobile model featured a questionable product architecture and product variation from an assembly work point of view (fewer engineering hours were consumed for the product design work compared with, e.g. some Japanese automobiles) (see Engström, 1982 who early was trying to estimate similar comparisons). Nevertheless, the best product quality for this automobile model within the Volvo Automobile Company was achieved at the end of this assembly plant's short life (some knowledgeable customers even asked for Uddevalla automobiles). However, the sales operation kept the origin of the automobiles sold confidential.

Furthermore, the assembly time reported here is accurate shop floor information, not secondary information (indirect observation) or information logged according to the company-stipulated routines.

*

Similar results regarding efficiency and product quality were also obtained at the assembly docks at the Volvo Tuve heavy truck chassis plant, which assembled the most complex heavy truck chassis for the most demanding Japanese market from 1992 until closing down in 2002 (see Engström, Blomquist & Holmström, 2004, for further details regarding this heavy truck chassis case). As reported elsewhere, similar effects were also gained earlier in the Volvo Truck Company's temporary Arendal workshop.

Other partly unforeseen results gained, i.e. requirements of tools and other manufacturing equipment and space utilisation

As mentioned above, one unforeseen result was the improvement in administrative flexibility, which mainly involved the internal administration of the KDP and PKIs and various already-explained matters coupled with this administration. Such results also involved other types of flexibilities of a more physical nature.

Administrative flexibility includes (1) the efficient introduction of new product variants, (2) ease of fulfilling frequent product change orders and (3) simplification of revisions required by central departments.

The physical results were (4) flexible work time scheduling and (5) the ability to work while being partially manned, i.e. only some of the small parallel workgroups had to be present if extra automobiles were required. Assembly line work presupposes that all workstations are manned to manufacture anything.

Moreover, (6) significantly more multi-purpose non-special tools and other manufacturing equipment were employed than was necessary for assembly lines and assembly line work. Some tools and other manufacturing equipment did not require modifications due to, e.g. new product variants.

Many of the often costly special tools and much of the manufacturing equipment demanded by the central Manufacturing Process Engineering Department proved unnecessary. Such central demands were made because established standards were often declared necessary for product quality. However, these central demands proved faulty or questionable in many respects.

In addition, (7) ergonomic benefits were achieved by replacing pneumatic-driven torque wrenches with power hand tools and manual torque wrenches. Allowances for individual preferences, such as adaptations for gender-specific human characteristics like smaller hands, could also be made more readily.

*

A rarely reported benefit was (8) the space efficiency, which is comparatively lower for an advanced and unorthodox parallel product flow construction; this resulted from a lower transport frequency of the products due to increased work cycle times, i.e. the products were moved fewer times. This reduction in shop floor space resulted from the automobiles being inaccessible for assembly work while moving along.

An assembly line like the Volvo Torslanda automobile plant requires approximately 1.5 times more space for the assembly work than an assembly plant constructed like the Volvo Uddevalla plant (see Table 2 in Ellegård et al., 1992 which contains valuable information). The difference in the areas for rectification work between these two assembly plants is approximately 50%. These figures were obtained by measuring the layouts of the Volvo Kalmar, Torslanda, and Uddevalla assembly plants (interestingly, the Kalmar assembly plant had similar space requirements as the Torslanda plant (the assembly plant in Kalmar featured a semi-parallel product flow design).

2.6.6 Further Details on the Assembly Work

In this section, the author will briefly detail the assembly work within the small parallel workgroups at the Volvo Uddevalla plant. The assembly was characterised by extensive work preparation, which is always conducted before initiating any assembly work.

The primary method of conducting the assembly work within the small parallel workgroups

Most of the ordinary assembly sessions were conducted as the operators (1) fetched a pair of material kitting fixtures that had to be divided and positioned on each side of the automobile bodies.

Next, they usually (2) brought a plastic container from the material kitting fixtures to one of the work positions along the automobiles before initiating the assembly. Subsequently, the operators (3) positioned small and medium-sized components within reaching distance for the forthcoming assembly work (on automobile bodies, extra tables, tool racks, etc.).

Then, the operators (4) opened the small transparent plastic bags and made other preparations, e.g. arranging particular tools and other manufacturing equipment orderly. (5) Both during and after the work, product quality assessment was performed by several different constellations of operators.

Also, in most cases, the large-sized components were fetched from the material kitting fixtures when required. The shop floor was also used for some components. Several such components, such as the long fragile brake line piping, were placed on the shop floor in preparation for fitting. During these work preparation activities, fragility, textures and sensitivity to scratches of the components were also considered and sometimes handled in various ways.

Such extensive work preparations are not possible for assembly line work. In an assembly line, the work cycles are short; components are fed differently, and walking distances along the material facades during material handling work consumes a significant proportion of the work cycle (Wänström & Medbo, 2009) (remember, the handling losses are one of the three different time losses).

In addition, these materials kitting fixtures progressively gained more empty areas after the removal of large-sized components and plastic containers with medium and small-sized components (as said previously, these plastic containers contained medium and small-sized components, while small-sized components were placed in small transparent plastic bags).

Group work within the small parallel workgroups

As emphasised repetitively, the compulsory activities created by the advanced material feeding technique generated collective and individual overviews of how the assembly work proceeded. Such detailed overviews helped provide continuous information to all operators within each small parallel workgroup on what was happening, meaning that the progressively emptied material kitting fixtures were mirrored by gradually completed automobiles, as did the changes in operators' work positions.

Another repetitive and obsessively emphasised fact is that the operators at the Volvo Uddevalla plant assisted one another regularly. Organising sub-groups of two operators proved to be a necessary and desirable measure that promoted cooperation.

These actions were possible, even though, in some cases, the operators could not directly oversee one another's work, such as when they were guided by the sound of co-operators' assembly work magnified by the automobile bodies. Moreover, most operators mastered **surplus competencies**, which means creating extra competencies inside such small parallel work groups beyond the minimum requirement (however, these extra competencies were only sometimes achievable at full work pace but still valuable because of how the work organisation functioned). As stated, the minimum requirement was one-fourth of an automobile.[20]

Two operators were required to work together to fit the complete sub-assembled dashboards inside the automobile bodies. Thus, such work tasks naturally necessitated cooperation between the two operators (when dealing with an awkward and large sub-assembled component). The presses used for the front window were another example of necessary cooperation procedures because these were shared among several automobiles or workgroups. In addition, the inevitable need to transfer automobiles and shift their positions during assembly work naturally required cooperation (yet another obsessively underlined fact).

It must also be mentioned that learning and training were provided in the sixth parallel assembly workshop for female operators on parental leave and male operators returning from military service. Operators who needed to establish their capabilities to reach the appropriate work speed

[20] The author and his closest research colleagues' analyses of the assembly work in the Volvo Uddevalla plant included an evaluation of the perceived work and work conditions through a closing-down questionnaire survey. This evaluation underlined the importance of surplus competencies while learning and training on the shop floor. The parallel assembly workshops with the highest product quality focused on learning and training surplus competencies among operators.

The counterpart, another assembly workshop, was equally highly efficient but prioritised individual learning and training, and it had a notably lower product quality (see Engström et al., 1995b, 1996a, 1996b for more details).

The first publication evaluates the five parallel assembly workshops through a closing-down questionnaire survey on the number of excellent and unsatisfactory workgroups, productivity and product quality (including the variation thereof), start time, total time, etc.

The two latter publications discuss and provide statistical evidence on several exciting findings, such as competence development and work satisfaction (the author challenges the reader to determine why surplus competency was crucial.

before returning to their original workgroup were also provided such opportunities.

*

In conclusion, both individual and collective work was performed differently from assembly line work (even if such assembly work is often referred to as group work, which sometimes is true, although this is group work of another sort).

The operators in these small parallel workgroups inevitably had to master their assembly work as they learned and trained, including gaining surplus competencies. However, the volume of such competencies differed depending on the learning and training principles interpreted and implemented within each parallel assembly workshop, as explained in a previous footnote.[21]

[21] To demonstrate how the engineered holistic learning approach functioned in practice, (1) a word-processing program on a personal computer may be presented. Such a program provides multiple ways to write and adjust any text. From the beginning, novices can already gain an idea of the program's merits.

Another analogy is (2) the time-consuming process of learning and training infants to read and write by putting them through school lectures in combination with an initial reading of children's books together with their parents, which culminates in such mastery after years of schooling (an analogy the author hands over to the reader to figure out). Both these analogies demonstrate protracted learning and training processes.

Yet another analogy is (3) the differences between reading a daily paper and using modern digital devices to accurately illustrate the possibility of gradually mastering complex knowledge based on possibilities to differentiate and integrate and exploration of, e.g. other mental mechanisms explained in Section 2.5.1.

Daily papers add depth and richness to the knowledge acquisition process with their established and predefined physical volumes. Editorial contributions are deliberately placed on the second page, while sports/athletic issues and television programmes are placed on the back pages. Text is woven together with illustrations and photos, following an established and partly implicit logic.

Conversely, the screens of mobile telephones or personal computers provide humans with far fewer opportunities to assimilate and process information (the author challenges the reader to consider these three analogies and how they are transferable to the configuration of the material kitting fixtures in the case discussed).

2.6.7 *The Rather Complex Hierarchy of the Intra-Workgroup Pattern that Defined the Division of Labour Within the Small Parallel Workgroups*

In addition to what has already been described regarding differentiation and integration, which includes references to at least three mental mechanisms for the engineering of the holistic learning approach, the assembly-oriented product structure also helps create a hierarchy that leads to differentiation and integration.

*

Figure 2.8 illustrates the rather complex hierarchy of the intra-workgroup pattern implemented at the Volvo Uddevalla plant. This pattern featured clusters of components and the corresponding assembly work to form a matrix of work modules.

The intra-work group pattern and the work modules describe how the division of labour within the small parallel workgroups was organised. This work pattern explicitly and implicitly provided the operators with a three-dimensional model of the product and the assembly work, which was expressed in several ways that magnified the relationships between details and wholes.

This model consisted of modules involving individual operators, pairs of operators and workgroups. The modules depended on both the work positions and orientation of the automobiles (an orientation resulting from the demands imposed by the product architecture and product variation).

This rather complex intra-workgroup pattern was even more far-reaching since it formed a general base with specific administrative, mental and physical initiatives or measures that were crucial for the Uddevalla plant's function.

*

This section revisits the assembly work on the shop floor discussed earlier in this chapter. It will also involve discussions of how the product information accessible in the KDP and the PKIs was treated. The focus this time is on the division of labour in the small parallel workgroups. Figure 2.8 illustrates such division of labour in assembly workshop nos. 1–3.

A simplified matrix of work modules represents this division of labour on the shop floor. This simplification concerns (1) the integrated subassembly work that should be included in the figure discussed, which is

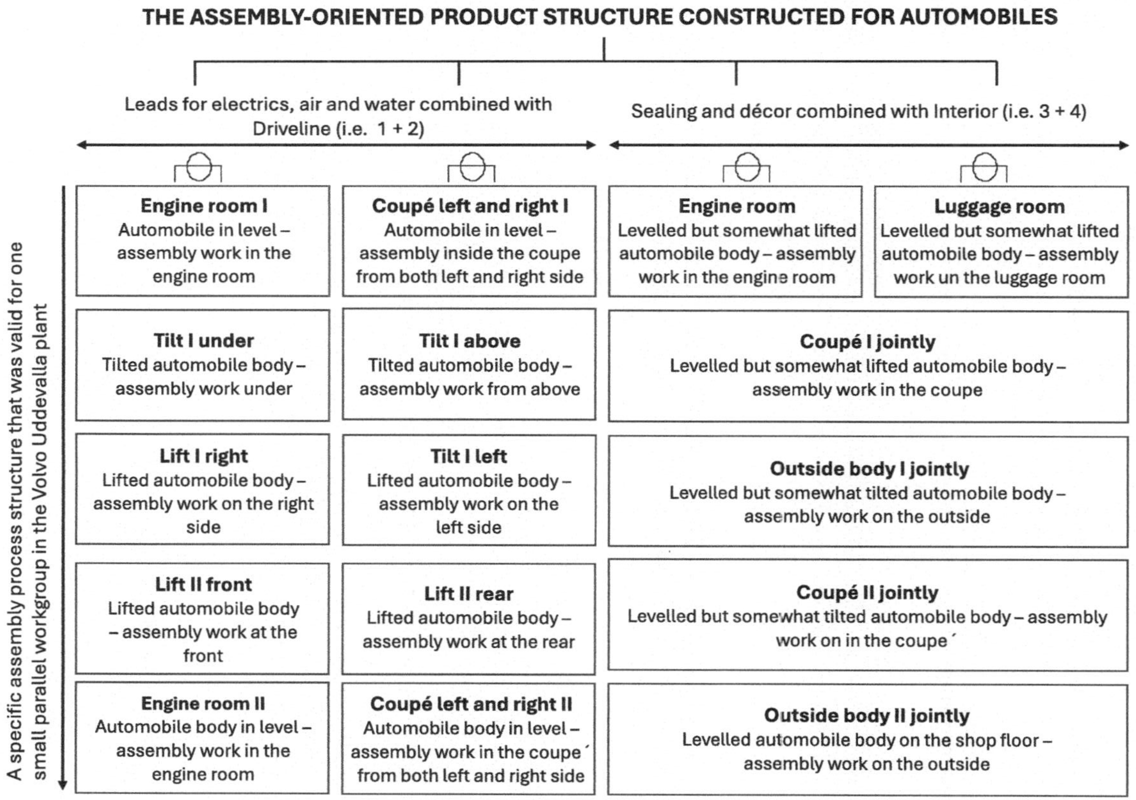

Fig. 2.8 The rather complex hierarchy of the intra-workgroup pattern of four operators was implemented in the Volvo Uddevalla plant (*Source* The author's drawing)

not the case, and (2) that all work modules (illustrated by rectangles) are inappropriately assigned the same assembly times.

These simplifications are necessary to clarify (streamline) the message conveyed to the reader. The latter action means that the heights of the work modules are identical. These work modules were permuted depending on layouts, operators' choices, workgroup sizes, etc.; operators in the small parallel workgroups could customise the division of labour in their workgroups.

Furthermore, the material kitting fixtures' configuration remained consistent regardless of the assembly workshop and workgroup because the same automatically guided vehicle system transported the automobiles and material kitting fixtures (remember, the interface in the form of identical roundabouts and the awkward manufacturing scheduling system inescapably imposed certain unwanted and unforeseen restrictions).

In addition, (3) only the assembly work of four out of seven operators is illustrated in the figure discussed as these operators worked in pairs on two automobiles, i.e. four automobiles were available (see Figure 1.1 in Medbo 1999 for an overview and further details on the layouts and the group work).

Figure 2.8 demonstrates the rather complex hierarchy of the intra-workgroup pattern using several rectangles and symbols, including the operators. The latter ones are inserted as 'crab symbols'.

Two coordinates frame the work module matrix. The horizontal coordinate from left to right above the matrix corresponds to the assembly-oriented product structure. The vertical coordinate running from the top left corner to he bottom left corner corresponds to the assembly time consumption (or to the extensively long assembly sequences), as defined by the construction of the advanced and unorthodox assembly systems.

This matrix consists of 16 work modules. The automobiles were transferred approximately mid-way through the assembly work in parallel assembly workshop nos. 1–3. They were moved from the first pair of operators (left) to the second pair (right), i.e., from the two tilting devices to the two lifting tables, as was evident by the names given to each work module.

Once again, the figure should be understood to express the fact that the first pair of operators on the far left uses components from the first two main groups of components in the assembly-oriented product structure, while the second pair deals with the remaining two main groups.

*

To exemplify the functions of this matrix, consider the work module in the upper left corner of Figure 2.8 ('Engine room I'), which is the first cluster of components, with the corresponding assembly work performed alone by one operator. This operator then worked individually with their co-operator down through the last four individual work modules (the rectangles are separated), thereby shifting both the work positions and the orientation of the automobile bodies.

The other pair of operators first work individually and then with the four subsequent collective work modules (the rectangles are oblong), i.e. 'Coupe' I jointly', 'Outside body I jointly', 'Coupe' II jointly and 'Outside body II jointly'. The operator who arrives first at a collective work module begins the assembly work until their co-operator arrives a little later. Finally, at the end of each pair's work, a checklist for product quality issues of the operators' previous assembly work was included (the operators were usually checking each other's work).

*

In conclusion, the construction of the layouts in the parallel assembly workshops and the advanced material feeding technique created an intriguing division of labour suited for the group work implemented.

2.6.8 *The Advanced Assembly Work Instructions*

The work modules were named after the operators' work positions along the automobiles, as these automobile bodies were reoriented in several ways during assembly work. The work modules had identical labels to one or several plastic containers with small and medium-sized components.

The advanced assembly work instructions used for the assembly work are briefly explained. These assembly work instructions were organised identically to other steps discussed and elucidated, i.e. by the rather complex hierarchy of intra-workgroup pattern. For example, these assembly work instructions were hierarchically organised with increasing magnification, establishing relationships between details and wholes (remember that advanced assembly work instructions constituted a model explicitly and implicitly provided to the operators).

*

To focus on one work module, the operators were able to progress level by level to gradually more elaborate extensively long assembly sequences (or times) as these became available below the surface of each work module, which had two levels underneath it.

These successively more detailed, extensively long assembly sequences were organised along the left side of the paper printouts or computer screen representations. That is, the two most detailed levels of the advanced assembly work instructions featured assembly sequences defined for the operators according to the division of labour as derived from the above-situated matrix of work modules (see Fig. 2.9), (remember each product variant was assigned its unique and extensively long assembly sequence that was of varying length, and these assembly sequences were, as said, gradually expanded as the hierarchy involved unfolded).

These levels, the two most detailed and extensively long assemblies, provided the operators with access to several columns of locally complemented, edited, and transformed product information.

Specified horizontally across the paper printouts or computer screen representations, information was presented on the next-to most detailed level in the following order: component denotations, torque required, socket size, and tools to use.

Edited and transformed product information on the most detailed level was expressed in the following order column per column: number of components required, component denotations, positions synonymous with plastic container labels, number of small transparent plastic bags and, finally, the four-digit plus four-digit number assigned to each PKI number. Therefore, these two most detailed levels of the advanced assembly work instructions contained four and five columns, respectively. The typographical techniques implemented increased the possibility of navigating between and on these two levels.

These advanced assembly work instructions were even more intriguingly constructed; however, the details cannot be explained in depth here. However, one critical example is that because of the advanced typography implemented, the components' denotations were complemented by nicknames and synonyms. Such nuances demonstrated how the KDP and PKIs were complemented, edited and transformed into suitable local information.

In brief, unique 'coastlines' were evident on the two most detailed levels. The advanced typography created unique figures, i.e. featuring various lines, styles, indentations, etc. For example, detailed work tasks (the smallest work elements in the assembly work instructions) without ingredients (components) were written in italics. In contrast, those that

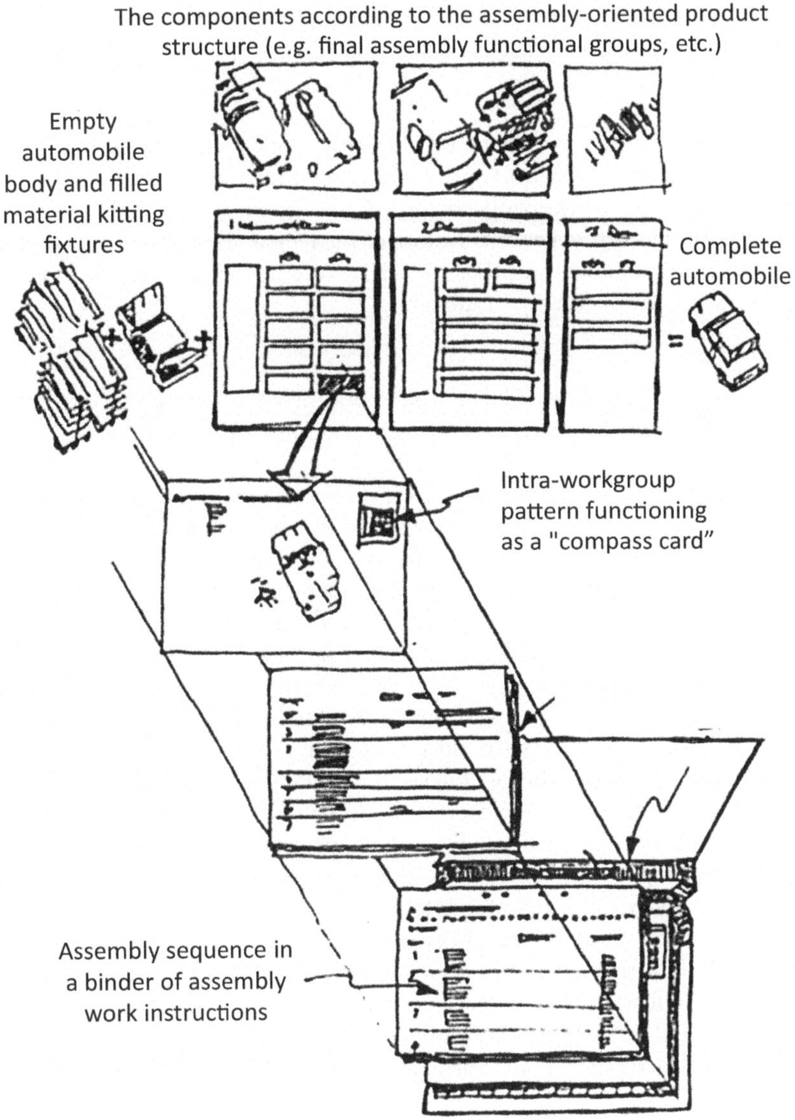

Fig. 2.9 An illustration of how the holistic learning approach was engineered at the Volvo Uddevalla plant, which involved employing an assembly-oriented product structure that gradually permitted access to the two most detailed and extensively long assembly sequences (ultimately to the fragmented PKIs). Some of the concepts or terms in the figure may be inaccurate since the figure was retrieved from an earlier publication (*Source* Engström & Medbo, 1991)

needed ingredients were printed in standard text (remember the 'dual perceptions capabilities' discussed previously).

*

One crucial detail in the advanced assembly work instructions was listing the PKI-numbers at the bottom of the (expanded) hierarchy of assembly work or the intra-workgroup pattern. Consequently, the PKIs from the central Manufacturing Process Engineering Department became valuable to the operators, explaining that the PKIs were not exclusively prospected by the local manufacturing engineers.

*

In conclusion, the advanced assembly work instructions were constructed consistently with the matrix of work modules (but with appropriate references to the configuration of the material kitting fixtures and the assembly-oriented product structure). This intra-workgroup pattern, in turn, depended on the layout construction in the parallel assembly workshops. Because of the enclosed PKI numbers (and ultimately also the PKIs), this approach was valuable for operators with expanded skills, i.e., operators who could constructively consult central documents and product information (quite handy for creating new assembly work instructions or updating old ones).

As stated, the local manufacturing engineers and the operators on the shop floor worked with and used the advanced assembly work instructions. However, the operators used these instructions sparsely and usually only when some matters called for extra action (during their daily work, they regularly used codified information supplied with the materials kitting fixtures, such as that describing the contents of the plastic containers, and relied on the fact that the correct components were always enclosed). Note, most professional work in other sectors of society benefits from access to relevant information, especially in matters of complex knowledge; however, a human cannot usually immediately recall all details, but they are still able to master complex knowledge (remember Marton´s explanation) (Marton, 1970).

2.7 Final Conclusions and a Brief Examination of Competitive Assembly Systems in Japan

The operators' actions and options are different in Japanese assembly systems. The fast and variable speed of the product flow sets the pace of the operators' assembly work, combined with a somewhat overwhelming work environment, as an uninitiated visitor might observe. Overall, what happens on the shop floor in Japan is similar to an amusement park full of bling and sounds, which obviously must be efficient, in contrast to the more serene work environment in the Swedish assembly plant discussed above.

However, the reader may have wondered why the latter assembly plant ultimately became constructed in the way described in this chapter. This topic will be cleared out below. After that, some conclusive comments concerning the work organisations used for assembly work in the two mentioned countries will be sketched based on the content previously presented.

2.7.1 The Final Choice of Layouts in the Semi-Completed Volvo Uddevalla Plant as a Consequence of a Deliberately Delayed Decision During a Somewhat Strained Period of the Projection Process

The author and his closest research colleagues became involved in constructing the layouts for the six parallel assembly workshops in the partly completed assembly plant in Uddevalla; this was the case when six blank areas were available inside the walls of each parallel assembly workshop. Constructing the layouts in these parallel assembly workshops was postponed until the very last segment of the projection process because of the choice of the length of the work cycle. This controversial question permeated this projection process.

This commitment to layouts was initially the sole responsibility of the manufacturing engineers employed within the projection process. However, by the summer of 1987, the projection process had reached a bothersome critical state.

*

After a busy and somewhat stressful summer in our experimental workshop in Gothenburg, the steering committee of the projection process was forced to end their summer vacation a week earlier to be briefed on our results. These results were based on our assessments of several product variants' appropriate and valid assembly time requirements.

Consequently, our proposal was based on information on the product variants assembled in Uddevalla and on central information applicable to other product variants. We needed to use insights from the construction of advanced assembly work instructions to obtain this information. We evaluated the product variants assembled in Uddevalla in terms of their assembly time requirements, including some variants that, perhaps in the future, might be assembled there. These assembly times were used to substantiate our layout proposal.

Similar information was impossible to obtain within the projection process or elsewhere within Volvo. This statement may seem perplexing, but it becomes understandable if the reader can understand the administrative idiosyncrasies discussed elsewhere.

The results of this engagement led to the rejection of a competing layout suggestion proposed by the automotive company's manufacturing engineers. This suggestion was far more traditional and proposed 'double docks', a layout comprised of a series of workgroups with two parallel workstations for automobiles and three operators who would alternate between the two workstations. The rejected proposal featured a 20-minute work cycle.

Furthermore, the far more traditional proposal was characterised by, among other things, the superfluous (non-assembly-active) shop floor space required for the automatically guided vehicles' high-frequency transports of automobiles. Moreover, the operators' proposed shifting between the two workstations meant that only one workstation was assembly-active, contributing to the extra need for space.

Generally speaking, as explained elsewhere, longer work cycles decrease the transport frequency of products, and the fewer times products are moved, the less shop floor space is required and vice versa (working on a moving product must be avoided).

Ultimately, the proposal was rejected because it could not fulfil the planned manufacturing volume of 40,000 automobiles annually. Based on the interaction with the automatically guided vehicle system and the fact that only half of the automobiles were accessible for assembly work, a meagre 28,000 automobiles per year was projected (and a problematic need for synchronising the movements of automobiles and workgroups inside each parallel assembly workshop would have been a direct consequence)

Conversely, the layout proposal chosen was characterised by extensively long work cycles and advanced work organisation, resulting in 56,000 automobiles per year being theoretically possible to manufacture. However, such an excessive manufacturing volume was not recommended or even sustainable because this proposed assembly system would then not have sufficient inherent flexibility to reduce time losses and achieve the required technical autonomy.

The difference between 40,000 and 56,000 automobiles per year represented, among other things, additional shop floor space available around the operators where space was most needed.

To repeat, extensively long work cycles reduce the transportation frequency, thereby saving shop floor space and creating possibilities for introducing various buffer mechanisms, such as in the case discussed. Note that this is only a brief explanation of a complex situation; see Engström, Jonsson & Medbo (1998) for an exhaustive description with the explanatory figures needed.

*

In sum, the 24 vacant areas in the three first started parallel assembly workshops were available inside the walls of the building facility for a short period in the summer of 1987. These blank areas were predestined to be automatically served with automobiles and material kitting fixtures. However, the automatically guided vehicle system restricted the use of these vacant areas. It limited the choice of layouts (thus, the need for the roundabouts in the middle of each parallel assembly workshop). In the end, the projection process became a journey of no return (the assembly system had to function correctly regardless of the non-traditional construction).

Reverting to shorter work cycles became impossible, a fact that generally positively impacted the operation during the assembly plant's short life. All of the employees had to ensure that the assembly plant functioned correctly.

As discussed in this chapter, this was necessary and unavoidable, even if some crucial underlying mechanisms were probably not fully grasped by and correctly communicated to stakeholders. As often reiterated, the situation exemplified by the awkward manufacturing scheduling system was an essential engineering idiosyncrasy that was never resolved.

Undeniably, it was a complex, intriguing, and sometimes stressful projection process that still needed to be completed, but the rather abrupt closing-down decision ended this.

2.7.2 Concluding Comments Regarding the Advanced Work Organisation Implemented in the Most Refined Versions of Japanese Assembly Systems and the Most Advanced and Unorthodox Assembly Systems

Apart from the apparent dissimilarities in product flows and work cycle lengths between the most refined versions of Japanese assembly systems and the most advanced and unorthodox assembly systems, these sophisticated assembly systems also incorporate several unique features. Group work and work organisation characteristics differ since particular features (e.g. intermediate buffers) are either non-existent or used differently. Furthermore, each assembly system has unique characteristic ways of operating, as discussed in several other anthology chapters.

The most refined versions of Japanese assembly systems have a more hierarchical and individualistic work organisation, with clearly defined tasks and relatively short career paths. Meanwhile, the most advanced, and unorthodox Swedish alternatives involve significantly more multidimensional and collective work. The skills encouraged and required are likewise different, as are the opportunities to gradually expand learning and training.

Such differences appeared because the construction of the Swedish assembly systems discussed harnessed various distinctively human skills. For example, it was necessary to grasp (1) the advanced assembly work instructions specified by a model of e.g. the division of labour explicitly and implicitly provided to the operators (see Figure 2.9 and the discussions in Sub-chapter 2.6).

Recall (2) how humans gradually master complex knowledge in specific ways, beginning with integration, which starts from a vague understanding of certain types of wholes and progressing towards differentiation, which leads to the mastery of yet more details (see Fig. 2.9 and the discussions in Sub-chapter 2.4). Furthermore, the author reminds the reader to reflect on a metaphor of a word-processing program described in a footnote.

In other words, while still metaphorically speaking and using the analogy of the word-processing programme, the most advanced and unorthodox assembly systems will challenge the operators in comparable but much broader senses since more multi-dimensional skills are an advantage in these systems. Such skills are learned and trained further in ways that allow operators to begin their shop floor careers from several different starting points other than those generally found within the international automotive industry.

These advanced and unorthodox Swedish assembly systems will benefit from an extensively broader recruitment area than the Japanese. In the Volvo Uddevalla case, a comprehensive recruitment strategy was implemented to reflect the characteristics of Swedish society.

However, the Japanese systems (or rivals) are no less sophisticated in such respects. These assembly systems are founded on almost 100 years of established tradition in scientific management, and has perfected the mode of operation based on these premises.

Such unorthodox and advanced initiatives as those described above are unlikely to occur today within the Volvo Automobile Company and other Swedish automotive companies, which today are convinced to choose more traditionally constructed assembly systems (today, in the 2020s, nothing of the original experiences and insights from the assembly plant discussed remains in inside Volvo since the few really knowledgeable individuals are either deceased or retired; only ill-willing gossips and rumours persist at the Volvo Torslanda automobile plant).

*

In the Japanese assembly systems, blue-collar employees working on assembly lines are recruited more selectively than was the case in Sweden. They use different practices that were and are possible in Japan's segregated labour market of ordinary and intentionally used temporary workers. No such labour market mechanisms existed in Sweden during the short life of the Volvo Uddevalla plant. However, nowadays, the Swedish automotive industry uses temporary workers combined with flexible worktime schedules, which is a different type of flexibility than what was created in the two advanced and unorthodox assembly systems in Tuve and Uddevalla.

Finally, the sheer scale of the most refined versions of Japanese assembly lines is overwhelming for an uninitiated visitor. The division of labour is severely restricted. The total work content of the automobiles is divided equally among numerous operators and several workgroups with a short work cycle time, around one minute. Simultaneously, the automobiles are moved rapidly along a long assembly line.

The scale of the Swedish assembly system discussed was comprehensible, and the work environment was more laid-back. The work cycle was extensively long and varied with the operators' skills, preferences, and the assembled product variants. Moreover, surplus competencies, workgroup sizes, and work paces differed.

References

Asterhall, H. & Åkesson, S. I. (1986). *Rum för arbete*. Industrial Architecture and Planning, Chalmers University of Technology (Master Thesis).

Berelson, B., & Steiner, G. A. (1964). *Human behaviour–An inventory of scientific findings*. Harcourt, Brace and Word Inc.

Blackler, F. H. M., & Brown, C. A. (1978). *Studies in Britih Leyland and Volvo*. Saxon House Westmead.

Chomsky, N. (1957). *Syntactic structures*. Mouton.

Chomsky, N. (1975). *The logical structure of linguistic theory*. Plenum.

Christmansson, M. (1997). *Production systems for manual repetitive jobs. Effects on autonomy and variety of work and prevalence of musculoskeletal disorders in upper limbs*. Department of Operation Management and Work Organisation. Chalmers University of Technology. Gothenburg (Ph.D. Thesis).

Dahlström, E., Gardell, B., Rundblad, B.G. & Wingård, B. (1970). *Teknisk förändring och arbetsanpassning. Ett sociologiskt bidrag till forsknings- och planeringsdebatten*. Prisma (in Swedish language).

Dudley, N. A. (1968). *Work measurement: Some research studies*. MacMillan.

Eckerström, G., & Södahl, L. (1981). *Ekonomisk analys av annorlunda fabriker*. Management Press.

Ellegård, K. (1983). *Människa – Produktion: Tidsbilder av ett produktionssystem*. Kulturgeografiska Institutionen Göteborgs Universitet, Göteborg (Ph.D. Thesis in Swedish language).

Ellegård, K. (1989). *Akrobatik i tidens väv. En dokumentation av projekteringen av Volvos bilfabrik i Uddevalla*. Kulturgeografiska Institutionen, Handelshögskolan vid Göteborgs Universitet. CHOROS 1989:2 Göteborg (in Swedish language).

Ellegård, K., Engström, T., Johansson, B., Johansson, M. I., Jonsson, D., & Medbo, L. (1992). Reflective production in the final assembly of motor vehicles – An emerging Swedish challenge. *International Journal of Operation & Production Management. 12*(7/8): 117–133 (originally published in the in proceedings of EIASM proceedings 1991 in Gothenburg).

Emery, F., & Thorsrud. E. (1976). *Democracy at work. The report of the Norwegian industrial democracy program*. International series on Quality of Working Life. Martinus Nijhoff Social Sciences Division. Leiden.

Engström, T., & Karlsson, U. (1981). *Alternative production system to assembly line – A problem concerning material supply*. Proceedings of VI International Conference on Production Research, D. M. Zelenovi'c (Ed.), Institute of Industrial System Engineering, University of Novi Sad, Novi Sad. Proceedings, pp. 267–270.

Engström, T., & Karlsson, U. (1982). *Alternativ montering*. Institute for Management for Innovation and Technology, Göteborg (in Swedish language).

Engström, T. (1983). *Materialflödessystem och serieproduktion*. Institutionen för Transportteknik. Chalmers Tekniska Högskola, Göteborg (Ph.D. Thesis in Swedish language).

Engström, T., & Medbo, L. (1990). Material flow analysis, sociotechnology and natural grouped assembly work for automobiles and trucks. In H. Porsnschlegel (Ed.), *Research and development in work and technology* (pp. 221–242). Physica Verlag, Heidelberg.

Engström, T. & Medbo, L. (1991). *Management and preconditions for long cycle time assembly: Some findings*. Management and new production systems, proceedings of the 3rd International Production Management Conference on Management and New Production Systems, European Institute for Advanced Studies in Management (EIASM), Gothenburg, pp. 31–36 (published later in International Journal of Operations & Production Management).

Engström, T., Hedin, H., & Medbo, L. (1992). *Design analysis by means of axonometric hand-drawn illustrations*. Proceedings of International Product Development Management Conference on New Approach to Development and Engineering, the European Institute for Advanced Studies in Management (EIASM), Brussels, pp. 147–157.

Engström, T., Jonsson, D., & Medbo, L. (1996a). *The Volvo Uddevalla Plant: Production Principles, Work Organzation, Human Resources and Performance Aspects: Some results from a decade's efforts towards reformation of assembly work*. Department of Transportation and Logistics, Chalmers University of Technology, Gothenburg (report on Work Environment Found Projects Dnr 93-0217 and 94-0516).

Engström, T., Jonsson, D., Johansson, B., & Medbo, L. (1996). Alternatives to line assembly: Some Swedish examples. *International Journal of Industrial Ergonomics, 17*: 235–245.

Engström, T., Jonsson, D., & Medbo, L. (2000). The method of successive assembly system design: Six cases studies within the Swedish automotive industry. In A. Gunarsekaran (Ed.), *Agile manufacturing: 21st century manufacturing strategy* (pp. 621–642). Elsevier Science Publishers.

Engström, T., Jonsson, D., & och Medbo, L. (2001). *Assembly work structuring based on restructuring and transformation of product information. Comparisons of management practices between Japan and Sweden*", Handelshögskolan vid Göteborgs Universitet, Göteborg (later translated into Japanese and published in an academic Japanese journal) (Journal of Humanities and Social Sciences, Nagoya City University, No. 12, March 2002, pp. 279–309.)

Engström, T., Blomquist, B., & Holmström, O. (2004). Reconstructing the History of the Main Volvo Tuve plant: Some general trends, reasons and consequences for different assembly system designs. *International Journal of Operation & Production Management, 24*(8), 820–839.

Engström, T., Johansson, J. Å., Jonsson, D., & Medbo, L. (1995a). Empirical evaluation of the reformed assembly work at the Volvo Uddevalla Plant – Psychosocial effects and performance aspects. *International Journal of Industrial Ergonomics, 16*(4–6), 293–308.

Engström, T., Jonsson, D., & Medbo, L. (1994). Extended work cycle assembly – A crucial learning experience. *International Journal of Human Factors in Manufacturing, 4*(3), 293–303.

Engström, T., Jonsson, D., & Medbo, L. (1996b). Production model discourse and experiences from the Swedish automotive industry. *International Journal of Operation and Production Management, 16*(2), 143–161.

Engström, T., Jonsson, D., & Medbo, L. (1998). The Volvo Uddevalla plant and interpretations of industrial design processes. *Integrated Manufacturing Systems, 9*(5), 279–295.

Engström, T., Jonsson, D., Medbo, P., & Medbo, L. (1995b). Inter-relation between product variant codification and assembly work for flexible manufacturing in autonomous groups. *Journal of Materials Processing Technology, 52*, 133–140.

Engström, T., & Medbo, P. (1997). Data collections and analysis of manual work using video recording and personal computer technique. *International Journal of Industrial Ergonomics., 19*, 291–298.

Granath, J.-Å. (1991). *Architecture, technology and human factors. Design in a socio-technical context*. Division for Industrial Architecture and Planning. Chalmers University of Technology. Gothenburg (Ph.D. Thesis).

Hart, H., & von Otter, C. (1972). *Lokal lönebildning: En sociologisk formulering av ekonomiska differentieringsprocesser*. Sociologiska Institutionen, Göteborgs Universitet. Göteborg (Ph.D. Thesis in Swedish language).

Hörte, S. Å. (1983). *Medbestämmande. En analys av kontaktmönster mellan parter vid arbetsställen*. Sociologiska Institutionen, Göteborgs Universitet. Göteborg (Ph.D. Thesis in Swedish language).

Johansson, M. I. (1989). *Product design and materials handling in mixed-model assembly*. Department of Transportation and Logistics. Chalmers University of Technology (Ph.D. Thesis).

Johansson, M. I., & Johansson, B. (1990). *High automated kitting system for small parts - a case study from the Volvo Uddevalla Plant*. The 23rd International Symposium on Automotive Technology & Automation, proceedings, Vol. 1, pp. 75–82.

Johansson, J. Å. (1996). *Psychosocial work environment and work-related Musculoskeletal Symptoms*. Department of Psychology. Gothenburg University. Gothenburg (Ph.D. Thesis).

Karlsson, U. (1979). *Alternativa produktionssystem till lineproduktion*. Sociologiska Institutionen. Göteborgs Universitet. Göteborg (Ph.D. thesis in Swedish language).

Marton, F. (1970). *Structural Dynamics of Learning. Gothenburg Studies in Educational Sciences 5*. Almquist & Wiksell Publishers, Stockholm (Ph.D. thesis).

Medbo, L. (1999). *Materials Supply and Product Descriptions for Assembly Systems*. Department of Transportation and Logistics, Chalmers University of Technology, Gothenburg (Ph.D. Thesis).

Nilsson, L. (1981). *Yrkesutbildning i nutidshistoriskt Perspektiv. Yrkesutbildningens utveckling från skråväsendets upphörande 1846 till 1980-talet samt tankar om framtida inriktning*. Göteborgs Studies in Educational Sciences 39, Göteborg (Ph.D. Thesis in Swedish language).

Nonås, K. (2005). *Vision versus reality in organisational change*. Department of Psychology. Gothenburg University (Ph.D. Thesis).

Thorsrud, E., & Emery, F. (1969). *Mot en ny bedriftsorganisation, Eksperimenter i industrielt demokrati*. Johan Grundt Tanum Forlag, Oslo (in Norwegian language).

Thorsrud, E., & Emery, F. (1964). *Industrielt demokrati. Representasjon på styreplan i bedriftene? Noen norske og utanlandske erfaringheter*. Universitetsforlaget, Oslo.

Wänström, C., & Medbo, L. (2009). The impact of materials feeding design on assembly process performance. *Journal of Manufacturing Technology Management, 20*(1), 30–51.

Wild, R. (1975). On the selection of mass production systems. *International Journal of Production Research, 13*(5), 443–461.

CHAPTER 3

Application of Certain Principles for Education and Shop-Floor Learning Practices for Assembly Work at the Volvo Uddevalla Plant—A Brief Overview

Lennart Nilsson and Tomas Engström

3.1 Overview

This chapter exploits particular knowledge regarding a selection of vocational learning principles and, to a certain extent, some engineering matters. This knowledge was crucial during the projection process of the Volvo Uddevalla plant and has proved necessary to make to practice by involved practitioners and the authors. Both authors were deeply engaged in this project process during the construction, start-up and full-scale manufacturing phases from 1985 to 1992.

L. Nilsson
Department of Pedagogics, Gothenburg University, Gothenburg, Sweden

T. Engström (✉)
Department of Technology Management and Economics, Chalmers University of Technology, Gothenburg, Sweden
e-mail: tomasengstr@gmail.com

H. Nohara and L. Medbo (eds.), *Technology Convergence and System Divergence*, https://doi.org/10.1007/978-981-96-1910-8_3

First, it must be stated that both authors deal with two approaches to perceiving the product: one mechanistic and one organic. Additionally, there are two perspectives on education and shop-floor learning: one additive and one holistic; this is one way to dichotomise the message conveyed, which may be puzzling for the reader. The intention is to differentiate between what may or may not appear as similar lines of reasoning; to some extent, these are two sides of the same coin.

In this context, the following four problem areas are treated:

> First, an introduction provides a brief background sketch of the earlier industrial context in the mid-1980s, including a short introduction to vocational learning principles.
>
> Second, Nilsson presents a more well-reasoned presentation of vocational learning principles by (1) elucidating four characteristics valid for education and shop-floor learning principles through active participation. Nilsson also (2) treats four characteristics of natural work, followed by (3) six educational elucidating concepts.
>
> These three presentations are based on one of the authors' earlier research and development work, which predates the projection process of the assembly plant discussed. In addition, these three presentations must be connected to the projection process discussed.
>
> Third, Nilsson further explores what is presented below by introducing five phases of what is denoted as expanded assembly work. That exemplifies how vocational learning principles were used more practically during the projection process, further developing Nilsson´s earlier experiences and insights.
>
> Fourth, Engström and Nilsson sketch how the selection of vocational learning principles proved possible to amalgamate with some engineering matters during the projection discussed. For example, the mechanistic versus the organic approach to perceiving the product is explained, including debating how individuals orient themselves in time and space.

In conclusion, some experiences related to and insights into the encounter between vocational learning principles and some engineering matters articulated.

The two associated appendices describe (1) the assembly work in the small parallel workgroups at the Volvo Uddevalla plant and (2) the operation of our experimental workshop in Gothenburg. This workshop was a window into the projection process organisation and towards the temporary education and shop-floor learning workshop in Uddevalla, which proved to be constructive and necessary since all participants in the projection process were heavily involved in other practical and urgent matters. At the same time, this workshop was, in many respects, a sanctuary for the forthcoming described encounter between vocational learning principles and some engineering matters.

*

These four problem areas just mentioned will make the reader aware of certain more common matters that are also valid for other sectors of society. After that, the two authors gradually started to grasp how assembly work should be carried out in the assembly plant, which in most respects was an antithesis to assembly lines and assembly line work featuring short work cycles and serial product flows.

The assembly plant treated featured extensively long work cycles, employing several small parallel workgroups that managed their product flow, etc., following a development line that refined group work, primarily in line with Scandinavian socio-technical research practices.

Due to the projection process, assembly work involved particular engineering matters. Several of these were crucial non-traditional features that could have been more carefully examined in this chapter. However, some of these essential features were innovative and unique, as more carefully explained in another chapter of this anthology. Among other things, the advanced material feeding technique practised and the particular ways of enhancing the perception of the product were distinctive, i.e. they were partly achieved by advanced assembly work instructions. Some of the features were already known and established, while other were unique.

Note that some of the non-traditional features practised are common within various other sectors of society. However, such initiatives still need to be practised for assembly lines and assembly line work within the automotive industry. Otherwise, where else in society is a correct and stringent vocabulary not an asset? Such disregard was previously the case for assembly line work.

3.1.1 Prologue

The point of view for the step-wise path of the Volvo Uddevalla plant projection process was developing socio-technical research practices, which were earlier tested for e.g. the Volvo Kalmar plant.

The Volvo Uddevalla plant featured a non-traditional assembly system without a conveyor belt and extensively long work cycles (substantially enlarged work content), requiring professionally skilled operators. It involved non-repetitive assembly work without machine pacing, i.e. assembly work on stationary products.

However, this regeneration and successive development of assembly work required additional education and shop-floor learning strategies, which emerged from the shop-floor reality created in Uddevalla. This implied that the work content needed to become the starting point for developing the education and shop-floor learning strategies.

The extensively long work cycles needed a radical reform of the assembly work compared to conventional assembly lines and assembly line work practices. That is, the assembly work had to be organised from the operators' point of view, comprising five phases: (1) knowledge of material and (2) structuring of material; (3) pure assembly, which in turn was followed by work tasks dealing with (4) inspection and (5) adjustment/inspection.

One consequence was that it was not only manual skills, i.e. manual dexterity and motor skills, that were important. Various intellectual competencies also had to be considered, some of which were new from the perspective of assembly line work. Education and shop-floor learning must focus on creating professional skilled work involving, e.g. operator communications. Such enhanced communication ought to be driven by straightforward questions related to the five phases of what is defined as expanded assembly work in the main text.

The result of such enhanced communication between humans differs from what was and still is practised at assembly lines and assembly work. Traditional assembly systems feature segregated work between blue- and white-collar workers, cultivating manual assembly work. Instead, in non-traditional assembly systems, the operators and workgroups on the shop floor have to master some of the white-collar employees' work tasks.

The goal is to create professional skilled identities focused on automotive assembly. This should be distinguished from aged and obsolete assumptions of craftsmanship. This questionable understanding has sometimes been used to dismiss some non-traditional assembly systems, mainly discussed in this chapter.

Describing the assembly work discussed solely as craftsmanship is a misunderstanding since, in the case discussed, it was instead a question of rather complex cross- and multidisciplinary engineering, as is especially true for this anti-thesis of traditional assembly line work, where restricted human skills are prospected.

Instead, we are concerned with a contemporary industrial manufacturing model that significantly differs from scientific management. Operators should not solely carry out fragmented, individualised and manual assembly work fitting components without designations within the limited frames constituted by the narrow time windows of the work cycles. At the same time, the operators involved in the group work in the non-traditional case must unavoidably also consider and help the nearest co-operators in several respects. However, at the same time, they should directly or gradually recognise how the product is designed, i.e. the product architecture and product variation.

Realising socio technical research practices also involves recruiting operators from various categories, i.e. recruiting both old and young operators, as well as of multiple ages and both men and women. Consequently, for the reasons thus far explained, the prominent manufacturing elements are small workgroups featuring complementary and overlapping work tasks, i.e. surplus competencies.[1]

These workgroups preferably consist of 4–9 individuals with enhanced communication possibilities. In the automobile case, such workgroups function correctly if each small workgroup's division of labour is organised into sub-groups of pairs of operators.

3.1.2 Some Selected Concepts or Terms

Due to the nature of this chapter, the authors have paid extra attention to vocabulary because, in many respects, it is a result of cross- and multidisciplinary research and development work based on intricate and long-term cooperation between the social and technical science disciplines.

Such approaches were relatively common within the industrial-oriented research context in the Gothenburg area around Gothenburg University in the 1950s and 1960s. Later, in the late 1970s, these approaches spread to Chalmers University of Technology. They are treated more exhaustively in another chapter of this anthology.

*

[1] Combined qualifications or complementarity are similar concepts or terms used in other contexts.

A selected number of (1) important concepts or terms from this chapter are presented below (see Table 3.1). Usually, these are marked in bold text and underlined in the main text the first time, but in some cases, the context explains their meaning.

The authors consider that such concepts or terms require extra attention from the reader since their origin may need clarification.

In addition, (2) a selected number of concepts or terms are presented in another table to inform the reader of their origin since they may appear strange for different reasons; however, these are not judged to require typographical marking (see Table 3.2). Finally, (3) a third table presents essential acronyms (see Table 3.3).

An editorial approach sometimes requires explaining the primary concepts or terms from what any author, consciously or otherwise, regards as not belonging to everyday language mastered by most readers (such an approach is useful for certain cross- and multidisciplinary efforts, such as co-authoring chapters in an anthology founded on intricate and long-term cooperation between the social and technical science disciplines).

3.2 Background and Introduction

Two references for what will be treated in the following are (1) the main Volvo Torslanda automobile plant in Gothenburg and (2) the Volvo Uddevalla plant. Moreover, this chapter is based on a revision of three chapters from a book dealing with the Volvo Uddevalla plant, i.e. Ellegård et al. (1992), which is in turn mainly a translation of Ellegård (1989).[2]

*

[2] This chapter is a product of a long joint venture research collaboration between Japan and Sweden. The initiative comprises several other perspectives than those treated here. The projection and full-pace manufacturing of assembly systems are related to external society, as treated elsewhere by other authors' contributions to this anthology. One such example is the way individual carriers are organised formally in the two countries.

Generally speaking, (1) in the Japanese automotive industry, committing to and passing an obligatory education before gaining higher-ranking jobs is standard practice. In contrast, in the Swedish automotive industry, you may apply for higher-ranking work and, if accepted, receive an education without any obligatory exam.

Moreover, (2) Sweden's loyal wage policy means that proportionately minor wage differences are offered within each employment category, while considerably higher wages are gained mainly by changing work and accepting responsibilities for more subordinates (success at a specific job has marginal effects on salaries).

Table 3.1 Selected less-common concepts or terms from this chapter clarify their origin in bold text and underlined. Note that some concepts or terms on central and local information have partly intentionally been synchronised with another chapter of this anthology authored by one of the two authors

Concept or Term	*Comment*
Additive perspective (on education and shop-floor learning)	Established
Advanced assembly work instructions	Not established, i.e. the authors′ choice
Advanced material feeding technique	Not established, i.e. the authors′ choice
Assembly-oriented product structure	Established in the Volvo Uddevalla plant; this product structure was practised locally to describe the product architecture and product variation from an assembly work point of view
Central assembly process information (the PKIs), i.e. 'Process/ Kontroll-Instruktioner'	Established in the Volvo Automobile Company case; the assembly work was defined centrally
Central product information, i.e. the design-oriented product structure (the KDP) and the central assembly process information (the PKIs)	Established in the Volvo Automobile Company case; the product architecture and the product variation was defined by such a product structure
Design-oriented product structure (KDP), i.e. 'Konstruktionsdata Volvo personvagnar'	Established in the Volvo Automobile Company case, meaning that the product architecture and product variation defined centrally by a design-oriented product structure, as communicated by the central information system KDP, i.e. 'Konstruktions Data Volvo Personvagnar'
Educational concept	Not established, i.e. the authors′ choice
Expanded assembly work	Not established, i.e. the authors′ choice
Function Group Register	Established, a generic description of an obsolete product used as a compulsory standard by the Swedish automotive industry, on which the design-oriented product structure used within, e.g. the Volvo Automobile Company, is based.
Holistic perspective (on education and shop-floor learning)	Established
Inner monologue	Established

(continued)

Table 3.1 (continued)

Concept or Term	*Comment*
Mechanistic approach to perceiving the product	Established
Natural work	Not established, i.e. the authors′ choice
Organic approach to perceiving the product	Established
Surplus competencies	Not established, i.e. the authors′ choice

Table 3.2 A table of selected concepts or terms from this anthology chapter clarifies their origin, which are not marked in specific ways

Concept or Term	*Comment*
Education and shop-floor learning principles (through active participation)	Established
Extensively long work cycles	Not established, i.e. the authors′ choice
Incomprehensible strings of cryptic codes	Not established, i.e. the authors′ choice
Large automotive products	Established
Product architecture and product variation	Not established, i.e. the authors′ choice
Spatial and verbal network	Not established, i.e. the authors′ choice
Specific coordinates and symmetries	Not established, i.e. the authors′ choice
Suitable wholes	Not established, i.e. the authors′ choice

Table 3.3 Acronyms used in this chapter

Acronym	*Comment*
KDP	Acronym for 'Konstruktions Data Volvo Personvagnar', which within the Swedish automotive industry is organised according to the Function Group Register; a Swedish standard for the automotive industry according to Volvo terminology, i.e. a central information system comprising the (overarching) design-oriented product structure, which in turn is based on each automotive company's interpretation
PKI	Acronym for 'Process-/Inspection Instructions ', according to Volvo terminology, i.e. central process engineering documents from the central Volvo process engineering department

The main Volvo plant is traditional and equipped with body, paint and assembly workshops, as well as assembly lines. It has a serial product flow layout with short work cycles, i.e. some minutes of assembly work. In contrast, the Volvo Uddevalla plant contained six parallel assembly workshops with very different assembly systems.

The reform of assembly lines and assembly line work to what is sometimes referred to in everyday language as assembly work on stationary products creates possibilities for reform from short work cycles to a skilled profession. Notably, the critical innovation for the non-traditional case discussed lies not in the fact that automobiles were standing still or that extensively long work cycles were practised but mainly in how assembly work was structured in specific advanced ways.

Briefly, small parallel workgroups operated simultaneously on several automobiles and subassembly workstations. Specific external and internal buffer mechanisms were intended to create internal flexibility, which, in turn, led to an advanced work organisation. Moreover, central product information was transformed locally in Uddevalla, both a necessity and a potential merit. This approach supported the operators in intriguing ways (see another chapter of this anthology, where this is elucidated).

*

The non-traditional Volvo Uddevalla plant featured six parallel assembly workshops containing several small parallel workgroups of different sizes (usually seven or nine operators depending on the layout) and various work cycles, i.e. the work cycles were very long and varied. During assembly work, the automobile bodies were height adjusted and tilted to allow assembly work underneath (see Fig. 3.1), and several subassembly workstations were integrated into each small parallel workgroup's regular work, i.e. doors, instrument panels, etc.

A separate common material store delivered material kits consisting of three pairs of intriguingly constructed materials kitting fixtures with components for one automobile. These material kitting fixtures were delivered to the small parallel workgroups by an automatically guided

Neither (3) are high-ranking academic skills worth much to the Swedish automotive industry. This fact was exemplified by the fact that during the Ford-owned period of the Volvo Automobile Company, the salary for the sole staff member with a doctoral degree increased substantially compared to earlier practice.

Fig. 3.1 An operator performing assembly work on a tilted automobile body in one of the five parallel assembly workshops used for assembly work, i.e. fitting components underneath (*Source* The Volvo Automobile Company, the Volvo Uddevalla plant)

vehicle system and, after delivery, were split up at the entrance of each small parallel workgroup.[3]

Each small parallel workgroup controlled its own separate product flow. The required components were successively delivered using an advanced material feeding technique through the mentioned material kits.

Since the assembly plant was organised according to specific Swedish socio-technical practices, it was an alternative to assembly lines and assembly line work that thereby in many respects was a refinement of socio-technical research practices.

The social and technical science disciplines merged during the projection process in ways explained in detail in other chapters of this anthology.

*

To sum up, the assembly plant in Uddevalla exhibited a substantially more advanced and refined example of socio-technical research practices

[3] The work cycles were regularly at least 80 and 100 minutes, depending on the layout and the preferences of small parallel workgroups. For example, two female operators regularly assembled complete automobiles themselves. Note that the work cycles varied due to the preferences and prerequisites of the small parallel workgroups.

than several other simultaneous Swedish initiatives. Conversely, the large-scale and older Volvo Torslanda automobile plant was substantially more traditional in all possible respects.

3.2.1 Sketching a Synoptic Background of an Early Industrial Context

Though not as radical concerning ambitions for totally new building facilities as the assembly plant discussed, an exception to assembly line work as in the Torslanda plant was the assembly docks at the Volvo Truck Company's main truck chassis plant in Tuve, which was projected approximately one and a half years after the assembly plant in Uddevalla projection process was initiated. It was also a non-traditional assembly system based on the same experiences and insights. This assembly system assembled heavy truck chassis from 1992 to 2002.[4]

The authors regard the large-scale Volvo Torslanda automobile plant as the antithesis of the small-scale Volvo Uddevalla plant and the assembly docks for heavy truck chassis. Thus, we refer to another chapter of this anthology that explains why and how this now closed-down assembly plant was non-traditional from an engineering and organisational standpoint.

In short, the Volvo Torslanda automobile plant featured two assembly lines and a material feeding technique based on traditional assembly line practices, e.g. those featuring assembly lines and assembly line work. Today, this plant is heavily influenced by the Japanese automotive industry's successes with assembly lines and assembly line work. Nevertheless, Japanese plants are still considered more refined than their Swedish counterparts.

[4] The assembly docks originated from research and development work carried out in our experimental workshop in Gothenburg by Engström and his nearest research colleagues. These docks assembled the most complex heavy truck chassis, which achieved superior efficiency and product quality compared to the performance and product quality at the two assembly lines. Similar effect was also observed in the Volvo Uddevalla plant, i.e. 2–4 hours less assembly time consumption and a substantially improved product quality.

That comparison concerned the Volvo Torslanda automobile plant, where, during the end of the non-traditional assembly plant's short life, the same automobile model, i.e. the Volvo 740 automobile model, which was a four-door rear-wheel driven sedan, was manufactured. However, the exciting fact is that these performance figures still must be explained and sufficiently treated in any publication. (See Engström, Jonsson and Medbo, 1996a one of several other publications that deal with this topic.).

As mentioned, parallel product flow was used in Uddevalla, where each small parallel workgroup managed its own product flow. The material was fed in the form of material kits carried by material kitting fixtures containing all needed components for one individual automobile, which is very different from assembly line practices.

Additionally, exactly how these material kitting fixtures should be configured was unclear at the onset of the projection process. It was initially questionable if this was possible due to several reasons not dealt with here in this chapter. The length of the work cycles was an even more controversial question for the projection process organisation.

Finally and notably, the length of work cycles differed, i.e. the large-scale assembly plant features work cycles that only last a few minutes. In contrast, the small-scale non-traditional assembly plant used extensively long work cycles with a minimum competence of approximately one-fourth of an automobile and above. Before closing down, five of the six parallel assembly workshops were at the end working at full speed, while the sixth assembly workshop was solely used for education and shop-floor learning purposes.

*

A common misunderstanding is that extensively long work cycles for assembly work are impossible due to individuals' lack of memory capacity, i.e. their inability to memorise longer assembly sequences. However, this assumption must be rectified. Instead, it is a matter of mastering generative and progressive skills within specific defined frames, i.e. one action successively triggers new actions.[5]

For example, in the Volvo Uddevalla plant, such mechanisms were purposely constructed and enhanced as the operators were trapped between the pairs of material kitting fixtures on each side of the automobiles. An assembly-oriented product structure functioned as a defined frame, as discussed exhaustively in another chapter of this anthology.

As observed elsewhere, these kitting fixtures were deliberately configured to support the assembly work. The step-wise reoriented automobile

[5] For example, this phenomenon can be seen as corresponding to individuals travelling along a familiar or partly familiar route; one scene at a block-out curve triggers the subsequent scene at the next curve. Conversely, the total voyage is impossible to imagine directly and simultaneously, but only through unfolding scene by scene exposure. During such adventures, actions and decisions are continuously perceived as taking place on a geographical map, representing one or several suitable wholes.

bodies and the co-operator's miscellaneous actions and changed work positions repeated the education and shop-floor learning process. Various skills were activated repeatedly, including cognitive, communicative, motor and verbal skills.

At the end of this assembly plant's short life, many operators could perform very long work cycles. Extensive surplus competencies were also achieved, i.e. extra competencies available within small parallel work-groups beyond what was required as a minimum (these were only sometimes obtainable at a full work pace but are still valuable due to how these work organisations function). See Engström, Johansson and Jonsson et al. (1995a) regarding the surplus competencies available at the end of this assembly plant's short life.

3.2.2 Introduction of a Selection of Vocational Learning Principles

The projection process discussed gave education and shop-floor learning their own identity. This was considered an essential means of achieving fundamental reforms in what may be summarised as current technology and tradition.

Such reforms involved creating something new, contrasting with the current technology and tradition of short work cycles that had hitherto been dominant in Volvo's automobile and heavy truck chassis manufacturing; this was especially the case because several participants of the projection process organisation wanted the Volvo Uddevalla plant to break radically with assembly lines and assembly line work. From the beginning, a more general effort to improve assembly work permeated the entire Volvo organisation of the projection process.

The skills needed for assembly work were commonly regarded as a consequence of the current technology and tradition involved during the period discussed, i.e. assembly lines and assembly line work, which, among other things, are derived from the scientific management view of manufacturing. The experiences and insights required regarding various types of work organisations practised earlier were essential attributes of already established knowledge that may or may not provide crucial inputs. (However, this was not proved to be the case; instead, such knowledge proved counterproductive.)

Generally speaking, education and shop-floor learning were considered subordinate to preconditions dictated by the current technology and tradition. Thus, education and shop-floor learning became compensatory measures, i.e. aiming to bridge the gap between (1) the demands put on operators by (2) the current technology and tradition combined and (3) early experiences and views that practitioners had about work. Because education and shop-floor learning were consequences of these factors in traditional assembly line work, a professional assembly work identity had not emerged in assembly line work. Nor was such an identity valued in this context.

One of the overarching aims of the projection process was to create an example of how to combine (1) a view of individuals as constructive and creative (i.e. partly regarded as available within the social science disciplines) with (2) a non-traditional assembly plant (i.e. partly regarded as available within the technical science disciplines). Both were assumed to produce a desired quality and a defined manufacturing volume. (Guiding questions for such thinking at this early period were, what happens when the aim is to reform the current technology and tradition, and how should this be achieved?).

*

During the projection process, the aim was to plan and implement fundamental reforms in the current technology and tradition, which were put into practice. Moreover, new habits must be created. However, much of the already established knowledge was related to the current technology and tradition. The expertise available within Volvo was, therefore, of limited value. Additionally, the time perspective imposed certain restrictions, especially since the projection process imposed a tight schedule (money was available through governmental funding, and some immediate action was undoubtedly in demand).[6]

The projection process became, to some extent, a revision procedure that included two rejected proposals on how the assembly plant in question ought to be constructed (Ellegård, 1989). Revisions gradually allowed several crucial non-traditional features to prove their merits.

This feat was achieved (1) partly due to the temporary education and shop-floor learning workshop in Uddevalla that started in the spring of 1986. This temporary workshop predated the final construction of the

[6] This was one of the reasons for the involvement of junior and senior researchers from Chalmers University of Technology and Gothenburg University, i.e. senior researchers dealing with internal material handling and vocational learning were engaged already from the start by some foresighted participants of the projection process organisation (an engagement that grew over time).

assembly plant in question and (2) partly used results gained in our experimental workshop in Gothenburg. Our experimental workshop was located close to the temporary office premises in Gothenburg, which the project process organisation hired before it was moved to Uddevalla, as a closed-down shipyard in Uddevalla was tidied up and successively made accessible for building and construction work.[7]

*

A more theoretical explanation of the vocational learning principles requires understanding the complex relations between (1) the current technology and tradition and (2) the possible preconditions and opportunities in education and shop-floor learning. This situation, leading to the projection process at an early stage, attributed an intrinsic value to education and shop-floor learning.

This statement means that measures necessary to create knowledge and promote long-term competence development were unrestricted by current technology and tradition, as these parameters were given in the autumn of 1985. Consequently, the two authors of this chapter became interested in developing something new and facilitating an extensive reform.

Therefore, practitioners in the projection process judged a reform necessary. For this and other reasons, various company and union representatives formed a local interest group that engaged Nilsson. The local interest group aimed to emphasise a more fundamental reform than usual when the Swedish automotive industry projected new assembly plants.

[7] The operators employed at the temporary education and shop-floor learning workshop in Uddevalla were at the start financed by a governmental subsidy. This grant covered not only education and shop-floor learning of assembly work but also, during this period, accounted for almost all blue-collar employees in Uddevalla, including former shipyard workers from the closed-down shipyard where the assembly plant was built. This subsidy was valid for approximately one year. However, this longer period has sometimes been assumed to represent the actual education and shop-floor learning in the literature.

Some of these initially financed blue-collar employees adapted and tidied up the shipyard premises and prepared the separate common material store housed in a remaining building ('Kärraverken'). This common material store had to accommodate the components delivered for the first fifty identical product variants assembled in the temporary education and shop-floor learning workshop, as was supplied by the spare parts operation in Gothenburg (Knocked Down Operation, KSO). These components had to be appropriately organised in the common material store in Uddevalla. They gradually managed to select and organise material to fill the material kiting fixtures supplied to the temporary education and shop-floor learning workshop.

Based on the knowledge available during the period discussed, including the experiences and insights successively gained during the projection process, possibilities and restrictions were identified in the various education and shop-floor learning initiatives discussed in this local interest group. Several vital attitudes were gradually identified as guidelines for this particular problem area (for example, the six educational concepts explained below were identified and anchored).

3.3 Expanding the Notion of a Selection of Vocational Learning Principles

This sub-chapter explains a selection of vocational learning principles more exhaustively. These vocational learning principles are based on experience and insights from earlier research and development work, predating the assembly plant's projection process that is in focus in this chapter. In many respects, these explanations and reasoning are still coupled with the projection process of the Volvo Uddevalla plant, i.e. such as included in the vocational learning principles communicated during one of the authors' direct involvement from 1986 to 1992. (See Appendix 1 regarding the assembly work in the small parallel workgroup in the Volvo Uddevalla plant.)

3.3.1 Guidelines for Education and Shop-Floor Learning Introduced During the Volvo Uddevalla Plant Projection Process

A more general and already established approach to creating the guidelines sought was for Nilsson to introduce and emphasise **natural work** for the local interest group members. This concept had been firmly and previously established in the diversified agriculture and handicraft tradition. (1) This concept implies that it is considered the most important to stress the work's unique content and capability to enrich life by focusing on the product itself.

This understanding, in combination with (2) the demands the product imposes on individuals during the process of manufacturing from the first components to complete products and (3) the demands that operators generate during work processes, all together forms the starting point. This is a platform for developing education and shop-floor learning that became practised in the Volvo Uddevalla plant.

A very keen discussion regarding the orientation of education and shop-floor learning was introduced within the projection process already in 1985. As suggested, education and shop-floor learning were thus not solely consequential phenomena derived from current technology and traditions. Still, such aspects should also be regarded as a means for renewing this by leaving a scientific management view on manufacturing.

3.3.2 *Four Characteristics of Natural Work as Explained by Nilsson During the Volvo Uddevalla Plant Projection Process*

To emphasise the new orientation discussed earlier, natural work should be related to the dominant aspects of the current technology and tradition influenced by scientific management. According to Nilsson (1985b), natural work is characterised by the following:

> Characteristic no. 1: Individuals must completely control the day's events.
> Characteristic no. 2: The work must encompass an entirety.
> Characteristic no. 3: The work must be meaningful to individuals and not be predetermined by the time factor.
> Characteristic no. 4: Knowledge transfer must occur from generation to generation within a profession.

*

These four characteristics of natural work are explained below in more depth and generalised. At the same time, the authors considered the non-traditional case in this chapter.

Characteristic no. 1: One characteristic of natural work is that individuals have complete control over the day's events and how they spread their time and various other events over a more extended period. This contrasts with the mechanistic attitude of scientific management views on manufacturing.

The entirety of the work, together with its specific content, forms the core of establishing other preconditions on the shop floor. This characteristic of natural work creates the necessary preconditions for a functional work rhythm, facilitating the ability to gain appropriate detailed overviews and contributing to periodicity between work and rest (including lunch and coffee breaks). Thus, the work constitutes points of reference, i.e. not narrow or predetermined time windows that are administratively divided into elements as small as possible. Here, crucial qualitative differences are found between natural work and assembly line work, which continues to dominate the automotive industry.

Characteristic no. 2: The entirety of the work (or at least functional shares of more prominent elements) implies that individuals can see contexts that should also be a point of view in education and shop-floor learning. This skill development results from newly recruited individuals working with professionally skilled operators while dealing with authentic products. Education and shop-floor learning are not secondary to manufacturing, administered as a distinct and separate activity; in other words, active participation is essential.

Characteristic no. 3: For work to be experienced as natural, it must contain meaningful qualities from an individual human being's point of view. If such attributes are lacking, these must be provided while broadening and enhancing work content. The work can thus be analysed with regard to whether it contains enhancing qualities. A pattern of different qualities emerges by studying the unique content of the work concerning the qualities that give it meaning, a new approach that forms the basis for the regeneration and successive development of everyday work and its routines, and creating initiatives to question traditions (Nilsson, 1981b).

Characteristic no. 4: Models or representations usually relate to specific periods. It is vital to relate to those of earlier periods to obtain new perspectives; this is why skills development will be emphasised in ways that include transferring knowledge from skilled to newly recruited workers. The point is to demonstrate the importance of such personal knowledge transfer from generation to generation within professions. However, such transfer must be carried out by education and shop-floor learning on authentic products. As stated previously, this was and is a powerful early instrument for creating an entirety of the work content.

Such attributes also embrace aspects of social life, including communication through facial expressions, gestures and speech, which enhance communication between individuals of all ages at different stages in their professional development. Most work carried out in society contains roles that relate to the identities of apprentice, journeyman and master (Nilsson, 1981a).

The discussions in the spring of 1985 between the members of the local interest group and other participants of the projection process organisation indicated a strong common desire to integrate education and shop-floor learning into the assembly work, i.e. active participation. This decision could be expected to be an essential factor for the regeneration and successive development of the assembly work (the starting-up and organising of the temporary education and shop-floor learning workshop in the ways practised certainly underlined this standpoint, but the authors are not detailing this topic further here for reasons of space).

*

The principal results of these discussions are summarised in the following six concepts, denoted as **educational concepts**. The term concept in this setting is a more general term for education and shop-floor learning principles, as well as for viewpoints on the necessary measures to create, transfer and further develop skills according to the authors' intentions, based on Nilsson (1986).

- Educational concept no. 1: Anchor education and shop-floor learning on assembly work with authentic products manufactured for the market.
- Educational concept no. 2: Education and shop-floor learning must be organised to allow the product and assembly work to be surveyed.
- Educational concept no. 3: Education and shop-floor learning must encompass assembly work in its entirety.
- Educational concept no. 4: A functional work rhythm should replace the administrative division of working time.
- Educational concept no. 5: Skills must be transferred from the professionally skilled to the less educated and learned.
- Educational concept no. 6: Practice tempo training while maintaining the quality level.

3.3.3 *The Six Educational Concepts*

The six educational concepts are detailed further below, connecting them to the discussion in the previous section.

Anchor education and shop-floor learning regarding assembly work with authentic products manufactured for the market, educational concept no. 1

As previously stipulated, natural work requires education and shop-floor learning to be established by manufacturing authentic products. This educational concept involves interrupting a separation between the school-type form of education and the education and shop-floor learning on the one hand. This fundamental precondition for selecting appropriate vocational learning principles provides models or representations of recommended new approaches for such initiatives.

Operators' demands for meaningfulness in their assembly work, relevant education and shop-floor learning can be met. As stated previously, this is when operators can review their assembly work, and the finished product can be used as demanded by the market.

New approaches entail using the product as an education and shop-floor learning instrument, avoiding the otherwise usual separation of education and shop-floor learning into small elements. From the point of view of education and shop-floor learning, such approaches lead to difficulties in seeing meaning in what is going on, as when the final results are removed and thus repositioned to be far outside the time perspectives that operators can generally overview (this applies at least during the period of education and shop-floor learning).

Education and shop-floor learning must be organised to allow the product and assembly work to be surveyed, educational concept No. 2.

As mentioned, education and shop-floor learning must be organised so that operators who have carried out particular assembly work can overview both the assembly work and products in a more extensive time perspective; this means that the point of view cannot be small elements that are sliced up into even smaller ones.

Instead, efforts must be made to organise education and shop-floor learning to maintain the context of the work content, which rejects assembly lines and assembly line work. The overarching aim was, therefore, to develop and organise assembly work according to the **organic approach to perceiving the product**, which is an antithesis to the **mechanistic approach to perceiving the product**, as is treated further in the forthcoming sub-chapters. These two approaches are included in a line of reasoning comprising two perspectives on shop-floor education and learning.

People responsible for organising education and shop-floor learning must grasp the qualitative differences between mechanistic and organic approaches; this must be the case since such experiences and insights provide a basis for education and shop-floor learning, which also develops surplus competencies. Such procedures mean operators can simultaneously assimilate knowledge and proficiency using an arsenal of tools.

*

In this context, it must be emphasised that the organic approach to perceiving the product also creates scope for handling corresponding complexity, as is achieved through the various ways individuals use their skills. At the same time, communication occurs, meaning a fertilisation of knowledge and memory triggered by recognising multiple relations between phenomena, as allowed by operators' usual intellectual processes. This desirable view is further supported if phenomena are given correct and meaningful denotations that reflect their work content. An accurate and varied vocabulary connected to diverse verbal communications facilitates recreating contexts for operators, which, in turn, gives them the confidence to handle complex work processes.

The mechanistic approach is essential in education, shop-floor learning and work involving short work cycles. At the same time, the organic approach to perceiving the product emphasises suitable thought patterns for the relations between elements, providing knowledge of a more stable nature and offering new possibilities to deal with product variants of the automobile model.[8]

*

A crucial challenge for the participants of the projection process organisation was to emphasise re-creation of the work process of complete automobiles during the projection process, that is, from a non-traditional point of view and understanding it according to the organic approach to perceiving the product.

This way of thinking generates suitable thought patterns regarding new approaches, which must be reflected on all levels of analysis required.

[8] For example, upholstery materials and colours are characteristics of some of the components fitted inside an automobile's coupe, which makes it evident that they are connected, even if they are not fitted to the automobiles on the same occasion. Nonetheless, various large subassemblies like engines and gearboxes within the same automobile model also create similar consequences.

The challenge in the non-traditional case discussed was that the **advanced assembly work instructions** and the advanced material feeding technique needed to be constructed and harmonised with such new situations on the shop floor. (This was an intriguing research and development work, especially considering the earlier assumption of the enormous number of product variants.)[9]

Education and shop-floor learning must encompass assembly work in its entirety, educational concept no. 3.

Education and shop-floor learning must encompass assembly work in its entirety. Competence must be increased primarily in the operators' field, meaning that different emotional, intellectual and manual demands will be placed on operators and groups working together. Therefore, education and shop-floor learning must be oriented towards (1) developing operator skills and (2) optimal operator cooperation. The collective results of the workgroup are crucial to achieving the desired quality and manufacturing volume.

Consequently, the primary manufacturing entities in the Volvo Uddevalla plant were the small parallel workgroups. Several very different work tasks were integrated into their work assignments. In other work organisations, they were spread out over different operators, with far more roles and work assignments located at various geographical positions on the shop floor.

One effect of broadened and enhanced work content aimed at the assembly plant was the need for an expanded arsenal of tools that operators had to be educated in and learn to master, including tools for planning work and performing assembly work, inspection and any necessary adjustments. The combination of emotional, intellectual and manual demands meant that education and shop-floor learning resembled forms of handicraft.

*

To successfully carry out more comprehensive education and shop-floor learning featuring extensively long work cycles, given that the projection process rejected two proposals on how the assembly plant ought to be

[9] Some engineering matters prototyped in the education and shop-floor learning emanated from research and development work carried out in the experimental workshop in Gothenburg, such as the advanced assembly work instructions and materials kitting fixtures.

constructed, it became necessary to create shop-floor preconditions for operators to develop an **inner monologue**.[10]

This means understanding (1) what assembly work is needed and (2) the execution of the work itself. These skills define two different sides of the total knowledge mastered. It was also (3) necessary to develop an arsenal of tools that allowed individual operators and workgroups to perform their assembly work, inspect it themselves and apply these experiences and insights while assembling different product variants.

By allowing education and shop-floor learning to encompass assembly work in its entirety, particular possibilities also presented themselves, i.e. what to do and why, would gradually emerge for all involved as time passed. Thus, the balance shifted in the usually significant division between blue and white-collar work assignments. Indeed, this balance throughout the temporary education and shop-floor learning workshop's life was somewhat partly out of balance because blue-collar employees were over-represented, and help from other resources within the Volvo Automobile Company was restricted.[11]

This implied that the current dominant digitalisation and unpredictable thinking must be abandoned in favour of a philosophy where the dominant tenets instead are operators' work patterns related to the assembly process of creating complete products.

A functional work rhythm should replace the administrative division of working time, educational concept no. 4.

[10] An inner monologue relates to the connection between the work of the hand and the mind. The aim is to link phenomena to operators at these levels. An inner monologue in this context refers to individuals' mental orientations through self-reflection and thoughts. While work is carried out, it aims to create contexts between individuals' outer and inner, between actions and individuals' ideas.

In such cases, it is a matter of fertilising the operator's mind by providing the correct ingredients. According to the main text, at the assembly plant considered, the correct and meaningful component designations can then be discussed. However, such fertilisation is increased further by an unavoidable need to communicate with various co-operators and local manufacturing engineers, especially during periods of massive product change orders and the introduction of new product variants (on occasions when it was unavoidable to consult the fragmented PKIs, or the local manufacturing engineers). The PKIs were sometimes conveniently provided by photocopy machines in the offices at the end of each parallel assembly workshop.

[11] In short, the task of the projection process organisation was to turn a dilapidated shipyard into the most modern Swedish assembly plant, complete with building facilities, information systems, manufacturing equipment, etc.

Maintaining a functional work rhythm is necessary, which also concerns awareness of creating habits that establish a prominent and periodic chain of assembly work while simultaneously considering both rest and sustenance. Such a work rhythm can be found through prominent events. For example, lunch can be a reasonable starting point for a functional work rhythm on ordinary working days.

Some phenomena are related to whether or not these occur before or after lunch. Morning and afternoon activities are scheduled before or after coffee breaks, meaning ordinary working days are divided into four periods of 1.5–2.0 hours each. Assembly work and its visible outcome, e.g. the number of products manufactured, are related to such periodicity.

The number of products to be manufactured before and after lunch is determined by the amount of work required, current technology and tradition, and the competencies of personnel responsible for planning, which must be carried out in cooperation with operators. The main point is to achieve a manageable and reasonable division of various activities to establish a functional work rhythm rooted in human life, which may create new habits.

Skills must be transferred from the professionally skilled to less educated and learned, educational concept no. 5

Skills must be transferred from already professionally skilled persons to those learning the profession. The choice of principles for education and shop-floor learning usually aims to achieve an appropriate balance between (1) the school-type form of education and (2) the shop-floor learning through active participation.

*

In the school-type form of education, there is the need for specific premises (lecture rooms, teaching media, etc.), which creates an extra distance between professional skilled operators and those newly recruited. However, as emphasised previously, education and shop-floor learning through active participation relate instead to the handicraft tradition.[12]

In such handicraft cases, knowledge transfer from generation to generation within a profession will be regarded as both a right and a duty. Such transfers mean that the role of an apprentice during an initial period is critical, ending with a final examination. The role of a journeyman,

[12] As had been developed during the period of the guilds system.

who followed different masters within the same guild or profession over a period, follows that of the apprentice. Such an earlier approach often entailed moving over geographical distances (sometimes even abroad). The journeyman period was and is still determined by an examination (based on demonstrating the skills mastered by an individual masterpiece). This success was a final precondition for becoming a master.

Fundamental to the handicraft tradition is that work is the starting point for education and shop-floor learning through active participation, i.e. there is no detour over the school-type form of education. Skills are transferred from experienced individuals in the profession to newcomers. The daily work is illustrated through facial expressions, gestures and speech. It is carried out in relatively small but manageable portions, and this is a standard education and shop-floor learning process. All this results in what is generally referred to as instructor roles. In the Volvo Uddevalla plant, in some cases, these are referred to as integrated work tasks within all more professionally skilled operators' spheres.[13]

Whatever strategy for education and shop-floor learning is chosen, as has been repeated, professionally skilled individuals must be available with experiences and insights into an arsenal of tools, work divisions, work organisation, etc. for educating and training the new generation.

*

For the most part, preconditions on the shop floor could be better; however, there is still an opportunity to create them in stages. The arsenal of tools could first be expanded for individuals within a particular selected field. After that, individuals can be models for a new generation in the role of instructors; this specific strategy was attempted during the initial phase of the projection process and start of the temporary education and shop-floor learning workshop.[14]

[13] The senior researcher with competence in vocational learning principles organised particular sessions during the assembly plant's short life.

For example, (1) categorising and sorting different components was one such session conducted in separate education facilities. It is essential to know that human understandings emanate from various points of departure (more generally, understanding complex phenomena from vague assumptions of wholes towards an expanding number of components, as is treated in another anthology chapter).

As suggested elsewhere, (2) the operators on the shop floor were learned to instruct one another through gestures and mimics, including verbal communication.

[14] An extensive mutual confidence between the members of the small parallel work-group existed. For example, the younger operators respected the older ones and vice

Several operators recruited from the automobile assembly plants in Kalmar and Torslanda were employed in Uddevalla. These operators were given opportunities for education and shop-floor learning to expand their toolkit. The intention was to explain, illustrate and present the work content and construction of assembly work for newly recruited operators.

These more experienced instructors functioned almost like journeymen for several months with the first operators recruited in Uddevalla. Their ambitions and mutual personal involvement enabled bridge-building between generations using knowledge despite shortcomings in the temporary education and shop-floor learning workshop.[15]

However, this proved to be the strength of this strategy. (Many of these shortcomings were generated by the projection process and later corrected, while others caused problems at the assembly plant until it was finally closed down. This was annoying since some of these shortcomings interfered with the early recruits' education and shop-floor learning.)[16]

*

versa in quite uncommon ways. In some cases, within some workgroups, valuable personal belongings were left on the tables in the multipurpose break rooms during ordinary working days.

These break rooms were positioned near the work groups' ordinary workstations, facing the shop floor through glass windows. Thus, it was always possible to observe what happened on the shop floor and during assembly work to see what was happening in the break rooms.

[15] Some of the first employed operators in the temporary education and shop-floor learning workshop became the original instructors responsible for a particular part of the assembly work on the complete automobiles, i.e. for one-fourth of an automobile. They all had obligatory agreements to help one another during the preceding and subsequent assembly work, meaning surplus competencies were also critical. The original instructors at the temporary education and shop-learning workshop in Uddevalla were four blue-collar employees from the Volvo Torslanda automobile plant who had all been involved in introducing the Volvo 740 automobile model.

[16] To be clear, no appropriate information system was available from the start of the temporary education and shop-floor learning workshop. The novel local information system in Uddevalla was required and first completed when the building facilities were completed. The operators were later transferred to the first started parallel assembly workshop, i.e. the parallel assembly workshops were started successively. Projecting the new local information system was an extensive task for the projection process organisation. It was a matter of designing and introducing a novel local information system much more advanced than its counterparts in other Volvo assembly plants (this information system was denoted PUMA).

The recruitment approach of the Volvo Uddevalla plant was based on the premise that individuals should be complementary. An aim when recruiting new operators was to recruit men and women, with at least 40% being women; this was a step-wise recruitment process since the parallel assembly workshops were started this way. They were predated by the temporary education and shop-floor learning workshop, which were finally closed down.[17]

Different ages were also recruited to achieve a diversity of young, middle-aged and older operators. This recruitment approach was also very constructive in developing the role of the instructor and the relations between the instructors and the instructed.[18]

*

Three factors were considered fundamental for the strategy for education and shop-floor learning at the assembly plant (as highlighted previously):

- First, assembly work with authentic products and work processes: The automobiles were an instrument for education and shop-floor learning, which, from the point of view of learners, was also a concrete and visible end. In contrast, from the automobile company's perspective, the completed product verified that such education and shop-floor learning had resulted in concrete outcomes and achieved a prescribed level of quality.

[17] As stated, the sixth parallel assembly workshop in Uddevalla, which was never used for full-scale manufacturing purposes, was instead used for women returning to work after giving birth, for male operators returning from military services who needed to do tempo trainng before returning to their original parallel workgroups, etc. (Other reasons for transferring some operators to the sixth parallel assembly workshop did also occur, such as an individual with particular difficulties in cooperating within the small parallel workgroups).

[18] Many younger female civil servants were applying for work at the Volvo Uddevalla plant, while former shipyard employees were less interested. Uddevalla was and is still a relatively small town in an industrialised region. For example, the operators frequently meet co-operators and superiors in their leisure time at local libraries, petrol stations and supermarkets; work hours and leisure time were, in some respects intermixed.

A common opinion was that the group work offered at the assembly plant was better suited for females. For example, females were not interested in a formal test on being able to complete automobiles single-handedly. Slightly more than 20 male operators voluntarily met this demand, but there was only one female. There are likely several other explanations, as well as some data, that support such common opinions.

- Second, there must be professionally skilled operators acting as instructors; there was a risk involved here for the unwary. For example, responsibilities for developing assembly work instructions may become an unnecessary side-track. Thus, such responsibilities may have been transformed into separated work assignments delegated to a limited number of operators, who probably would have transformed this work assignment into their professional skilled niche. If this happens, will the rather obvious connection between (1) a regeneration and successive development of the assembly work and (2) the work content harboured in assembly work instructions be lost? Consequently, this may result in adopting the school-type form of education, but without the advantages of a school system.
- Third, instructors must be competent and professionally skilled in maintaining education and shop-floor learning. They explain and demonstrate through facial expressions, gestures and speech. They must simultaneously carry out assembly work, following specific models or representations.

Fulfilling these three factors requires a comprehensible language; it must be created if it does not exist. The component designations and names of tools and other manufacturing equipment must be familiar to instructors and instructed operators. The component designations required should be divided hierarchically and, if relevant, turned into groups of components based on product functions, or sub-functions thereof or on other criteria, as discussed elsewhere.

Human communication, knowledge and memory cannot be developed only by employing information defined in numerical or verbally disfigured forms. Developing knowledge demands a preservation of what is uniquely human. With a comprehensible language that allows or generates a nuanced perception of the product, assembly work becomes more developed, i.e. it is not turned into far too digitalised and unpredictable thinking.

There are extensive opportunities for qualitative advances in assembly work and developing skills in a comprehensible language, which can enhance those elements and build different patterns and relations among themselves. However, this requires that such patterns and relations be identified. (To some extent, it became a task to detect and formalise the research and development work carried out in our experimental workshop in Gothenburg since the perception of the product through the administrative functions of the automotive company was severely distorted.)

*

It is also necessary for those who receive instruction (i.e. the newly recruited operators) to develop knowledge of how education and shop-floor learning can and should be advanced. The dialogue between professional generations is crucial for mutual understanding of something essential and shared. Verbal and face-to-face communication, supplemented by written content and numerical information, is critical for the regeneration and successive development of the assembly work.

If the demands for the three fundamental factors explained above are met, preconditions will be created for newly recruited operators to feel well-received. The instructors' guidance will provide opportunities to succeed with essential work tasks, thereby building a foundation for a constructive, skilled professional identity. Moreover, newly recruited operators can make friends on the work site since the instructor and the learners are mutually interdependent.

Practice tempo training with a maintained quality level, educational concept no. 6

Tempo training is an essential instrument in education and shop-floor learning in such a direction that assembly work corresponds to what operators experience as meaningful. Tempo training involves deciding beforehand on a desired quality and a defined manufacturing volume. The entire assembly must then be carried out to address the set goals.

During a tempo training session, work patterns must also be observed; this must be achieved concerning relations between operators and other operators as well as various positions of material and tools, etc. The aim is to accomplish work patterns that avoid human injury and damage to materials and tools.[19]

Tempo training should occur individually in workgroups and cover periods from a half to a whole day, including initiatives on subsequent days. The tempo training should then be expanded to include several other operators. Such collective tempo training should also correct various

[19] For large automotive products and small workgroups performing assembly work, it is crucial to have extra work positions available to avoid generating unnecessary waiting times, which Wild (1975) denoted as system losses; this is essential both for assembly lines and for non-traditional assembly systems.

oversights in the construction of the assembly plant and achieve smooth flows of products and materials. The product path through manufacturing is the main point of view; this applies the entire way, from structuring the first components required to delivering complete products.

All tempo training should should involve comparison with the desired quality and a defined manufacturing volume as well as an analysis of the reasons for any deviations in progress before and during tempo training sessions to avoid creating misunderstandings, then interpreting any differences between planned and actual results as preparation for the next tempo training. All individuals in tempo training should actively participate in the required analyses, comparisons and conclusions. Thus, each training session is preparation for the subsequent session.

3.3.4 Five Phases of the Expanded Assembly Work

Until now, the traditions and everyday routines that have dominated the automotive industry, as far as education and shop-floor learning are concerned, are such that these benefit short work cycles. As highlighted above, different models or representations are necessary for this and other reasons to achieve fundamental reforms in the current technology and tradition towards long or extensively long work cycles.

Short work cycles and related education and shop-floor learning principles are characterised by extensive repetition. From individual operators' points of view, work content continues to be severely restricted.

If assembly work is to be developed further, a relatively low degree of repetition is required. A far more broadened and enhanced work content allows the operator to extend their arsenal of tools to master such work content. The variety of tools combined with the operator's mental activity—an inner monologue about the work tasks in question—is critical for developing a direction towards professional skills.

*

To achieve professional skills, a learner must be given guidance to bring together competencies relating to the work. In contrast, in assembly lines and assembly line work, practical work, including education and shop learning, is divided into elements that are as small as possible. A regeneration and successive development of assembly work, as proposed here, cannot be based on a division into small elements. (Remember that education and shop-floor learning were given their own identity within the projection process discussed.)

Long or extensively long work cycles place more demands on individual operators and operators in workgroups than short ones. The former employs prospecting in all aspects of human capabilities, i.e. emotional, intellectual and manual abilities. It has been proven possible to break away from the prevailing separation between intellectual and manual work tasks and between blue- and white-collar work assignments and tasks.

Questioning established and partially standardised ways of managing and organising blue and white-collar work disrupt assembly lines and assembly line work. (A shift in the balance between blue- and white-collar work assignments is one example of a consequence considered previously. At the same time, it provides opportunities for something contemporary and constructive for future generations (creating something new, contrasting with the current technology and tradition, which is another example mentioned earlier).

*

Numerous initial discussions within the projection process regarding the possibilities for long or extensively long work cycles were primarily dominated by prevailing conceptions rooted in a tradition of short work cycles. A frequent and typical question posed was:

- How long must work cycles be for operators to learn and train?

It gradually became clear that this question was unanswerable. It is impossible to use a dimension of time as a point of view; this and similar problem areas must instead be approached from a more content-related perspective. This way of approaching this and similar questions leads to other questions, such as:

- (1) What does such assembly work consist of? (2) Are the expectations on operators functionally and logically related to the total work content of their assembly work so that operators' qualifications are exploited in all possible aspects?

The following discussions generally address the perspective of new models or representations transformed into principles that lead to assembly work. Operators should plan and carry out assembly work and evaluate the results they have achieved with help from co-operators. Thus, in Uddevalla, broadening and deepening such problem areas concerning the specific aspects of assembly work was unavoidable.

Assembly work was therefore broadened to include not only the assembly work itself but also the preceding and subsequent phases, comprising work tasks related to (1) knowledge of the material, (2) structuring of material as carried out before assembly and (3) pure assembly. These were, in turn, followed by work tasks dealing with (4) inspection and (5) adjustment/inspection.

See Fig. 3.2 for an overview and examples of these particular aspects. The initial phases of the pure assembly work comprise activities such as quality control of components and pre-structuring thereof. In contrast, the following phases include the final inspection and any necessary adjustments, including a further inspection. One may use the concept or term of expanded assembly work, hinting that assembly work may be more far-reaching than solely assembly activities.

*

Work tasks constituting an extensively long work cycle of assembly work thus put many and varied demands on operators. It is worth repeating that accomplishing assembly work along the above reasoning requires that operators be allowed to develop an inner monologue. Such developments should be related to an awareness and foresight of consistency in the assembly work.

To aid the development of skills required to develop an inner monologue and also to enhance and communicate with other operators (including those at the beginning of their education and shop-floor learning), several guiding questions formulated for each phase were required to exemplify what newly recruited operators were expected to learn and to serve as guidelines for operators sufficiently skilled in tutoring.

*

To sum up, the ability to pose answerable and relevant questions is fundamental to developing knowledge. For example, the guiding questions indicate the direction of education and shop-floor learning, stimulating

Model of the different phases in the expanded assembly work				
Knowledge of materials	**Structuring**	**Assembly**	**Inspection**	**Adjustment/inspection (action)**
Questions such as: Which components are to be found in the assembly zone? How are the components fixed? Are the components the correct ones? What quality demand are there on this component?	Questions such as: In which function are the components included? How is information about the assembly obtained through documents? What tools are to be used? What demands on precision are there for assembly/fit? How should the work pattern look as regards assembly work?	Questions such as: How should the work patterns look as regards relations between individual/s - tools - components - car body? How can variation be achieved in movements to prevent injury to people and damage to materials and tools and attain desired quality and work intensity? Performing assembly tasks. How should worker-inspection be carried out during and after the work process?	Questions such as: How should the completed assembly look when it has been correctly carried out? What instruments are used to determine whether the work has been correctly carried out? What routines are to be followed if defects are detected? How is the completed inspection documented?	Questions such as: How are the various types of defect rectified? How are rectified defects and desired quality attainment checked? How are rectifications and new inspection documented? How are results reported, and to whom, to prevent recurrence of defects?

Fig. 3.2 The figure illustrates the five phases of the expanded assembly work. The mutual communication between operators in the roles of instructors and newly recruited operators should gain some degree of stability regarding the broadened and enhanced work content through guiding questions, as illustrated in the figure above and also contribute to reinforcing professional skilled identity through such work content in assembly work. Assembly, therefore, involves something more than the act of assembling. Note that this is an old version of this table published earlier, i.e. it could be more coherent with the main text (*Source* Ellegård et al., 1992)

intellectual and manual work tasks. Such guiding questions place demands on the newly recruited operators, the instructor and the availability of constructive ways to support involved operators. By relating these guiding questions to the different phases of assembly work, the intention is to develop operator work patterns, contributing to a desired quality and a defined manufacturing volume.

3.4 The Encounter Between a Selection of Vocational Learning Principles and Some Engineering Matters

As stated before, several of the participants of the projection process wanted the assembly plant to break radically with assembly lines and assembly line work. One non-traditional and somewhat Utopian initial idea was that several small workgroups should assemble complete automobiles on a few workstations. The automobiles would be kept inside small workgroups during assembly work and moved from operators' workstations only after completion.

3.4.1 Contextual Background Regarding the Encounter Analysed

At the start of the projection of the Volvo Uddevalla plant, there was no experience with or insights available into the crucial non-traditional features required if an assembly system of the kind envisioned was chosen, such as how to construct the advanced material feeding technique required for such assembly plants. As hinted before, due to almost unavoidable circumstances, this was a step-wise process during the projection process, leading towards an increasingly non-traditional assembly plant.

In the light of what was assumed and known in this early period, the Swedish automotive companies' general opinion was that such constructions would have (1) resulted in an unmanageable and far too complex material feeding technique and (2) the space needed to display the material on the workstations would be far too extensive.[20]

[20] Engström (1983) provided possible clues by challenging this assumption by proving the insight that the components in one individual automobile or heavy truck chassis were manageable, an insight gained during a governmental research project financed by the National Swedish Board for Technical Development (STU) and denoted 'Alternative assembly' (Engström & Karlsson, 1982).

Karlsson (1979) was the first scientist to prove the merits of socio-technical research practices in a PhD thesis using an intriguingly constructed manufacturing system. The Saab–Scania body shop ascertained that efficiency and humanisation were possible simultaneously. Advanced and correctly regulated small workgroups, long or extensively work cycles and parallel product flows were the keys to this achievement during the mid-1970s. However, in this case, it was a matter of grinding and welding work on the empty automobile bodies, not performing assembly work.

(3) The length of the work cycles was another controversial issue, which also permeated the projection process as discussed in this and other chapters, including the work planned to be carried out in the temporary education and shop-floor learning workshop. Constructive answers to these and several other critical difficulties required urgent attention.

Examples of overarching questions handled by the projection process organisation and our experimental workshop

A general assumption was that one way to address these difficulties was to construct a material kit containing all the components required for one individual automobile and, simultaneously, group components. Almost from the beginning of the projection process, it was assumed that such material kits were possible. However, an open question was how these kits should be constructed and configured.

Such advanced material feeding techniques may or may not work in symbiosis with appropriately constructed assembly work instructions during the period treated. These facts implied that the grouping of the material on the shop floor had to be logically organised according to established vocational learning knowledge and other demands that also made sense for the researchers working in our experimental workshop. However, such groupings needed to be accepted and clarified for the projection process participants.

It may be very difficult for 'uninitiated' readers to understand why the assembly system aspects discussed represents a breakthrough in industrial history. Therefore, we want to make readers aware of this point.

According to the senior researcher concerned with internal material handling matters, these thoughts implied that several pairs of material kitting fixtures should harbour components that had to be organised according to the always predictable product architecture and product variation. However, such an approach presupposed that the product architecture and product variation was defined from an assembly work point of view. In principle, this information is predictable from the perspective of the central Volvo Product Design Department. It can also be so on the shop floor, i.e. provided the central product information is dealt with correctly (this was definitely not the case from the start).

Consequently, one critical question was crystallised:

- Exactly what approach and what methods could ensure that the product architecture and product variation was described from an extensively long assembly sequences point of view?

The central Volvo Product Design Department's point of view had to be considered but extensively reformed in one way or another. Such thinking proved possible after intensive research and development work in our experimental workshop in Gothenburg, where the material kitting fixtures were constructed and then further fine-tuned in the temporary education and shop-floor learning workshop in Uddevalla. (See Appendix 2 for more comprehensive explanations of the research and development work carried out in our experimental workshop.)

*

In conclusion, the following questions were initially found to be essential during the projection process:

- How should components be identified and groups of components be formalised computer-wise and formed physically, and moreover how should the assembly work instructions be constructed regarding product variants?
- Which education and shop-floor learning principles should be applied early in the temporary education and shop-floor learning workshop, bearing in mind that they should be appropriate also later during the full-scale production phase of the completed assembly plant?

To answer these and other similar questions, it gradually became necessary to question the assumptions already established by central product information systems and old shop-floor realities.

Current technology and traditions required to be questioned and reformed in ways that take advantage of phenomena that have meaningful relationships and are learned more efficiently; this in contrast to phenomena that lack functional relationships that are sufficiently significant and intrinsic

Finally, these were examples of overarching scientific questions that could not be directly addressed according to established and standardised Volvo practices involving external consultants. Consequently, our experimental workshop in Gothenburg and the senior researcher dealing with internal material handling were engaged in the projection process through trust based on long-time personal relationships with particular persons within Volvo, which, as hinted at previously, proved essential.

Examples of specific questions relevant to the research of vocational learning

As mentioned previously, senior researchers involved included research colleagues from two scientific disciplines, internal material handling and vocational learning, with experiences and insights that gradually proved essential for the projection process. Due to these circumstances, the two senior researchers began a continuous, constructive collaboration. This collaboration was achieved through a mutual interest in the required reorganisation of the projection process. (Three of the four Swedish authors of chapters of this anthology are still working together after more than four decades.)

*

An example of a more fundamental question regarding vocational learning principles in this context relates to assembly work and operators' control thereof:

- How should assembly work be managed and reformed, so operators can gain more detailed overviews and understand contexts better? At the same time, how can they see meaning in assembly work within a time perspective that they are active in?

This projection process provided a critical opportunity for the senior researchers available already at the initiation of the projection process, to create a more general foundation for non-traditional assembly plants. Constructive dialogue between practitioners and researchers was also unavoidable and necessary.[21]

[21] To be clear, the organic approach to perceiving the product also includes organising ideas using purely technical science disciplines (assembly work, assembly work instructions, product flows, etc.). It is also concerned with individuals' understanding, i.e. mostly professionally skilled operators. The education and shop-floor learning principles successively became clear also to the responsible practitioners.

Realising the principles for education and shop-floor learning through active participation facilitates and promotes the development of individuals' mental worlds, vitalising their emotional and intellectual lives. Only then can those who perform assembly work master the current technology and traditions, not be its subordinate servants. Specifically, this requires applying various relevant social and technical science disciplines to phenomena on the shop floor.

The only way to resolve the dilemma confronted at the start of the projection process for one of the senior researchers was to obtain complete automobiles and organise systematically disassembly procedures.

In brief, it was a matter of disassembling shiny new automobiles and spreading out the removed components on a shop floor. At the same time, all relevant automotive information was gathered and used, e.g. constructing and identification of suitable wholes. (Why this was required is something the authors challenge the reader to successively figure out or remember by himself or herself.)

3.4.2 The mechanistic approach to Perceiving the Product on the Shop Floor

As suggested previously, there are two different ways to define assembly work on the shop floor in combination with education and shop-floor learning through active participation: (1) the dominating mechanistic approach to perceiving the product and (2) the organic approach to perceiving the product. Let us examine some essential aspects of each, starting with the mechanistic approach.[22]

In the mechanistic approach to perceiving the product, the assumption is that the final result is the total of all elements. Generally speaking,

Indeed, education and shop-floor learning by active participation were also crucial, as this was introduced in the temporary education and shop-floor learning workshop and, then, on the shop floor of the step-wise completed assembly plant.

[22] As explained above, novel local information system in Uddevalla was projected and introduced in the projection process; not only physical artifacts such as building facilities, tools and other manufacturing equipment were projected. Indeed, it was a crucial initiative because various unfortunate effects could be corrected locally. For example, the need for an appropriate information system hampered our experimental workshop in Gothenburg and the temporary education and shop-floor learning workshop in Uddevalla. (Recall that the central information that reached the shop floor of any local assembly line plant was fragmented and unavoidably distorted.).

in dealing with the Swedish automotive industry, demands for standardisation and rationalisation have resulted in substituting correct and meaningful component designations with component numbers and mutilated component designations. For Volvo, this situation was aggravated by selections of cryptic codes based on how the product variants were codified, created at the central Volvo Product Design Department.[23]

*

Thus, components were and still are assigned incomprehensible strings of cryptic codes adapted to machines and mutilated component designations suited to central information systems.

As highlighted previously, this was problematic for all the participants of the projection process organisation. This situation was initially created through earlier assumptions, such as an enormous number of different product variants and an extensive number of components that were supposed to be required to complete each automobile. Such imagined preconditions created an almost impossible dilemma for the projection process organisation.

Detailing the mechanistic approach to perceiving the product

In the Volvo Torslanda plant, central information systems and old shop floor pratices favoured the mechanistic approach to perceiving the product, which in turn blocked learning and training of long assembly sequences, an impediment generally assumed to be caused by operators' difficulties in memorising long assembly sequences. Thus, prominent component characteristics, such as colour, fragility, shape and size, were irrelevant on the shop floor for the operators. Consequently, various central ways to define components were converted or reduced to component numbers and mutilated component designations, sometimes complemented by selections of incomprehensible strings of cryptic codes.

These facts increased the complexity of the shop-floor work and negatively affected the work of the local manufacturing engineers. These unfortunate situations were gradually not caused by the components but

[23] Frequent rebalances, new product variants and other unavoidable matters are always required in assembly lines and assembly line work. Ensuring that the operators are fully occupied and that the product flow is smooth is also always a struggle. This situation aggravates further since the components and work tasks constantly need to be reformed, which is usually uncontrolled from the operators' perspectives.

rather by how operators on the shop floor (and the manufacturing engineers) treated them due to how the central product information was supplied in assumed coherence with the components.

Specifically, we are talking about **the central information system (the KDP)** ('Konstruktions Data Volvo Personvagnar') and the **central assembly process information** (the PKIs) (''Process/Kontroll-Instruktioner'). Such facts were discovered and became evident during our research and development work in our experimental workshop.

Consequently, for assembly lines and assembly line work at the automotive company discussed, it was impossible to associate the mutilated component designations that reach the shop floors together with the physical components. As was just said, however, it was far more than just the components that were difficult to grasp but this included understanding how the underlying administrative mechanisms functioned.

Later in the completed assembly plant, only appropriate and reformed central product information was allowed to reach the shop floor at the full-pace manufacturing phase. Picking components while filling the material kitting fixtures with proper components in Uddevalla's separate common material store was relatively easy. As discussed earlier, the assembly work also proved to be not particularly challenging in terms of the material structuring phase of the expanded assembly work.

Thus, the traditions and everyday routines that have dominated the automotive industry's various earlier initiatives made things difficult for assembly line operators, making the assembly work problematic. Such difficulties lay in finding meaning in unconnected information based on selections of incomprehensible strings of cryptic codes adapted to machines and mutilated component designations. When things are made difficult this way, the idea quickly arises that operators must be given limited work content to cope with the assembly work that has to be accommodated within a narrow time window created by short work cycles.

However, impediments created by the traditional assembly work itself become arguments for preserving short work cycles for any assembly work when the mechanistic approach to perceiving the product is formalised in central information systems and further consolidated by multiple administrative routines. In this case, such impediments will be aggravated and entrenched throughout the entire organisation.

Clearly, inappropriate information systems should be reformed or replaced. Computer technologies can become counterproductive, especially when traditions and everyday routines are followed.

3.4.3 The Organic Approach to Perceiving the Product on the Shop Floor

The organic approach to perceiving the product is based on suitable wholes. Such wholes contain phenomena that have various purposes, qualities and mutual relationships, and depend on the quality of such relationships. Purposes, qualities and mutual relationships determine whether elements can be organised into, e.g. discernible and hierarchically organised groups of components. That is, the components must be grouped according to choices of material, product functions and sub-functions thereof, positions in and on the automobile bodies, etc., as well as the most prominent component characteristics such as colour, fragility, shape and size.

Correctly carried out, the large, medium and small-sized components will unavoidably create new main groups and sub-groups of components, i.e. an assembly-oriented product structure. As suggested, recognising such prominent component characteristics and hierarchies is crucial for grasping suitable wholes.

Overview of the organic approach to perceiving the product

Transferring the idea of the organic approach to perceiving the product to assembly and material handling work in the non-traditional case discussed makes it evident that the basis (or point of view) must be the complete automobile, which forms a suitable whole. This thinking constitutes the value of discernible and hierarchical groups of components. However, such phenomena must be identified, i.e. grasping the always predictable product architecture and product variation in different ways was and is still the key here. (Incidentally, the authors mean grasping in both administrative and physical senses.)

In practice, this proved unavoidable and necessary for operators performing extensively long work cycles, and was also needed for the operators involved in material handling. The latter case refers to the operators in the separate common material store responsible for filling the material kitting fixtures with appropriate components. In their practical roles, the latter operators must also be professional at choosing precisely the correct components.

*

As stated previously, one crucial point for the reader to understand is that it is possible to define large automotive products, such as automobiles, buses and heavy truck chassis, as organised according to established vocational learning knowledge.

In short, this means that appropriately organised components are based on several prominent component characteristics that must be considered simultaneously. Consequently, components can be arranged into different categories (similar to families of individuals), which, in turn, can be arranged into even larger groups (identical to kinship groups of individual humans).

Components and component hierarchies, assembly and material handling work, which becomes outputs, will be much more straightforward because components are assigned identities if these elements are required to be assigned correct and meaningful designations. Memberships regarding families and kinship groups are thus far more transparent, and the components are more or less evident as they belong to more extensive and smaller groups of components.

Detailing the organic approach to perceiving the product

Such mechanisms as those explained above call for constructing groups of components; Engström and Nilsson practised this during later research and development projects within the Swedish automotive industry, i.e. after the experiences gained during the projection process in focus in this chapter.

As stated previously, such groups of components are synonymous with discernible and hierarchically organised groups of components, a more general concept or term used previously in this chapter. In practice, however, in the context discussed, the **assembly-oriented product structure** is also used to position these identical concepts or terms on the same analytical level as the central **design-oriented product structure**. The critical issue was that the central information system, the KDP, was annoying for the researchers' purposes and could not provide any suitable wholes as treated elsewhere. (See Appendix 2.)

Consequently, the assembly-oriented product structure is a concept or term with at least two synonyms just mentioned; even more synonyms exist, as will be touched upon later. In the non-traditional case discussed, such a product structure proved possible to construct for the automobile model assembled in Uddevalla, namely for the Volvo 740 automobile model. Such product structures can be constructed for any large automotive product assembled, but also for other automotive products for different analytical purposes, as Engström et al. (2000) reported, e.g. for clarifying the product variants for a large diesel engine at Scania in Södertälje.

The subgroups in an assembly-oriented product structure correspond to families, and these are, in turn, forming even larger groups of discernible and hierarchically organised groups of components, i.e. the main groups of components in an assembly-oriented product structure. (See Appendix 2 for more comprehensive explanations of the research and development work in our experimental workshop.)[24]

The automobile's path through any work process can be grasped as an organic growth pattern instead of a far more modest addition of components within a narrow time window from the point of view of assembly lines and assembly line work. Consequently, such assembly processes can be viewed as organic (biological) processes rather than isolated mechanistic additions.

As explained previously, discernible and hierarchically organised groups of components and the expanded assembly work represent essential qualities from the point of view of education and shop-floor learning. Such qualities vitalise operators' emotional and intellectual lives. In contrast, the mechanistic approach to perceiving the product that today dominates working life within the automotive industry most certainly limits the possibilities of further development since it is linguistically stunted and more numerically oriented.

3.4.4 Two Perspectives on Education and Shop-Floor Learning

Yet another way to explain some essential matters of the lines of reasoning that have been treated so far is to discuss and distinguish between two perspectives.

This was achieved by applying a **holistic perspective on education and shop-floor learning**, practised at the assembly plant until it was closed in the spring of 1993. At the same time, Japanese-inspired assembly lines and assembly line work are internationally accepted methods for assembling large automotive products. Such assembly systems represent

[24] An assembly-geographical atlas may be the best synonym or metaphor since such artifacts contain numerous interrelated geographical maps with different purposes, resolutions and scales. Most individuals can still interpret these maps. Moreover, some of these somewhat implicitly interconnected maps are optional. (The author challenges the reader to figure out or remember why this is the case and why it is essential.).

an antithesis coined as applying an **additive perspective on education and shop-floor learning**, which has been successful within the automotive industry worldwide. (However, this perspective emerged from another way of regarding humans, a fact which gradually became clear and prominent during the projection process.)

Today, in the 2020s, the Swedish labour market is different. Customer-oriented work time scheduling has been introduced, and temporary workers are allowed in ways that were not permissible or even possible, as mentioned briefly in a footnote (a new type of segmentation of the Swedish labour market was introduced).

*

Figure 3.3 illustrates the holistic perspective, showing the most prominent large-sized components required to assemble one-fourth of the assembly work carried out in and on the automobile bodies in the Volvo Uddevalla plant, i.e. the interior.

The last main group of components included in the assembly-oriented product structure used at this assembly plant dealt with carpets, left-hand and right-hand panels, seats, etc. If these components had instead been positioned on a shop floor according to an assembly sequence, the reader would certainly not have been able to comprehend them with such detailed overviews.

The authors challenge the reader to imagine that only a few minutes of a work cycle on an assembly line are available before the semi-complete automobile bodies are transferred to the subsequent workstation. At the same time, new semi-complete automobile bodies will unavoidably arrive some minutes later at the reader's imagined workstation.

Consequently, an important distinction and analogy, emanate between (1) interrelated geographical maps, similar to an assembly-geographical atlas, and assembly functional groups. These are two other synonyms for an assembly-oriented product structure composed of discernible and hierarchically organised groups of components, as highlighted above in this sub-chapter.

The image in Fig. 3.3 of one-fourth of the assembly work in and on the automobile bodies contrasts to (2) routes corresponding to assembly sequences, representing an additive perspective. These analogies (on one hand, the geographical maps and on the other hand, the routes) are embodied in the two perspectives discussed.[25]

[25] As mentioned, the realisation of various conceptions in time and space is sometimes denoted as phenomenography.

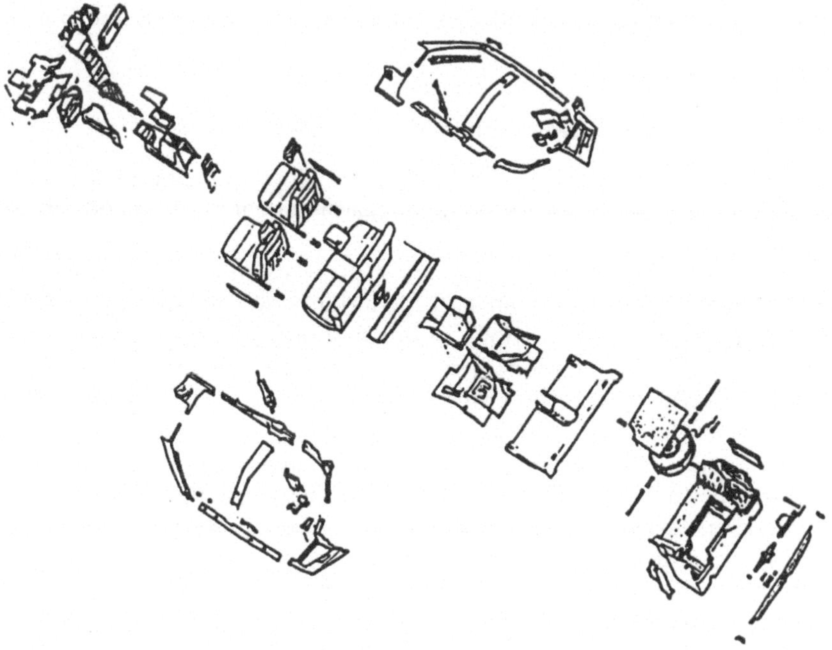

Fig. 3.3 Selection of the most prominent components needed for the assembly work of one-fourth of the work in and on the automobile bodies of the Volvo 740 automobile model. That is the fourth and last of the four main groups of the most prominent components comprised in the assembly-oriented product structure for this automobile model, the Interior. The illustration in the figure is an example of the hierarchical system of illustrations constructed in our experimental workshop in Gothenburg (*Source* Ellegård et al., 1992, an original illustration from our experimental workshop)

Notably, since the components illustrated in the figure are arranged to fit in and on the automobile bodies, it also becomes evident that the automobile features specific coordinates and symmetries, such as the fact that the components are fitted in pairs around the longitudinal centre line of the automobile bodies. Components whose positions deviate from this symmetry have an evident position, such as the steering wheel, battery and spare wheel. The components occurring in pairs are identical and can be confused since they fit at each position. Conversely, the components mentioned may also be impossible to fit at the wrong position. This symmetry resembles how animal or human bodies are configured out of components.

3.4.5 Orientation During Work Processes

All orientation is related to time and space. Therefore, the phenomena treated in this section must be grasped regarding their position and spread in time and space.

Orientation during work processes such as assembly work is defined in various ways, such as front-back, left–right, over-under, etc., similar to the geographical orientations of north–south, west–east, etc. In this context, as already mentioned, it is essential to emphasise that automobiles display specific coordinates and symmetries like many other biological and mechanical objects.

Such objects may feature a front and back, left and right side, etc., and a centre line that explicitly and implicitly defines a system of coordinates. (Remember the assembly-geographical atlas, which was synonymous with an assembly-oriented product structure or the concept of grouping components following the principles of discernible and hierarchically organised groups of components.)

Individuals' orientation while considering the perception of space dimensions

The specific coordinates and symmetries mentioned above considerably facilitate operators' perceptions of the always predictable product architecture and product variation. These specific coordinates and symmetries contribute to the creation of appropriate detailed overviews. Therefore, the position of phenomena in the space dimension is vital in orientation contexts.

As suggested, for operators, the relations of the components to the automobile bodies, including their cardinal points, principally give both explicit and implicit definitions in space. (Remember the specific coordinates and symmetries discussed previously and that each component required to be assigned correct and meaningful designations, sometimes including the attributes like those mentioned previously.)

*

As mentioned earlier, the material handling work carried out in the separate common material store in Uddevalla demanded extensive precision in orientation. Complete automobiles could not be observed on the shop floor during work in the separate common material store. They could only be observed at the end of the ordinary course of the work, i.e. while entering and leaving the ordinary work site in the morning and late afternoon.

It is crucial to sustain the idea that the phenomena treated by various appropriate and coherent descriptions involve several appropriately denoted elements (i.e. with correct and meaningful designations). It is not a question of constructing various descriptions based solely on unidentified elements.[26]

Phenomena that differ from other phenomena are essential to note and substantiate. In this context, reference can again be made explicit in several ways. The hierarchies of different groups of elements are defined by explicitly and implicitly emphasising suitable thought patterns.

In the case discussed, such hierarchies were formalised using advanced typography in the advanced assembly work instructions or by the hierarchical system of illustrations constructed in our experimental workshop in Gothenburg featuring colour codifications, i.e. each main group of components in the assembly-oriented product structure was assigned a specific colour, as well as by various three-dimensional models. (See the so-called intra-workgroup pattern described more exhaustively in another chapter of this anthology.)

Thus, the assembly-oriented product structure was constructed by moving around various-sized components from complete but disassembled automobiles on the shop floor of our experimental workshop until the hierarchies made sense from an assembly work point of view. The necessity to depict the components while constructing the hierarchical system of illustrations meant we had sufficient time to contemplate and discuss various underlying reasons for choosing a product structure; this was a somewhat intriguing and time-consuming task. (This research and development work was even more complex and cannot be detailed here for space reasons.)

[26] While discussing the local administration of central product information, it must be noted that for the non-traditional plant discussed, as well as for all other traditional local assembly plants, there was an extensive number of troublesome product change orders, i.e. frequent administrative and physical revisions of details (some real, others administrative). Such revisions are trademarks for Volvo and other Swedish automotive manufacturers; the exact reason for this is unclear to the authors.

*

It is somewhat difficult to elucidate how such phenomena were identified and constructed in practice and theory in this sub-chapter (the materials available in our experimental workshop are likely required as illustrations). As hinted before, these procedures comprised activities such as identifying various discernible and hierarchically organised groups of components.

When individuals have attained orientation in such contexts, various new appropriate and coherent descriptions of elements can be approached. The meaning is that aspects of phenomena build different patterns and relations among themselves, in which, in turn, various functions, relations, structures, etc., can gradually be discovered. Phenomena can then, in turn, be related to yet other phenomena. (Remember the value of attributes like left–right, over-under, front-back, etc.)

*

Alternatively, from another perspective, while specifically treating the automotive industry, it must be highlighted that the central product information from the central information systems expected to reach the local shop floor is already available beforehand at central operations, i.e. sometimes even during the design and pre-manufacturing phases of large automotive products.

This is before any assembly system is constructed, and thus before the assembly work on the shop floor has been initiated, as was the case for the projection process discussed. It was also naturally possible during the full-scale manufacturing phase. However, this fact needs to be recognised and thus utilised during any projecting process of an assembly plant. Some sort of temporary analysis database would then have been necessary to construct, as was achieved some years later during the reopening of the Volvo Uddevalla plant, which explains how one of the authors and his nearest research colleagues computerised the systematically organised disassembly procedures and printed out intriguingly constructed labels in connection with the projection of Autonova Automobile Plant (see Engström et al., 2000 for more details).

Fortunately, in both the Autonova and Uddevalla cases now discussed, novel local information system were used for these completed assembly plants. These information systems, among other things, used correct and meaningful component designations fetched from computer files discovered in the central Volvo Product Design Department, which proved possible to explore. These designations were often required to be complemented with appropriate attributes if they were to be used for assembly work instructions.

This means that other types of extra information were often needed to be added locally to, e.g. the unique information system in Uddevalla, such as attributes like the front-back, left–right and under-up information, as well as the choice of colour, material, shape, size, etc. As was essential, such additional information was also added locally in Uddevalla to create the required and appropriate discrimination on the shop floor while, e.g. fitting several different plugs simultaneously in and on the empty automobile bodies. Extensively long assembly sequences unavoidably require additional component information that is not always found in the central computer files mentioned previously.

To some extent, the terminology used centrally also determined whether particular components have more or less apparent relations with other components; in other words, if these were relevant for the assembly work, these usually had to be respected. However, in many cases, this is not possible or even relevant because such relations, while available, are only sometimes recognisable to the shop-floor operators.

*

For example, some cryptic codes dealing with the central codification of product variants must either be eliminated or revised for the operators. Extra cryptic codes, with no relevance for the assembly work, usually (1) concerned chassis settings, i.e. specific reciprocal interacting components created these codes, while others (2) were related to different markets. These are just two examples of such extra cryptic codes that obviously are useless for operators on the shop floor.

This is the case since operators on a shop floor can neither drive a large automotive product under various conditions nor master the knowledge of markets. Nevertheless, such experiences and insights are still essential for central operations, where the product variants are selected and appointed to be assembled at each local assembly plant. However, identifying and considering the appropriate selection of the relevant codes concerning their influences on the assembly work proved essential. This became possible due to our research and development work in our experimental workshop in Gothenburg.

For example, such knowledge was needed while constructing the paper printouts that followed each product variant during assembly work, i.e.

the 'product variant specification'. Consequently, both the blue- and white-collar employees in Uddevalla used the experiences and insights gained in our experimental workshop. (By the way, frequent visitors to our experimental workshop from the Volvo sphere were quite common, and this meant that particular topics of interest, like those exemplified above, were possible to identify and discuss, but such occasions were sometimes also encouraging opportuneties for us, i.e. new inputs were welcome.)

*

However, phenomena such as those treated above were almost impossible to grasp within the Volvo organisation especially on the shop floor during assembly line work, because the internal logic behind how product variants were codified had been lost totally. (As hinted, such matters were possible to use constructively, although treating such topics goes beyond the scope of this chapter and are elucidated better in Engström et al., 1995a, 1995b.)

*

As suggested several times, developing the experiences and insights advocated in this chapter has proven possible. The effects or results of such education and shop-floor learning principles are less subject to the influence of the ravages of forgetfulness in individuals' mental worlds.

Employing such experiences and insights presented for the non-traditional case discussed can yield more lasting knowledge; however, these insights are also more general. Although this may seem non-traditional or strange for many practitioners within the automotive industry, it is common elsewhere and, indeed, is expected in other parts of society.

For example, in automotive matters, the specific coordinates and symmetries mentioned above exist in any large automotive product, and they work in symbiosis with correct and meaningful component designations. In the non-traditional case discussed, the assembly work on the shop floor was reformed, but this was hampered by preexisting notions about assembly work. An imaginary vaccination or immunisation of all blue and white collar employees in the Uddevalla plant against preconceived notions of assembly work would have helped, provided that it had been efficient against negative opinions about non-traditional assembly work in society and in particular among employees of the Torslanda plant.

For example, during some periods, it was necessary to turn away presumptuous visitors to the Uddevalla plant from central operations to maintain internal calm.

*

As reported in another chapter of this anthology, the comparable, more cost-efficient facilitation of introducing new automobile models and product variants was a surprise (Engström, Jonsson & Medbo, 1996a) (see Appendix 2). However, such processes had partly already begun in the temporary education and shop-floor learning workshop, where no assembly work instructions or computer connections to central information systems were available.

Thereby, the operators were directed to use the fragmented PKIs, the sole assembly work instructions available and partly unsuitable for the projected assembly work. However, this contributed to creating self-confidence for many of them (see another chapter of this anthology where this is discussed in detail). Such knowledge is relatively stable for individuals. This more stable knowledge was crucial for several reasons and contributed to increased flexibility and reduced work and worry.

*

After humans have gained orientation, detailed perceptions successively emerge because phenomena are localised spatially. For this and other reasons, it is relevant for humans to penetrate more deeply into selected aspects of particular phenomena. Different depths in perceptions, in combination with the possibility of shifting mentally between diverse levels with various focuses, surely help.

As was previously addressed, all the mental mechanisms explained have similarities with the changes of scale between different interrelated geographical maps of the world, which contain explicit and implicit hierarchies. For example, major land masses are separated, creating continents with countries with borders, regions and towns. Hierarchies become explicitly and implicitly evident (if the maps are used repetitively in practice, they are unavoidably internalised in the human mind). (see Fig. 3.4).

An analogy that implies that operators performing assembly work will gradually gain experiences and insights into moving mentally between various perceptions of reality grasped as interrelated geographical maps

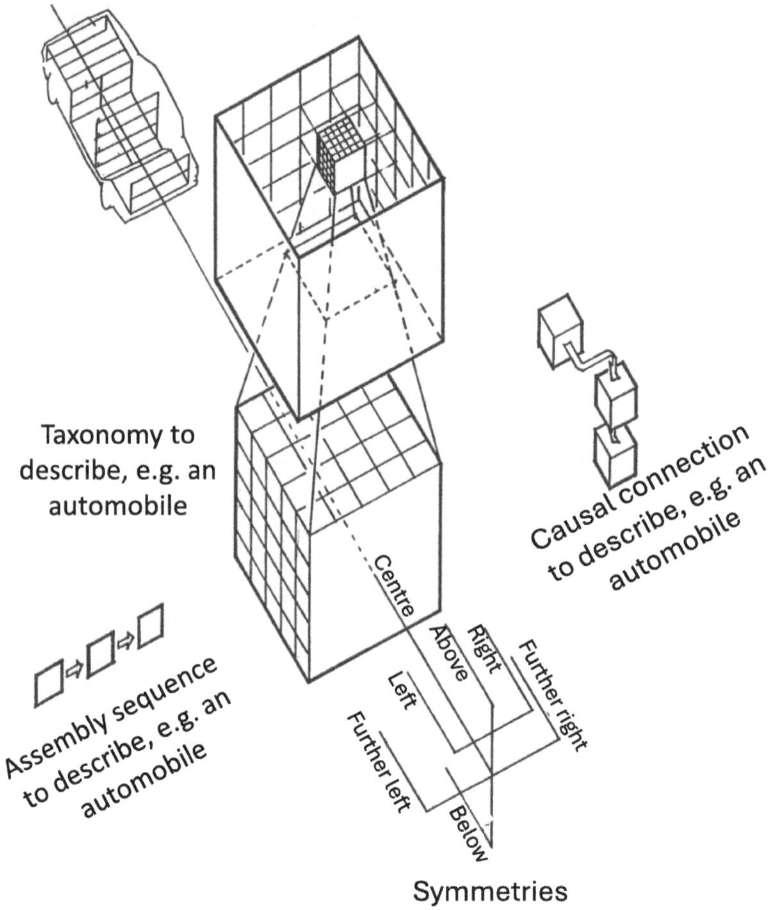

Fig. 3.4 Above is a schematic or symbolic illustration of how the product should be perceived according to various explanations and reasoning presented in this and earlier sub-chapters, namely as one short assembly sequence and as one specific relation (these are in the figure denoted 'sequence' and 'casual connection'). The illustration suggests that these phenomena derive from hierarchies of a partly more taxonomic nature, which define, for example, product architectures and variations of the automobile. Examples of the specific coordinates and symmetries discussed are also illustrated (remember the explicitly and implicitly defined specific coordinates and symmetries) (*Source* Ellegård et al., 1992)

has thus proved to be of great value. Fortunately, they are helped individually and collectively during assembly work. For example, in the non-traditional case discussed this was achieved by the configuration of material kitting fixtures. Even with these tools, communication between operators was still required, and the layouts were purposely constructed to promote face-to-face interaction.

Individual orientation while considering the perception of the time dimension

Another essential topic to explain is the time dimension. Clearly, individuals' perceptions of phenomena that appear before and after the present are necessary so that working life is not experienced as comprising many small moments of the present Thus, consideration, forethought and reflection assume that operators are also oriented in the dimension of time.

How individuals understand the time dimension is connected to (1) their activities' emotional and intellectual content during a given period. The experience of time is also (2) closely related to the experiences and insights of the purpose and meaningfulness of what they do. Individuals' work tasks must contain certain qualities that contribute to understanding the meaning from their point of view. The difficulties in observing lie in the fact (3) that individuals, for various reasons, have different time perspectives.

*

As described by Nilsson (1981a), some individuals have very short time perspectives, while others have substantially longer ones. Research suggests that individuals' time perspectives can be reformed if they perform work tasks, see a purpose and succeed. They must also experience themselves as successful, happy about this and appreciated by others who acknowledge it. The operators' time perspectives are thus expanded, and possibilities are created to see a purpose in things that lie further ahead. (Remember the expanded assembly work previously treated in connection with Fig. 3.2 in Sect. 3.3.4.)

Preparedness, realisation and understanding create farsightedness, which has advantages and qualities that only individuals can possess. The surrounding work must, therefore, be so constructed that such qualities are rewarded and possible for operators to attain. Appropriate shop-floor preconditions are thereby created so that they can acquire an expanded time perspective.

Conclusions regarding orientation in the space and time dimensions during work processes

To further summarise and develop what has been explained here and elsewhere, it became evident that the central and local information systems and other information communicated to the shop floor must be developed further for the non-traditional assembly work discussed. Such measures are required to help operators handle the complexity of their work and simultaneously experience success.

Thus, it is in the future required to continue to develop assembly work successfully beyond the experiences and insights reported in this chapter, i.e. to turn fragmented assembly line work into a skilled profession without the adverse effects exhaustively explained here due to short work cycles. However, other publications have yet to better explain how to deal with such effects.

Creating professional pride related to self-realisation is especially important, i.e. achieving orientation in space and time during work processes. (Another chapter of this anthology explains how this was possible by using an 'engineering' approach at the Volvo Uddevalla plant.)

However, since the closing of the Volvo Uddevalla plant in 1993, including the assembly docks in the Volvo Truck Company's main truck chassis plant in Tuve in 2022, the future could be brighter. Both these assembly systems were successful industrial full-scale initiatives that should form a platform for the future, not be brushed aside due to being closed down.

3.5 Experiences and Insights Regarding the Encounter Between Vocational Learning Principles and Some Engineering Matters

This sub-chapter summarises the encounter between some vocational learning principles and some engineering matters, as these research areas proved possible to define and relate in this chapter. The vocational learning aspects of education and shop-floor learning principles have been previously covered. However, some engineering matters remain to be better clarified. Consequently, two appendices also briefly describe some of the engineering matters required for the reader's understanding.

Summary of the Encounter Discussed

What vocational learning recognises as phenomenography refers to the assembly work discussed, which, in various ways, connects to assembly work derived from, e.g. an assembly-oriented product structure. Its realisation means that, in this particular context, operators involved in assembly and material handling work become owners of spatial and verbal networks connected to and integrated into their work. Such networks explain the relations between details and wholes in contexts involving varying depths, heights and breadths. Consequently, specific coordinates and symmetries appear.

Due to such spatial and verbal networks and several other measures and mechanisms already treated, it is still possible to handle both predictable and unpredictable aspects of the work, including various management interventions. These were often frustrating during the projection process discussed; as stated previously, the temporary education and shop-floor learning workshop was started featuring a cadre of selected blue-collar employees, and this workshop's operation was from the start embryonic. (Remember the interest group's purpose and the initial dilemmas for the projection process organisation mentioned previously.)

At the same time, the completed assembly plant was still being projected, i.e. unfinished. The work cycles were extended step-by-step, i.e. the temporary education and shop-floor learning workshop was an integrated part of the projection process. The assembly plant underwent the starting-up, running-in and full-scale manufacturing phases before it was closed down with at least one critical unresolved issues (a far too rigid and traditional manufacturing scheduling system treated in another chapter of this anthology). The projection process was, undoubtedly, demanding for all categories of employees. (Among other things, work cycles were extended as experiences and insights gradually were gained, but annoyingly, an appropriate information system required to be accessible from the beginning.

*

As suggested above, the education and shop-floor learning principles discussed meant that, in the completed assembly plant, central product information was supplied to the local manufacturing engineers both by gradually employing the central information system, the KDP, and somewhat by a vast number of paper printouts of the PKIs.

That is, there was an intrinsic value in the KDP defining the product architecture and product variation and the fragmented PKIs that proved possible to interpret directly on the shop floor in the temporary education and shop-floor learning workshop since, as already hinted, these documents expressed the assembly work as defined in detail from the central Volvo Manufacturing Process Engineering Department, even if this was not the original intention.

As stated elsewhere, the Volvo Uddevalla plant's fully educated and learned operators gradually became accustomed to the fragmented PKIs, which awkwardly defined the assembly work.

Nonetheless, they could have benefited from being more familiar with the original assembly line practices and routines, which, for assembly line work, is almost solely handled by the local manufacturing engineers. This context could have helped the operators to understand the PKIs better.

Thus, nuances and precision in the world of imagination and reality have a more profound meaning, even in the case of the fragmented PKIs. A richly nuanced and varied reality demands diverse perceptions, which call for appropriate and coherent descriptions. Such approaches also provide an opportunity to formalise individuals' experiences and insights successively and, in addition, consider the knowledge established earlier. (Remember the more stable knowledge mentioned earlier, while discussing the human being's perception of the space dimensions.)

In other words, the components in the assembled automobiles initially dictated how assembly work was to be carried out. However, various appropriate and coherent descriptions suddenly or gradually proved possible to be transformed into new models or representations, as has already been touched upon.

*

One conclusion drawn from the authors' and others' analyses is that right up until the results of the crucial aspects generated from the projection process became apparent for the practitioners involved, the non-traditional assembly systems were unthinkable due to the choice of serial product flows in existing assembly plants and the mechanistic approach to perceiving the product combined with the existing management of automotive companies. Such questionable restrictions generated most of the within-the-box thinking.

Although Volvo's interest in fair evaluations and more public communication regarding the non-traditional case discussed could have been more significant in most respects, such initiatives were partly allowed to be carried through due to the decision to close down. One of the authors and his closest research colleagues from Chalmers University of Technology carried out a cross- and multidisciplinary initiative to understand the assembly work in connection with the closing down period. (See some of the other chapters of this anthology or Engström et al. 1995a, 1995b.)

Conclusions and future initiatives, i.e. suggestions for the future

To conclude, assembly work possesses a generic logic provided that the product is defined with the always predictable product architecture and product variation as a starting point, determined from an assembly work point of view. This feat was possible since the central product information defined and supplied by central information systems could be transformed and edited locally in even more general senses in line with the methods developed during the reopening of the Volvo Uddevalla plant (see Engström, Jonsson and Medbo, 2000 for further details). Such achievements prove feasible since the automobiles were still described in detail during the product design and pre-manufacturing phases.

Such exploitation of the facts in the form of constructing assembly-oriented product structures still provides particular critical possibilities. However, this chapter could not treat the exact and practical procedures required to construct such product structures sufficiently; this was impossible for space reasons. Nevertheless, the approach and methods discussed were crucial for the assembly plant in Uddevalla, and constitute an undeniable starting point for further development of assembly and material handling work in line with socio-technical research practices in the future.[27]

*

To repeat, but from a slightly different angle, this means that products with similar assembly work should be identical in the perspectives

[27] In the case discussed, it was not a matter of solely employing healthy young males for assembly line work and simultaneously using other categories of individuals in different operations with other types of employment contracts, where for example, subcontractors with less favourable benefits carry out assembly work in the subassembly areas.

However, some years after the Volvo Uddevalla plant closed down, the legal preconditions for the Swedish labour market were changed. Customer-oriented work time scheduling and temporary workers were then introduced, providing another type of flexibility. At the same time, other options for employers became available with a more segregated workforce, making other mechnisms available to discipline employees in the absence of win-win employment relations.

of assembly and material handling work. From both perspectives, it proved possible to reform the central product information, as detailed elsewhere.[28] This allowed for the establishment of spatial and verbal networks, a vital potential hidden in the product design and pre-manufacturing phases. (By the way, the costs of the manual work to assemble large automotive products is marginal in relation to the costs of product design and pre-manufacturing activities, i.e. enhanced cooperation between these operations and the assembly operation along the lines sketched here and in another chapter has a great potential for the future.)

Detailed knowledge of the two different approaches to perceiving the product makes it possible to organise the work in the desired way. The mechanistic and organic approaches must be mastered, or at least their existence should be recognised. Consequently, the automotive industries Japan and Sweden or indeed any open-minded automotive company must realise that it is possible to construct assembly systems that combine efficiency and humanisation in the automotive industry, which may or may not lead to future initiatives for new socio-technical research practices.

Seeking alternatives to assembly lines and assembly line work was suddenly no longer necessary. Moreover, the recession in the automotive industryforced the termination of the smaller Volvo assembly plants in Kalmar and Uddevalla. Consequently, the large and traditional assembly plant in Torslanda, nowadays with just one with an assembly line was the sole survivor.

The giant Japanese automotive companies and their internationally reported successes were also crucial sources of inspiration for the Swedish automotive industry. (Today, at the Volvo Torslanda automobile plant, no knowledgeable individuals remain who were deeply involved in the experiences reported in this chapter. In the best case, these individuals are retired, but most are deceased. Only negative rumours and assumptions exist today.)

[28] In the main text, the authors explain how an assembly-oriented product structure was constructed based on discernible and hierarchically organised groups of components. This construction and the following research and development work has primarily nothing to do with functions or product functions, claims to the contrary notwithstanding. (It is impossible to divide the large automotive products considered evenly on the shop floor according to product functions; besides, this criterion is unstable and unsuitable for constructing an assembly-oriented product structure). However, the product variants had to be re-codified into levels according to generativity, which to some extent refers to product functions, as treated elsewhere (see Engström et al., 1995a, 1995b, which explains these matters in depth).

Appendix 1: The Assembly Work in the Small Parallel Workgroups at the Volvo Uddevalla Plant

Briefly summarised, it was a matter of organising a generative and progressive education and shop-floor learning processes on authentic products carried out during comparatively long periods for the newly recruited operators. As a result of the Volvo Uddevalla projection process, the operators in the completed assembly plant had to master the competencies of approximately one-fourth of an automobile as a platform to become an accepted member of the workgroup.

As mentioned, it was also important to master skills that complemented the minimum assembly competencies. For some operators, these education and shop-floor learning processes seemed almost endless, far past the minimum required to gain sufficient self-confidence and the ability to communicate and help co-operators.

Such accomplishments were feasible because the operators were educated and the shop floor work learned by a division of labour consisting of clusters of components (so-called work modules) following how their own and their co-operators' actions were related to the requirements for the automobile bodies to be shifted around during the assembly work. The manually handled tilting devices reoriented the automobile bodies by rotation and height adjustment. In some workgroups, lifting tables were used to adjust the height of the automobile bodies.

The assembly work was organised in pairs of operators and consistent with how the material kitting fixtures were configured and related to the product architecture and product variation from an assembly work point of view.

Moreover, the workstations were intentionally arranged to promote readings of facial expressions. Three or four automobiles were always accessible in various stages of completeness; the number of automobile bodies depended on the layouts. These automobiles served as physical assembly instructions that defined the preceding and subsequent assembly work. Combined with peripheral vision, the sound of the nearest co-operator's tools and other manufacturing tools, provided split-second information on the ongoing situation. (A division of labour involved so-called work modules that, in turn, formed a so-called intra-workgroup pattern, as is explained in another chapter of this anthology.)

Each reorientation of the automobile bodies called attention to particular large-sized components and plastic containers, which were evident by the configuration of the material kitting fixtures. These components were identified step-wise, as they were harboured on shelves, hooks and fittings. However, these reorientations also called attention to particular plastic containers containing specific medium-sized components and small transparent plastic bags with small-sized ones.

As stated previously, the more fully educated and shop-floor-experienced operators could, for these and other reasons, gradually expand their knowledge beyond the minimum by successively carrying out their co-operators' assembly work (albeit only sometimes at full work pace). The assembly work was organised in pairs of operators and clusters of components. Some of these clusters were collective (i.e. the first operator who arrived at such a cluster started to work until their co-operators caught up and joined).

Generatively and progressively, this assembly work promoted an understanding of the product architecture and product variation. This feat was made achievable due to the correct and meaningful component designations available in the advanced assembly work instructions; there, it was possible to check these assembly work instructions if an operator was unsure or suspicious for one reason or another (however, the configuration of the material kit was the main source of information).

The assembly work instructions defined an extensively long and unique assembly sequence for each product variant, using advanced typography and several other features, as is more exhaustively treated in another chapter of this anthology.

The automobiles' various product functions and other characteristics gradually became apparent from another angle of approach rather than solely by configuring the material kitting fixtures. Moreover, communication with the local manufacturing engineers, including prospecting the central product information, was necessary. This feat was achieved by prospecting the central information system, the KDP, and the fragmented PKIs, with the latter supplied as paper printouts (these two streams of product information were usually used by the local manufacturing engineers, e.g. for the construction of the local assembly work instructions suited for each particular choice of assembly system).

Appendix 2: Example of the Research and Development Work Carried Out in Our Experimental Workshop in Gothenburg

Several crucial non-traditional features were constructed for the assembly plant in Uddevalla, some of which emerged from our experimental workshop in Gothenburg, while others were initiatives of others (consultants and Volvo personnel). This workshop started approximately half a year after the projection process was initiated, and this operation was gradually integrated into the Volvo Uddevalla plant projection process. It continued until the end of the non-traditional assembly plant's short life.

This workshop was filled with automotive literature and all accessible and relevant Volvo and automotive documents (customised computer printouts, manufacturing engineering documents, service and repair manuals, spare part catalogues, etc.). Additionally, scraped prototypes of equipment roughly put together for trying various ideas, equipment prototyped for feeding small-sized components in small transparent plastic bags, material kitting fixtures, wagons for tools prototyped, etc., were available.

At the start of the projection process, two parallel product flow assembly workstations were preliminarily prototyped to establish several critical measures for the so-called double docks, as a necessary input for the early stages of the projection process. Similar preliminary initiatives were launched to evaluate the possibilities of tilting the automobile body during assembly work (at this early period, a controversial option, questioned by specialists at the Volvo Torslanda automobile plant).

The most noteworthy initiative in this workshop was protracted procedures to systematically disassemble several complete automobiles and position the removed components on the shop floor. These components and the systematically organised disassembly procedures achieved at least three purposes, as explained below.

The research and development work in our experimental workshop in Gothenburg

The systematically organised disassembly procedures were necessary (purpose no. 1) to construct the assembly-oriented product structure for the automobile model assembled in the Volvo Uddevalla plant. As stated in the main text, this product structure was required locally in Uddevalla to reform the central product information that, in one way or another, defined the always predictable product architecture product variation from the central Volvo Product Design Department to an assembly-oriented product structure, which, in turn, gave birth to a division of labour for small parallel workgroups. (The division of labour, i.e. the so-called intra-workgroup pattern, is a concept or term used in another chapter.)

In addition, the removed components (purpose no. 2) were depicted as we constructed the hierarchical system of illustrations of the assembly-oriented product structure, with illustrations that gradually became more aggregated at the top. At the same time, the fragmented PKIs were found at the bottom of this hierarchy.

These disassembly procedures (purpose no. 3) also provided an opportunity to grasp the product architecture and variation together with the central product information in ways not encountered earlier in the Swedish automotive industry.

Research and development work was carried out by prospecting completed and disassembled automobiles and gradually developing the systematically organised disassembly procedures. As mentioned earlier, these initiatives were supported by the automotive literature and all possible Volvo documents. Such cross-referenced procedures helped in understanding the central information.

These procedures were also substantiated by various types of expertise, popular automotive literature and selected expertise from the Volvo Torslanda automobile plant. In some cases, it proved necessary to compile several collections of documents, organised alphabetically. There was, for example, approximately 80 pages of compiled information on the different components and product systems for the Volvo 740 automobile model.

The implied activities were carried out, forming a base of knowledge that enhanced understanding. Numerous anomalies and discrepancies regarding the central information as well as information from other sources became apparent, many of which could not be detected in any other way. (As stated in the main text, removing unidentified components from the new and shiny automobiles was not an option for us.)

*

Another essential initiative was the start of the temporary education and shop-floor learning workshop in Uddevalla in the spring of 1986 before the first parallel assembly workshop was started up in the completed assembly plant when the buildings were completed and the gradual starting up of five of the six parallel assembly workshops began.

Detailing the research and development work further

This subsection further details what has been explained while paraphrasing the main text's explanations, reasoning and topics. The experiences and insights received from the systematically organised disassembly procedures discussed were, as stated, utilised to grasp the central information system, the KDP, which communicated the central design-oriented product structure. Naturally, this was also a way to understand the fragmented PKIs.

The central design-oriented product structure was and is still based on what is known as the Function Group Register, which results in a product structure not suited for assembly work since it emerged from the Volvo Automobile Company's interpretation of this register. This register is a compulsory standard for automotive vehicles in the Swedish automotive industry. Nevertheless, this register generated several unfortunate effects for the always predictable product architecture and product variation. This was most annoying for the researchers because the information based on this register could not provide suitable wholes since it is based on an obsolete vehicle. (See another anthology chapter for more details.)

As stated previously, the senior researcher within the internal material handling matters and his nearest research colleagues working in our experimental workshop were required to gradually comprehend how central product information was defined and supplied by various central operations and then provided to multiple local operations internationally.

Thus, it became evident that various unfortunate effects were created because of how the central product information was treated before it reached the shop floor in any local assembly line plant, an operation that, by necessity, comprised local selections of the correct central product information valid for the product variants assembled (excessive central product information was always applied to the local assembly plants).

These experiences and insights were initially unknown to most participants in the projection process, as building a new assembly plant in Sweden was rare. Still, these intricate effects were gradually recognised as the projection process progressively approached a step-wise completed, non-traditional assembly plant.

*

The already mentioned hierarchical system of illustrations is exemplified in Fig. 3.3 in the main text. The figure illustrates the locations of an automobile's most prominent components as these were fitted in and on the complete automobile. The large-sized components described represent one-fourth of the automobile model assembled in Uddevalla, i.e. the last fourth of the main groups of components of the assembly-oriented product structure. Note that the automobile bodies are invisible in this figure.

As suggested, the complete illustration system is even more intriguing. Among other things, on the less aggregated levels, the components are also illustrated as positioned in and on the automobile bodies but exploded outside imaginary automobile bodies. See also Fig. 3.5 below, which shows an illustration from a similar illustration system developed for heavy truck chassis. (See Engström, Hedin and Medbo, 1992, which instead presents the hierarchical system of illustrations for the Volvo 740 automobile model.)

This system of hierarchical illustrations was one example of methods used by the senior researcher within internal material handling and his closest research colleagues, which inevitably was one of several other initiatives needed for our analyses and communication with the practitioners involved.

Briefly, it was a matter of preventing the mechanistic perception of the product already established from dominating the projection process by spreading these hierarchical illustrations. We wanted to 'prevent some of the participants from regressing to their traditional experiences and knowledge'.

Note also that a crucial and later falsified assumption permeating the Swedish automotive industry was the enormous number of different product variants. This hypothetical postulate was an assumed dilemma

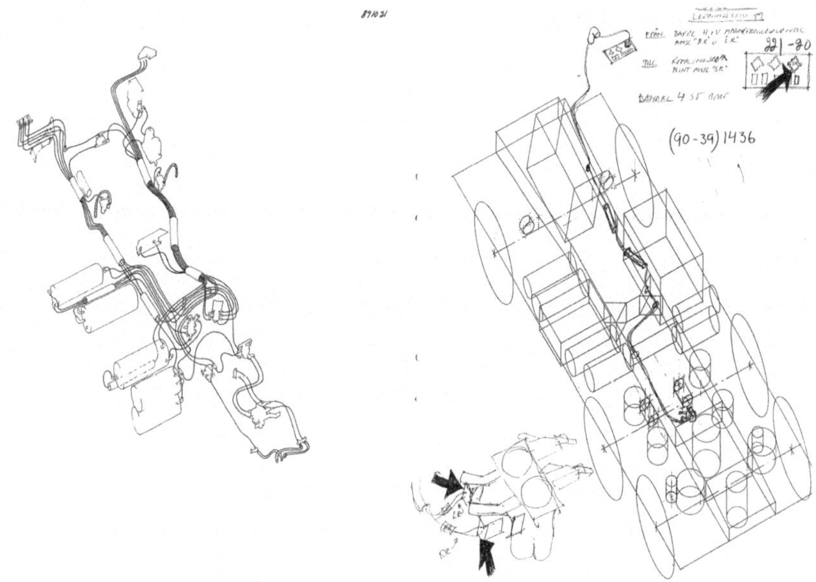

Fig. 3.5 This is an example of how the hierarchical illustration systems were constructed in our experimental workshop in Gothenburg. In the case of the figure above, the components were successfully removed from a heavy truck chassis and documented. At the same time, the outline was used to document their original positions and other purposes, as was the case during similar research and development work for the Volvo Truck Company that followed the Volvo Uddevalla plant projection process by approximately two years (*Source* Original illustrations from the experimental workshop)

that was cleared out in a hands-on manner due to the research and development work in our experimental workshop. (See Engström et al., 1995a, 1995b, which explains this matter more carefully.)

Examples of problem areas treated and comments on the hierarchical system of illustrations

The research and development work in our experimental workshop, substantiated by the experiences and insights gained in the temporary education and shop-floor learning workshop, gradually brought forward several crucial matters. This was essential since the central information systems had destroyed the understanding of product architecture and product variation of large automotive products assembled by the Swedish automotive industry.

*

The need for the hierarchical illustrations explained above was dictated by the fact that no appropriate illustrations were available within the Volvo Automobile Company, i.e. such illustrations needed to be constructed. These illustrations were required for long or extensively long work cycle assembly work. Consequently, such hierarchical illustrations were not required earlier or likely to be constructed for several reasons. (The authors challenge the reader to figure out or remember precisely why.)

The illustrations included in the fragmented PKIs delivered from the central Volvo Manufacturing Process Engineering department as paper printouts, had to be far more cohesive and normalised; moreover, they needed to be less sketchy. On the other hand, the KDP from the central Volvo product design department was available in digitalised form as communicated through this central information system and had its own peculiarities.

The local assembly plants had to cross-reference the PKIs with the KDP. Computer terminals were accessed in each local assembly plant. Additionally, the PKIs were numerous and physical and delivered as paper printouts that required 15 binders of paper printouts in A4 format that had to be accessible and continuously updated.

Each of these paper printouts covered approximately half a minute of assembly work, which usually specified one to four components. According to the central codification of product variants, these components were addressed to a large number of different product variants expressed in cryptic codes, i.e. an immense and confusing mass of fragmented information.

*

Following this line of thought, this automobile company's assembly lines and assembly line work demanded extensive administrative efforts. These and similar facts became apparent for the researchers working in our experimental workshop in Gothenburg, who had to make sense of a gigantic jigsaw puzzle of their own making, i.e. one consisting of paper printouts of the fragmented PKIs and the removed components from the disassembled automobiles. The temporary education and shop-floor learning workshop in Uddevalla also had to struggle with these paper printouts since, as stated previously, no computer connection was available, nor were there any assembly work instructions.

Effects gained during the period of closing down the Volvo Uddevalla plant

Even though a far too rigid and traditional manufacturing scheduling system, not fit for purpose, hampered both the planned function and the performance of the assembly plant in Uddevalla, its performance surpassed in several respects that of the assembly line in Torslanda for the same automobile model. However, how much upper management understood this and several other intriguing and underlying mechanisms is still an open question. (See another chapter of this anthology where such matters are elaborated.)

Direct observations based on the researchers' video recordings of nine different product variants revealed 2–4 fewer hours of assembly time than what was required for the same Volvo 740 automobile model at the Volvo Torslanda automobile plant. According to the automotive company's investigations, these automobiles were of the highest quality in the Volvo Automobile Company (Engström et al. 1995a, 1996b for further details).

*

Notably, the Volvo Uddevalla plant was the most cost-efficient Volvo assembly plant in introducing new product variants and making yearly model changes at the end of its short life in the spring of 1993. These achievements surprised most of the managers in Gothenburg.

This was one of several noteworthy positive effects gained due to reasons not fully explained or sufficiently treated in this chapter. It should be added to the more expected merits of a non-traditional assembly system, which proved to be possible to realise for full-scale manufacturing purposes.

Furthermore, surplus competencies were also created. The extensive competencies mastered by operators during the assembly plant's final year were, undoubtedly, a contributing factor to the administrative efficiency achieved since the traditional border between blue- and white-collar employees' work was dissolved and because both categories of employees drew on the same sources of information.

References

** = The publications marked with an asterisk (*) are not explicitly mentioned in the main text and the appendixes but are still related to the topics treated.*

Ellegård, K. (1989). *Akrobatik i tidens väv. En dokumentation av projekteringen av Volvos bilfabrik i Uddevalla*. Kulturgeografiska Institutionen, Handelshögskolan vid Göteborgs Universitet. CHOROS 1989:2 Göteborg (in Swedish language).

Engström, T., & Karlsson, U. (1982). *Alternativ montering*. Institute for Management of Innovation and Technology, Göteborg (report in Swedish language).

Engström, T., Hedin, H., & Medbo, L. (1992). *Design Analysis by Means of Axonometric Hand-drawn Illustrations*. Proceedings of International Product Development Management Conference on New Approaches to Development and Engineering, the European Institute for Advanced Studies in Management (EIASM), Brussels, pp. 147–157 (conference paper).

Ellegård. K., Engström, T., & Nilsson, L. (1992). *Reforming industrial work—Principles and realities in the planning of Volvo's car assembly plant in Uddevalla*. Arbetsmiljöfonden. Kommentus Förlag AB, Stockholm.

Engström, T. (1983). *Materialflödessystem och serieproduktion*. Institutionen för Transportteknik. Chalmers Tekniska Högskola, Göteborg (PhD thesis in Swedish language).

Engström, T., Jonsson, D., & Medbo, L. (1996a). *The Volvo Uddevalla plant: Production principles, work organization, human resources and performance aspects: Some results from a decade's efforts towards reformation of assembly work*" Department of Transportation and Logistics, Chalmers University of Technology, Gothenburg (report on Work Environment Found Projects Dnr 93-0217 and 94-0516).

Engström, T., Jonsson, D., & Medbo, L. (2000). The method of successive assembly system design: Six case studies within the Swedish automotive industry. In A. Gunarsekaran (Ed.), *Agile manufacturing: 21st century manufacturing strategy* (pp. 621–642), Elsevier Science Publishers.

Engström, T., Johansson, J. Å., Jonsson, D., & Medbo, L. (1995a). Empirical evaluation of the reformed assembly work at the Volvo Uddevalla plant – Psychosocial effects and performance aspects. *International Journal of Industrial Ergonomics., 16*(4–6), 293–308.

Engström, T., Jonsson, D., & Medbo, L. (1996b). Production model discourse and experiences from the Swedish automotive industry. *International Journal of Operations and Production Management., 16*(2), 143–161.

Engström, T., Jonsson, D., Medbo, L., & Medbo, P. (1995b). Inter-relation between product variant codification and assembly work for flexible manufacturing in autonomous workgroups. *Journal of Material Processing Technology, 52*, 133–140.
Karlsson, U. (1979). *Alternativa produktionssystem till lineproduktion*. Sociologiska Institutionen. Göteborgs Universitet. Göteborg (PhD thesis in Swedish language).
Nilsson, L. (1981a). *Yrkesutbildning i nutidshistoriskt Perspektiv. Yrkesutbildningens utveckling från skråväsendets upphörande 1846 till 1980-talet samt tankar om framtida inriktning*. Gothenburg Studies in Educational Sciences 39, Göteborg (Ph.D. thesis in Swedish language).
Nilsson, L. (1981b). *Yrkesutbildning i ljuset av produktionslivets reflexer*. Liber, Utbildningsforskning FoU, Rapport 42, ISBN01–40–70607–9 (report in Swedish language).
Nilsson, L. (1985a). *Utbildning för arbetslivet och lärande på arbetsplatsen*. Institutionen för Pedagogik, Göteborgs Universitet, Göteborg (report in Swedish language).*
Nilsson, L. (1985b). *Naturligt arbete och industriellt lärande*. Institutionen för Pedagogik. Göteborgs Universitet, Göteborg (report in Swedish language).
Nilsson, L. (1986). *Utbildningsstrategi för blivande bilbyggare: Uddevallaprojektet*. Göteborgs Universitet, Göteborg (mimeo in Swedish language).
Nilsson, L. (1989). *Förnyelse av industriellt lärande: Exemplet Vovlo Uddevallaverken*. Rapport inom projektet Dokumentattion av utvecklingsprocessen för Volvos Uddevallafabrik. Institutionen för Pedagogik. Göteborgs Universitet, Göteborg (report in Swedish language).*
Nilsson, L. (1992). "Lärandeprinciper som främjar långcykligt arbete" and "Förändrimg av lärande under 1900-talet." *Reflektiv produktion. Industriell verksamhet i förändring* (kapitel 2 och 4). Volvo Media Volvo. AB Volvo (in Swedish language).*
Nilsson, L. (1995). The Uddevcalla plant: Why did it succeed with a holistic approach and why did it come to an end? In Å. Sandberg (Ed.), *Enriching Production. Perspectives on Volvo's Uddevalla plant as an alternative to lean production* (pp. 75–86). Avebury Aldershot, Brookfield USA.*
Nilsson, L. (1998). *Vocational didactics; Results from research and practical applications*. Institutionen för Pedagogik, Göteborgs Universitet, Göteborg (report in Swedish language).*
Nilsson, L. (2000). *Strategies in order to support different kinds of production concepts for assembly work*. Osaka University, Faculty of Information Management, 22–25 November (conference paper).*
Nilsson, L. (2004). *Flexibility, control and learning strategies in the frame of different production systems. Conclusions from 20 years of participation in*

research and renewal of production systems in the car industry. The 12th GERPISA International Colloquium, Paris (conference paper).*

Wild, R. (1975). On the selection of mass production systems. *International Journal of Production Research.*, *13*(5), 443–461.

CHAPTER 4

Characteristics of the New Production Concept in Sweden Based on the Uddevalla Plant: Cognition and Action

Yutaka Tamura

4.1 Introduction: Aims and Issues

This paper discusses the Uddevalla Production System (UPS), a result of production reforms carried out in Sweden from the 1960s to the 1990s. This examination clarifies the position of UPS in the evolutionary process of production systems, with a particular focus on its relationship with Japan's representative production system, the Toyota Production System (TPS) and its ideal model, the Lean Production System (LPS). To conduct this examination, this paper focuses on the 'interface' between work design and execution. This direction of analysis is intended to clarify the relationship between cognition and production action, which has not been sufficiently demonstrated in previous UPS studies. The paper pays special attention to applying cognitive functions to production actions, as shown by UPS, which is presumed to have common ground with the

Y. Tamura (✉)
Faculty of Business Administration, Aichi Toho University, Nagoya, Japan
e-mail: tamura.yutaka@aichi-toho.ac.jp

H. Nohara and L. Medbo (eds.), *Technology Convergence and System Divergence*, https://doi.org/10.1007/978-981-96-1910-8_4

directions of TPS and LPS. Maintaining the comparison between UPS and TPS, we examine the commonalities and differences which UPS, TPS and LPS show regarding the use of cognition. Moreover, we clarify the significance of UPS's cognitive utilisation for future production actions.

This paper specifically examines three areas of the UPS. First, from the Socio Technical System (STS) perspective of the interrelationship between technology and society, how do we position UPS's achievements in Sweden? Second, what are the characteristics of the unique work design method adopted at the Uddevalla plant, known as UPS? Third, what similarities and differences can be observed with the Lean Production System? Let us examine the historical context of the STS job re-design in Sweden, which led to the establishment of the Kalmar and Uddevalla plants.

4.2 STS Development in Sweden

4.2.1 Why Is UPS Necessary?: STS Experiences in Sweden; The 'New Factories' Concept Proposal, SAF Engineering Department

Sweden has historically experienced a constant strain on the labour supply. As a result, absenteeism in production sites oriented towards mass production using conveyor systems faced difficulties, where criticism of conveyor production became a social issue. For example, in the late 1960s, workplaces in Volvo, a car company epitomising mass production, became dominated by immigrants as Swedish and Finnish workers became absent (Sandberg, 1995, pp. 88–89). By 1969, the turnover rate had increased to 52% and high long-term absenteeism was perceived as a social problem (Gyllenhammar, 1977, p. 53). Consequently, in 1974, Volvo initiated operations at its Kalmar plant, which had abolished line production, drawing global attention as an example of job re-design. The distinctive feature in Sweden was that job re-design was carried out using STS approach. The development of work re-design began in the early 1960s, influenced by organisational reforms in Norway. In Norway and Sweden, STS theory influenced work re-design, where SAF (Svenska Arbetsgivareföreningen: The Swedish Employers' Confederation, now Svenskt Näringsliv) considered experiences in Norway with interesting reform theory.

SAF established the Tekniska avdelningen (Technical Department) in 1966 to begin work re-design based on STS. In 1969, they established a new project, called the 'New Factories' project by which SAF intended to seriously undergo technical and organisational reform. The projects absorbed several progressive practical results and examined experiences at the plant level. At that time, project members Stefan Agurén and Jan Edgren indicated that their mission was developing technical reform to react against an autonomous work model. They had negative feelings towards autonomous work coming from Norway. The political attention regarding autonomous work in the Technical Department began at the start, they suggested. Why did the Department have such negative feelings? They mentioned that the wariness of autonomous work groups was an issue since the beginning of the Technical Department and that 'the work groups' intention to disengage from management authority' in Norway was a significant issue.[1]

As explained by Agurén and Edgren, the most distinguished project features aimed to solve 'production problems'. Based on this purpose, the project intended to expand technical flexibility. To obtain targets, the project had introduced short line assembly as being a 'plant in small plant', to realise cumulative production as 'production unit formation by product' while addressing the expansion of product variants as 'parallel flow'. From a management perspective, introducing autonomous organisation limited by management acceptance could retreat hesitating work at assembly production and reduce heavy-duty work since the 1960s. At the same time, management began discussing co-determination decision-making with unions. The union also responded to demands for a labour management's 'joint decision', which had been a significant demand issue on the labour union side. The 'New Factories' was undertaken as an empirical study by the SAF's Technical Department, though it was insufficient from the trade union side based on Landsorganisationen i Sverige: LO's perspective (Ishihara, 2003, p. 54). The project was a reform primarily focused on the technical areas, considering labour-management trends.

4.2.2 *Achievements and Challenges of the 'New Factories': Examples from the Kalmar Factory*

This section summarises key features of the Kalmar plant to compare with UPS. First, regarding technological aspects, the Kalmar plant abolished

[1] Interviewed Stefan Agurén and Jan Edgren in September 2012, in Stockholm.

the conveyor system and introduced AGVs (Automatically Guided Vehicles) on the line. By using AGVs, each body could be controlled to stop and move. Operators could also control the moving speed. Furthermore, installing buffers at the entrance of and between work process positions formed a layout structure where the assembly progress was restricted for balancing to adjust assembly operational speed. Then, by using the docking area removed from main line, the synchronisation between the main line and dock would become weak. Thus, by introducing AGVs, the buffers and the docking area at the Kalmar plant could be used simultaneously. The plant could hold high technical regulation capability against the synchronised movement of Ford line production.

The second key feature is that four or five stations were allocated to each work group. In a dock where each part of the assembly was performed, the work cycle became a maximum of 20 minutes, including the work responsibilities and work procedure of each group. In comparison, the work cycle at Torslanda Plant, Volvo's main plant, was 1 minute based on the traditional Ford assembly line. Therefore, the 20-minute work cycle at the Kalmar plant indicated a radical time extension. At the Kalmar plant, each group had a training responsibility for new employees; groups supported learning specifically through on-the-job training. Additionally, groups were authorised to select team leaders themselves. The organisational group's common intention was supported by social conditions (Akaoka, 1989, Chap. 6). The Kalmar plant thus attracted international attention as a plant that embodied the parallel concept of being a 'new factories', exemplifying representing group work with AGVs and dock assembly. Ultimately, the plant became known for eliminating the conveyor belt.

After 1977 and moving into the 1980s, however, the Kalmar plant faced challenges in expanding product variety. Some inefficiencies based on technical adjustability became evident. The use of AGVs and the dock system, which represented a technological innovation at the Kalmar plant, gave operators better freedom at work. However, the balance between the movement of the main line and AGVs was considered difficult. Thus, it was highly stressful to work with them and the docks were eventually removed as a result (Agurén et al., 1984, pp. 41, 44). Therefore, the 'New Factories' illustrated several accomplishments but required improvement.

4.2.3 *Lessons Learned from the 'New Factories' and the New Search: Focusing on the Japanese Approach*

Agurén and Edgren, who participated in the 'New Factories' project, evaluated how the project was considered a 'failure'. They suggested that the project had technical limitations in its pursuit of flexible production control dependent on mechanical technology. Moreover, increased investment costs in the production system became a managerial issue. Consequently, after the New Factories project's conclusion, the SAF Technical Department ended its activities; the Swedish STS characteristic management-driven technological reforms practically ceased in the 1980s.

What lessons have been learned from the New Factories project? The most important lessons concern self-awareness regarding the limitation of technical-led regulation. Therefore, project members sought to determine another route. Thus, Agurén and Edgren focused on the effectiveness of the quality control movement (QC circle activities) in Japan. Soon, years after their finished project in 1982, Agurén, Edgren and other members of SAF visited Japan and met Kaoru Ishikawa, a theoretical researcher of QC management. They studied Japan's QC control method. This taught them the importance of introducing Kaizen activities and cooperative union activities within a company. This led the SAF to pivot towards absorbing a Japanese management style. Such a management direction referred to as 'new managers' elicited a boom of Japanese management (Sandberg, 1992, pp. 274–279).

4.3 The Shifting of Sweden's STS: Uddevalla Plant

4.3.1 *Situation in the 1980s*

In 1982, the year the 'New Factories' project ended, significant agreements were reached between Swedish employers and employees. This was referred to as the Development Agreement (UVA: Utvecklingavtalet), which allowing for the application of the Co-Determination Act (MBL, Medbestämmandelagen) established in 1976, to private sector and was accepted by the management side. This agreement was concluded between the SAF and LO (Svenska Landsorganisationen), as well as the PTK (Federation of Private Salaried Employees in Industry and Services) recognising the unions' right to participate in work design at the factory level. Unions, including LO and Metall (Svenska metallindustriarbetareförbundtet, Swedish Metal Workers' Union, now IF Metall), were underpinned by the theory of STS and sought workplace reforms through projects such as DEMOS, UTOPIA and LOM.[2]

Examining the management environment in Sweden during the early 1980s, the labour market experienced a significant shortage of workers. For companies such as Volvo and other manufacturing firms, it became crucial to design attractive jobs and improve working conditions.

In summary, three factors were crucial for the realisation of the Uddevalla plant. First, management, including Volvo's CEO, P. G. Gyllenhammar's initiatives, dealt with the labour market situation. Second, the background of STS experiences. Third, the societal conditions that supported co-determination.

4.3.2 Uddevalla Project: 'Complete Stationary Car Assembly' Proposal

This paper refers to the production system introduced at the Uddevalla plant as the Uddevalla Production System (UPS). What characteristics does UPS exhibit? Examining the formation process of UPS, its initiation can be traced to the 'Uddevalla Project', launched in the fall of 1982. Uddevalla project activities began based on UVA that made it possible to share between employers and employees, involving Volvo (now Volvo Cars), Metall and the clerical workers' union SIF (Svenska industritjänstemannaförbundet, now Unionen) and central organization of salaried employees, TCO (Tjänstemannaförbundet).

Concerning discussion content regarding the Uddevalla project, we detail the processes of the assembly layout that was presented (Ellegård, 1989; Granath, 1991). For example, the first phase presented a design where car bodies flow along the line using AGVs, as in the Kalmar plant. For the Uddevalla project, however, discussions led to reducing the movement of car bodies on the line flow. The proposal discussions finally

[2] For example, the UTOPIA project, which started in 1982. This project advanced new technological configurations and labour organisation reforms, using the newspaper editing process as a case study. LOM is, from 1985 -1990, a project name, 'Ledning, Organisationen, Medbestämmande'; Leadership, Organization, Cooperation. The purpose of the project is to realise the expansion of worker participation and create a more democratic organisation, as well as to expand the utilisation of women in the workforce, see Sandberg [1992] p.242–258, Yuasa, Y. [1996].

reached stationary assembly, where bodies stayed in one or two positions across four stations within a dock; assembly was performed by 7–9 operators in a stationary group. This was now termed 'the stationary assembly' layout. Based on the work cycle time for each operator, allocated 1/4 of precisely the smallest 1/9, assembly became 120 minutes (Ellegård, 1995). This stationary assembly adopted by the Uddevalla plant received worldwide attention as being the first plant to completely abolish line flow. The layout was also formally authorised by Volvo management.

4.4 What Are the New Challenges Faced by the Uddevalla Plant?

As mentioned above, the Uddevalla plant had reduced the car body movement while realising stationary assembly. Therefore, opinions comparing the Uddevalla plant with the Kalmar plant followed; both plants had abolished production line assembly and expanded general work cycle time. Thus, determining the plants' shared similarities ensued. Now, if the Kalmar plant expanded its work contents in a dock, would it then be possible to achieve a 2-hour work cycle time at the Uddevalla plant?

The answer was definitely not. The reason is that work and task formation at the Kalmar plant was collected for each task based on a method of traditional work design. As a result, an extended 20 minutes was difficult to achieve. Therefore, to realise the 2-hour work cycle at UPS, new elements were required in the work design, which were derived from the technical division of labour. The next challenge was to determine what the added elements would be. This was significant in identifying UPS's uniqueness.

4.4.1 New Element of Control: Cognitive Domain

This paper hypothesises that the newly added element to the engineering division of labour was related to the realm of cognition. Being integrated for the first time, it enabled the extension of task execution. The utilisation of the cognitive domain was positioned as a novel approach in the development of STS in Sweden.

At UPS, the embodiment of line flow, however, did not emerge in front of operators. Therefore, operators could not be properly equipped with information guides on technological equipment to conduct their work tasks. From an operator's perspective, the order of assembly and how to attach parts were unclear. If operators did not have information about the part types on the assembly flow in advance, they would not understand the relationship between the body and parts fetched. A lack of understanding regarding general assembly meant that operators could not assemble for long work cycles effectively.

For these reasons, strengthening the relationship between assembly objects and the cognitive understanding and interface between human behaviour and technological equipment, including the cognitive activities of operators, became a significant practical issue. These matters of cognitive and behavioural design building that UPS developed had not been considered in the perspectives of Tayloristic and Fordism work design. Thus, these fields became new epochs for STS development.

4.4.2 *Construction of Different Human–Machine Interfaces on the Production System*

Plants are established in response to the conditions at the time of production set up; the arrangement of unique machinery equipment is followed by the corresponding deployment of the human workforce. Generally, in line production that utilises the conveyor system, the labour force is introduced based on the technological system of production equipment and its technical organisation. Technical and human factors stand parallel as two pillars supporting production. In essence, they form an interdependent relationship. UPS, as well as TPS and the ideal LPS model, each has a production model with its own technological and human roles. Each has its own interdependent situation, and method, or interface, which creates differences in production models. The interface process and characteristics, based on relationship development between technology and humans, was a solution selected for the UPS project. Each system shows its own unique combination.

However, the issue of interface building became a new challenge to consider. Building a relationship between technology and human cognition has been organised based on various management conditions in companies across each country and society. Each company possesses their own unique technological and human relations, factors that form various competitiveness advantages.

4.4.3 *Systematisation of Human Interaction*

When Ikujiro Nonaka, a renowned academic researcher of knowledge formation and organisational knowledge creation, visited the Uddevalla plant in 1990, he characterised the Uddevalla production as a 'human-centred production system'. He commended it for systematising 'human activities' within the production site and suggested it formalised tacit knowledge in production behaviour at the Uddevalla plant. Regrettably, Nonaka had not detailed the Uddevalla plant's production principles. Nonaka's evaluation of Uddevalla's characteristics in terms of tacit knowledge and explicit knowledge became a highly suggestive point in evaluating the development of STS theory in Sweden, as well as for making comparisons with TPS (Nonaka, 1990, pp. 250–251). While Nonaka noted that 'the establishment of an interface between the mechanical and human systems is essential' (Nonaka, 1990, p. 251), how should we think about 'establishing an interface between machines and humans' and 'systematising human activities'?

4.5 Conventional Line Work and UPS Work: Cognition and Action

Now, let us determine the characteristics of UPS. This discussion begins in terms of work action. The reason is that in UPS, where the line is eliminated, work action becomes a common axis of comparison with line production. When focusing on the work, the actual task of attaching a particular part at a designated location in a specified way is no different from line work at UPS. However, when we expand our view to include the situation surrounding the actor, the two have significant differences.

First, the steps on the line determine the attachment parts and location during line work. The line's technical environment limits the work because the bodies on the line move in one direction. Thus, the work location and position are constrained to either side, front, or back of the line's body and interior. Parts to be attached are arranged according to the work location, and work speed is controlled by the speed of the line's movement. Average work cycles are about 1 minute. Examining action and cognition, e.g. Toyota has a 'fixed-position stop line' (Tei-ichi teishi sen) where the range of movement within the operator's step is set according to each part (mirror, handle, etc.) to be attached (Monden, 2006, p. 263), the maximum movement is set to 5 metres (Tanaka, 2005, pp. 185–186). This indicates that the parts and the line determine the range of variation in operators' actions. Based on this set up, we can consider that the range of the operators' cognition also changes and is restricted by the 'fixed-position stop line'.

4.5.1 *The Role of Cognition in UPS*

In contrast, since UPS does not use a line system, devices such as conveyors cannot determine the position and orientation of the bodies as they do on a line. Therefore, UPS has an advantage where work and movements are not constrained by technological devices such as the line. However, since UPS does not use a line, tasks, processes and the range of movement of operator expand and vary as they assemble around the body during their work. Moreover, the positional relationship between the operator and the body changes, as required, while operators move around the front, back, left and right of the body to perform their tasks, depending on where and how the parts are to be attached. Consequently, UPS work created a new domain of relationships in the work set-up as standard operation, i.e. its cognitional standard, which involves choosing actions during assembly work. At the same time, recognising tasks, such as how to visually identify the car and how to proceed with work tasks while confirming the car and work position, need to be considered. These factors relating to recognition raise discussion points on work re-design: how to recognise work object and workspace by non-skilled assembly operators.

4.5.2 *How to Set Position: How to See and Do*

This aspect of relationship building between the cognition of acting and task selection has not been considered in line work where work cycles are short. This is because tasks are divided according to process and collections of small work pieces to adjust cycle time; the body's and operators' orientations are predetermined for each process during line work. Moreover, with a work cycle of around 1 minute and assembly points limited to a few places, actions can be transmitted to the operators as 'work procedures'. However, when designing work for a longer cycle, the relationship between the object of action, cognition and behaviour at the operator's position for action changes continuously over time. The action position also changes depending on the assembly point. Therefore, it becomes necessary to construct and organise work information that links 'cognitive design'. This defines how operators visualise three elements—the object of action, cognition and behaviour—to the operator. Combined with 'action design', these elements enable concrete task performance (Engström & Medbo, 1992, p. 135) in terms of the 'alignment' of (1) how work progresses, (2) how parts are displayed and (3) how work is described.

Therefore, relationship building in the operator's 'cognitive' and 'action' design domains is essential to realise UPS. It signifies the organisation of a new technology and human interface in the work design field. Compared to the line system of the TPS, the UPS interface set-up reveals several areas for examination, such as the role of work cognition as a starting point for task execution and the role of context for action performance.[3]

4.5.3 Layout and Cognition

The production layout at the Uddevalla plan was stationary, meaning that the body, i.e. the object of work, stayed in one fixed location. This allowed operators to grasp the entire body, i.e. the subject of assembly action, which improved the overall visualisation of component parts and the relationship between them and the operators' actions (Ellegård et al., 1992, p. 45). This layout facilitated the creation of a new mental image regarding the work object. At the same time, the stationary layout created a new assembly work-oriented function derived from product, which eliminated the partiality of individual assembly actions in the actual work.

In other words, even if actions were divided, each small activity of work could lead the operators to understand and reposition their actions regarding the overall image of the work object. Specifically, the body's position (the work object) would be set in a location that made it cognitively easy for the work group and the operators to visibly perceive it throughout their entire work performance. These layouts were set

[3] From the STS viewpoint on the 'interrelationship' between mechanical technology and humans, but the Kalmar plant was explained using the term 'symbiosis' (Karlsson 1995). Regarding Toyota, 'how to see and do' is a crucial consideration in designing assembly lines. At the Guangzhou plant in China, car bodies are moved to the assembly line for plate installation. The assembly area are divided into 8 segments, with each operator responsible for two segments. This segmentation was introduced to maintain a 'stable eye angle' during work, as explained in 2012 interview.

after considering 'eye contact' among operators, taking into account the importance of view point coordination (Engström et al., 1998, p. 292). This design policy, it is deeply related to technical aspect, to set the body position and layout based on the operator's perspective intended to create a cognitive and psychological environment that enabled operators to understand the significance of partial and localised tasks (Ellegård et al., 1992, p. 122).

The relationship between overall cognition and action is similar to the process of assembling a plastic model. A model assembler recognises what the object of assembly is before starting the act of assembly and proceeds to complete the model. During model assembly, all work is an act of partial assembly; however, showing the parts to be assembled and the overall image of the model in the design drawings, etc., assists in the recognition of partial assembly. This is because part of the overall assembly act and the overall image of the assembly object serve as guiding information for understanding the actions of the assembly process. Therefore, at UPS, starting the work with the overall image of the action object supported an important cognitive function regarding the interrelationship between technical structure, organisation and human behaviour. In traditional division of labour work design theory, the meaningfulness of work is limited to the organisational division of labour and the role of work dimensions.[4]

4.5.4 *Cognitive Organisation: Holistic Learning Theory*

Viewing the work object as a whole was an essential element for performing work at UPS from both cognitive and execution aspects. This holistic cognitive formation was implemented as a 'holistic learning theory' by L. Nilsson, an expert in vocational education. The characteristic of holistic learning is that it generates the act of assembly by learning to overview and cognise the whole object of action. In other words, it

[4] Discussions on the division of labour correspond to the opposition between machines and humans in line production. However, as indicated by the analysis of machine-cognition relationships by Suchman (1987) and Rasmussen (1986) since the 1980s. The issue of the interactive relationship between technology and humans intersects was argued in the concept of 'design for participation' in technology design which reflected in the DEMOS and UTOPIA projects (Ehn 1993).

relies on the overall visualisation of an object and forms hierarchical cognition starting with the totality of the object. It transitions from a partial, localised understanding to a holistic perspective. This transition enables the acquisition of a coherent impression of an object, making it possible to obtain an impression of the whole encompassed by partial cognition.

Holistic learning helps actors acquire an overall image of the object of action. At the same time, it helps them to understand this object more clearly by recognising the interrelationship between the whole and its parts, and between its parts and in-between parts. Furthermore, holistic learning provides actors with a cognitive process that enables them to grasp the meaning of the partial act regarding the entire object while finding meaning in the partial act; even though the local acts of the moment by moment is a part of the totality of the assembly act.

It was a challenge for UPS, which does not utilise line equipment, during production design and task organisation to ensure that operators could perform their tasks without error, even without line machinery. Therefore, it was necessary to construct a system that allowed each operator to judge their own actions and carry out their assembly work. The theory of holistic learning allowed for a cognitive framework (structure) that enabled the execution of actions and organised work cognition. So, how exactly was the theory of holistic learning applied to the actual process of car assembly? Next, look at the content of relationship building between technical organisation and the cognitive domain.

4.6 Cognitive Structure and Work Organisation Connecting Object Cognition to Work Process

4.6.1 Bridging Cognition and Work Action

In the process design at UPS, 'function' as assembly work design became a key element, just like 'holistic view'. During the transition from line assembly to stationary assembly, Engström was responsible for technical design. He recognised the importance of the existence and role of 'function' as part of the production system through the disassembly of completed vehicles. The goal of disassembly was to determine the assembly sequence while keeping the car body stationary, based on the theory of holistic learning. This involved showing operators how the body and parts relate while connecting this idea to assembly tasks. Under a stationary assembly layout where the body remains still, the challenge was determining how to set the core process as a structure of the operators' cognition while connecting it to task execution.

4.6.2 Disassembly of the Completed Car and Discovery of 'Assembly Function' for Assembly Work

During the disassembly process, parts lists (PKI) previously used by Volvo were referenced to trace back and confirm the connections between parts from a task assembly perspective. As a result, it became evident that the design-oriented method of classifying parts used by Volvo did not consider actual assembly work.

Therefore, an analysis of the correlation and relationship between components and parts was attempted at UPS. The result of this investigation revealed that components and parts were organised into functional units for assembly work point of view (assembly-oriented function groups) based on product functions. These represented the interrelationships between components and the parts themselves. Specifically, the collaboration and relationship between components and parts for performing assembly functions were clarified based on identifying the hierarchical relationships between components and parts, and as well as between parts. These functional groups for assembly were referred to as 'kin' and 'family' and subsequently named as such (Ellegård et al., 1992, p. 44).

4.6.3 Designing Processes and Work Organised by 'Assembly Function'

The 'Assembly functions'[5] discovered through disassembly of complete cars indicates the roles designated to artificial objects (components and parts) within a complete finished product. Specifically, an entire car could be divided into eight functions: door, electrical, air conditioning, water systems, joining, decor, powertrain and interior. These functions are common to all cars regardless of type or use and are inherent to cars as products. Based on these eight functional divisions and considering the assembly function relationship and work efficiency of the actual

[5] The 'assembly function' reflects two aspects: the first is the function that is derived from the components and parts, and the second is the related group of assemblies on which this product is based.

assembly process, actual assembly was organised into four groups: < electrical, air conditioning, water systems > , < decor > , < powertrain > and < interior > . Thus, the assembly was designed as a process in which eight functions were grouped around two functions and assembled in a one-quarter cohesion (1/4 car assembly). The discovery of these assembly-oriented function groups inherent within the assembly objects and the organisation into four groups based on workability led to an organised 'assembly-oriented structure'.[6]

In traditional line production, the 'functions' concerning assembly inherent in the assembly object product itself are separated on the line due to the division of line production, isolating them from the whole product. Therefore, for the operators, the set of assemblies that the product retains is reflected only to a limited extent in line production. In contrast, by eliminating line production, UPS made the product's internal structure, as assembly-oriented functions and functions groups, the basis of work design. By giving a central position to the connection between process, cognition and motion, these assembly functions reflected on product, became the fundamental framework of UPS.

4.6.4 Creating Work Cognition Through Specific Part Names

From an operator's perspective, 'assembly functions' alone provide insufficient information for assembly work. Even if the work is divided according to assembly function or division of each functional area of assembly, it is unclear how to perform the assembly work. To specifically carry out assembly actions, it is necessary to cognitively recognise which parts fit with which components while visualising the order and procedure. Traditional line assembly work determines the content of the operating actions according to the body's position on the line. However, in UPS, which did not use a line system, the operators (actors) had to create a memory or mental image of the information required to perform their actions, e.g. the location of the assembly work, the order of actions and the tools to be used.

[6] The word 'function' does not mean to go, stop or turn as in driving a car. In the following sections, the word 'function' is used to refer to the function for assembling work that derived from and depends on product, as shown by assembly function, assembly-oriented function.

Therefore, there was a shift to managing parts according to their specific concrete names. Up until that time, parts management was based on the Parts Number List (PKI) created by the Volvo design department. Therefore, for operators (the work performers), parts were remembered by numbers. This number-based management is efficient when handling the narrowly specialised processes of line assembly, where the parts used are limited. However, for UPS, comprising multiple line processes using work indications based on part functions, the numbers system could not concretely illustrate the relationship between the body and parts, or between the parts themselves. Therefore, UPS switched from numbers to specific or real names for parts.

Specific indications of part names altered the cognitive level of the objects of action, forming the basis for operators' understanding of work performance. This constructed a bridge that transferred operators' work cognition to specific actions. This naming of parts, along with the use of illustrated work expansion maps (described later), was called the creation of a 'new language' (Berggren, 1992, pp. 158–159) of the basic understanding of work design. The 'new language' implies that the structure of meaning interpretation of the work performed by operators was altered by making the words (part names) and grammar (based on the relationships between parts, and components) more explicit, enabling systematic grasp of the technical system. These changes of cognitive objects were inherently related to the cognitive structure, creating a room, new opportunities for the structural interpretation and understanding of tasks among the operators. This formed the basis of their interpretation of tasks performed, which was necessary to advance their work.

4.7 Formalising Interface Between Cognition and Action Shown Through Mapping

As mentioned above, Engström formed four assembly groups: < Electrical/Water > , < Decor > , < Drive System > and < Interior > , based on the eight functions established after car disassembly: < electrical equipment > , < air conditioning > , < joint > , < water system > , < joint/decoration > , < drive system > , < interior > and < door > where two imposition groups were established. The operators used these eight functions as a classification unit for the assembly objects. So, how can we communicate the work content based on the relationships between these individual parts and components to be assembled? Engström developed several types of maps as unique work display methods to clarify the cognitive work route.

These maps were broadly divided into two types: Maps detailing the overall structure of the vehicle being worked on (Fig. 4.1) and maps detailing the assembly process (Fig. 4.2).

First, Map 1 (see Fig. 4.1) begins with a finished product, the completed car, as its starting point. It divides and categorises components and parts into eight systems based on assembly functional area and describes the main components of each assembly functional area, and their parts branch off in stages. This map is called the 'logical product structure' (Engström & Medbo, 1992, p. 140). The map illustrates the components and parts in each assembly functional area, as well as the functional connections between the parts hierarchically from higher to lower levels. Visually portraying the completed product (the end result of the assembly process) provided operators with several vantage points: from the whole to the parts of the product's final form. Using this display method, the map allowed operators to grasp the product from a perspective of hierarchical positional relationships, i.e. from the whole car to the specific work area they were responsible for. With this complete product and hierarchical display, each operator could gain a mental image of the location of the parts to be attached as part of their work actions. Acquiring this mental image allowed them to understand their work as a series of actions for product completion while positioning each work action within the complete work process.

Therefore, Fig. 4.1's entire product display serves as a 'cognitive interface' that connects the actor's partial cognition regarding their part of the action to the entire product. The map is an informational tool of cognitive formation that uses a holistic display method, which depicts the product from the perspective of totality and shows operators the overall image of the work action objects, along with the relationship of each work location.

Figure 4.2 consists of two maps. First, the assembly proecss showing by allow A, and the section in the top right corner of Fig. 4.2 (enclosed by B) is 'the intra-group work-pattern'. The boxes marked ('work modules') indicate which part of the car is being handled and whether the work assignment is for an individual or a pair. Next, a series of A (arrows extending from the 'work modules') show the assembly of parts groups that comprise a series of assembly functions. These are displayed in four levels from highest to lowest, indicating what parts are assembled with

Fig. 4.1 The role of maps: Overview and visualisation of cognitive processes

which screws and tools. The work is shown in order of assembly, serving as the work instructions.

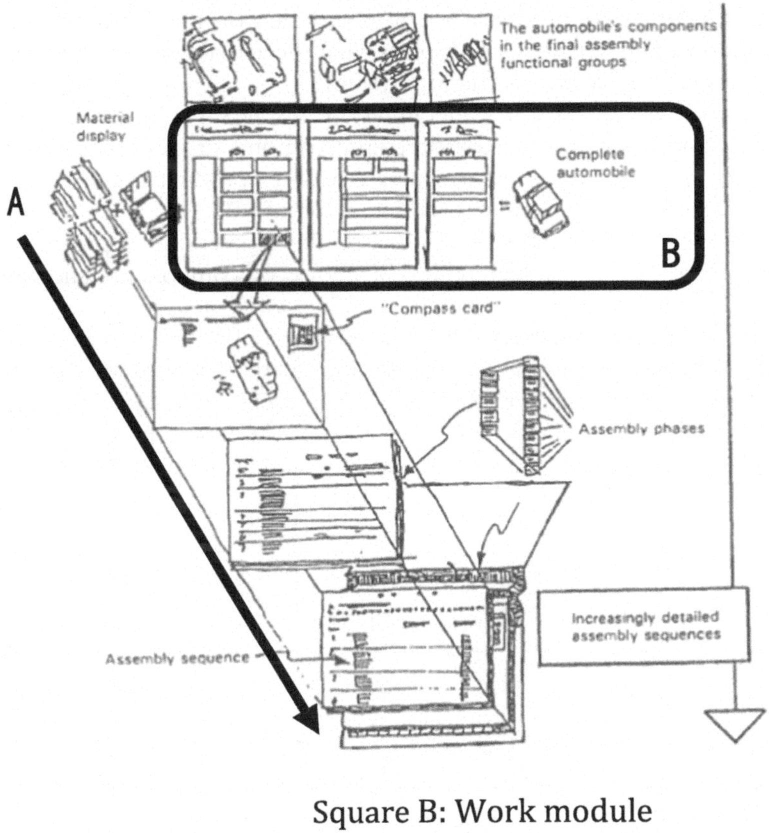

Fig. 4.2 The role of maps: Overview and visualisation of cognitive processes

4.7.1 Internal Logic of the Product Information in Assembly and Cognitive Function: Assembly-Oriented Design Depend on Product

Why was it necessary to create such a map, especially one that shows the entire work area, as shown in Fig. 4.1? We can say the following.

Figures 4.1A and 4.2B present the relationships between components and parts (the objects of action) from a holistic and a partial, process-oriented perspective. Figure 4.1 shows the hierarchical division of parts viewed as the entire product. In contrast, based on the positional information from Fig. 4.1, Fig. 4.2 displays what the actual work requirements, and is called the 'Work Instruction System'.

In other words, Fig. 4.1 hierarchically displays the functional logic within a whole car, giving holistic cognitive point of view. The product itself (the work object) shows the hierarchy of the product as a whole from the perspective of the entire product. In contrast, Fig. 4.2A indicated by the arrow A shows the order of assembly operations displayed with illustrations that depict hierarchical structure of components and untill the smallest parts should be assembled. The 'work modules' in Fig. 4.2B (minimum units of the assembly process) display the installation of parts locations regarding positional relationships with the body as the work progresses. Thus, Figs. 4.1A and 4.2A present two types of cognitive information for actions: the assembly of the whole product and its parts. They serve as guides for cognition and action during assembly (Engström & Medbo, 1992, pp. 138–141).

4.7.2 Ohno's Focus on 'Process of Assembly Work'

When we examine the relationship between cognition and action in traditional line production, its characteristics indicate that processes are organised by groups based on the division of parts. The work within these processes is allocated based on the engineering connections of parts and the balance efficiency of line production. This simplifies tasks through process division and subdivision, with work cycles divided into approximately 1 minute. However, Fujimoto indicates that Ohno reviewed the method of process division at Toyota, which lead to a shift from a 'Line-focused layout' to a 'Process-focused layout', and, further, to a 'Product-focused layout' (Fujimoto, 1999, p. 60). Ohno did not merely view the line as a connection of work locations but emphasised 'processes' and transformed them into 'processes divided by the product'.

From Ohno's perspective, the idea was to gradually expand the range of assembly work. This started with limited areas and expanding the scope of responsibilities of the production process, advancing the concept of 'multi-skilled development'. This encapsulates the characteristics of TPS (Fujimoto, 1999, p. 69). Regarding cognition, when processes are established, the cognitive range of operators consolidates and expands according to the unit of process. However, the cognition used for work would remain disconnected from the relationship with the complete finished car (as a product/the object of assembly). Since Ohno's challenge was to develop 'multi-skilled workers' who handle multiple processes, the expansion of the cognitive domain did not progress beyond the process. In contrast, at UPS, the relationship between body, parts and work cognition was established as a pathway of cognition. This cognitive path was generated by the assembly function derived from the product function.

What is the difference between Ohno and UPS? In Ohno's case, the focus was on part of one assembly line process. On the other hand, the axis of work design at UPS was the assembly-oriented functional group derived from product, enabling operators to recognise the entire product. The concept was to use this overall cognition to facilitate work on partial processes. Notably, the difference between UPS and TPS was that the change in the production system's design origin from process to product orientation was what created the difference between traditional line production and UPS production. The fact that the transition from a process-oriented work design to a product-oriented approach has not been thoroughly evaluated.

4.7.3 Overlooked by Womack

Womack and colleagues observed the situation only from a line production standpoint. They overlooked elements that replaced UPS's technological devices as a production method. Consequently, they failed to grasp how cognition, suppressed in line production at TPS had transformed into a new driving force for action. Womack understood UPS to be a mere retreat from line production and at the same time did not take advantage or evaluate of 'creative tension', derived from 'thinking' tool for management (Womack et al., 1990, p.102) used in LPS. Womack did not fully recognise the role of cognition in the production actions at UPS, which is also strategically employed in organisation setting in LPS.

As a result, they might have overlooked the importance of expanding the use of human cognition in both TPS and LPS which Womack called 'tensions', indicating a reversal from the traditional Ford system. The evaluation of the cognitive domain in human action by Womack was lacking in even LPS. This lacking of understanding the cognitive domain, and lacking of using cognition for support factor for technical factor organisationally is reason why Womack called Uddevalla plant 'neocraftsmanship' (Womack et al., 1990, p. 101).

This lacking and misunderstanding was compensated for primarily by Liker and others by 'Toyota Way' after Womack. Liker's perspective analysed the systematic use of cognitive domains that underlie the unique 'tensions' created by LPS, which Womack were unable to accurately capture. Liker's analysis examines how the organisational and conscious aspects of LPS are connected to production behaviour. Thus, we must consider an evolution in the perspective itself to grasp TPS and LPS.

4.8 Correlation Between Cognitive Functions and Concrete Acts: Acquisition of Cognition and Work Autonomy

Examining the actual content of work execution, this section focuses on kits while discussing the relationship-building features between the technical elements of UPS and human behaviour. The examination focuses on how flexibility was exhibited in the UPS production system design.

Specifically, how were changes in work speed due to changes in assembly numbers, the switching of products, or shifts in responsibility during work handled? The movement of each operator and workgroup must be considered when examining this issue. Specifically, in the case of UPS, when changes in production speed or product or part switches occurred, coordination within the workgroup and between operator pairs was essential. UPS's flexibility, which was not bound by technical equipment of the production process like a production line, depended largely on how flexibly operators could respond to work changes. Therefore, the work execution content of operators is a crucial point of examination for understanding UPS. Let us begin by examining the work modules.

4.8.1 The Function of Organising Work Actions and Cognition Based on Work Modules

As mentioned in the previous section, Fig. 4.1 displays the overall work, and Fig. 4.2A allows for hierarchical recognition of the subject from a work action perspective. How were instructions and displays made for the actual execution of work based on this cognitive organisation? The 'work module' in Fig. 4.2A played an essential role in presenting work execution.

Specifically, the 'work module' in Fig. 4.2B uses squares to indicate the smallest work execution unit. A single square represents an individual task, while a larger square indicates a paired task to be performed by two operators. In other words, the 'work module' presented the operator and the smallest unit of the assembly action. The flow of squares within the 'work module' indicates the sequence of tasks. This organises the 'work module' as a framework for recognising the order and person responsible for work execution. Thus, the 'work module' bridged cognition and action and formed the basis for organising work actions.

Therefore, the cognition of the entire work subject is presented in Fig. 4.1A. The actual actions are indicated in Fig. 4.2A, which illustrates each component and part; the assembly of each part was organised within each 'work module'. Therefore, the kit was a collection of parts used within the squares, shown in Fig. 4.2A. It was organised as the series of parts required to complete the 'work module' tasks, which was packed into a box (kit) and brought to the operators. Now, let us look at the main subject, i.e. the kit.

4.8.2 Formation and Characteristics of Behavioural Cognition in Kits and Tasks

As previously mentioned, the contents of a kit were set to correspond to the 'work module', a collection of tasks, as shown in the squares in Fig. 4.2B. Each work module was based on a standard time of about 15 minutes of work. The parts in the kit were also arranged as 'a set of parts' corresponding to this 15-minute work module. It included necessary nuts and bolts, which were carried to the work execution site on an AGV (Engström et al., 2008, pp. 124–126). The timing of the AGV bringing parts to the work position was adjusted based on work execution.

From the operators' perspective, the physical objects (parts) for assembly would flow to their hands, reducing the need for operators to fetch parts while maintaining task concentration. Furthermore, by organising these parts into a kit and parts themselves in a kit, it became easier to identify individual parts. For example, enhancing visibility and the flow of the kit helped operators recognise the sequence of work execution, supporting work efficiency. Medbo was primarily in charge of organising the parts kits at UPS and set up 'signal components' within the kits to examine whether they affected workability and whether they served as information to guide the operator's actions. He suggested that using kit boxes improved the visibility during work action, and therefore improved understanding (Medbo, 2003, pp. 263–281). Such kit-based part organisation is similar to what is known today as SPS (Set Parts Supply), which involves supplying parts in a box in the quantity needed for the corresponding work to the line.

4.8.3 *Differences in the Use of the 'Kit'*

There are notable differences between UPS and SPS, however, even in similar part kits, as follows. First, specific product information indicated by the physical parts composition at UPS was understood in combination with the cognitive structure (Fig. 4.1), showing the entire product, and Fig. 4.2A, organising elements that support the assembly action. By providing parts in the form of kits based on this cognitive structure, the parts formed a cognitive structure that aided in understanding the work cognitively. These parts, as physical and engineering entities, through the coordination of individual parts and parts among themselves, resulted in the individual parts playing a 'signal' role in providing a cognitive structure regarding their individual information. This organised a medium of information that helped operators understand the work execution more clearly.

In contrast, SPS exhibits a weaker connection to cognition. Instead, an operator's range of cognition is limited in the process; they are cognitively and physically separated from the overall product structure. Therefore, the SPS arrangement uses a group of parts supplied near the operator for easy pick up (temoto-ka) for each assembly operation along the process line. This is in an effort to improve individual task-related work performance, such as reducing incorrect part selection.

The role of parts and their kits in UPS and SPS, and the relationship between the kits and the operator's cognition during work, is directly related to work efficiency. In UPS, the kits as a cognitive organisational point of view supports human actions from cognitive domain, which shows a cognitive support technology (Tamura, 2022). Therefore, the organisation of kit parts, operators' work cognition and the method of work instructions must be examined from the perspective of cognitive engineering. This can be viewed in terms of communication and context between people and artefacts, as embedded in Suchman's (1987) view of machine-human communication.

4.9 The 'Point of View' of the Operator

We have discussed the relationship between work modules as units of work execution, the cognitive structure and organisation produced by kits and the act of working. Next, we examine the act of working based on the operator's 'perspectives'.

Why should we mention and focus on perspective? The reason is that unlike line work, the work positions at UPS were not physically determined. The work positions and postures of execution were left to the choice of each operator. If operators' choices of actions were inaccurate, work efficiency could not be maintained. Operators needed to develop their own perspective for work execution to suit the tasks they were responsible for. Therefore, understanding UPS requires examining the problem of relationship between action position and perspective selection during the operators' assembly actions.

The notion of perspective selection resulted from a technical choice made when the production line, as a technical condition, was abandoned in favour of a stationary layout. This included a shift in the driving force from the line to the operator's side. To explain the flexibility demonstrated in UPS production, it is essential to examine how an operator's perspective in part assembly action choices was determined and how it responded to changes and modifications in work execution. To understand the meaning of these changes more clearly, the focus of analysis shifts to the level of the operator's perspective.

4.9.1 Connecting the Whole to Act

Let us consider the operator's perspective as a 'cognitive perspective' and a 'perspective in action'. First, the cognitive perspective is used for recognising the target, as shown in Fig. 4.1A. This viewpoint visually identifies the object of action. On the other hand, perspective in action arises during work execution, which is required to act. Specifically, the perspective in action refers to a perspective occurring when an operator selects the correct parts; the assembly position changes during the act of assembly. Previous UPS discussions explain that when an operator performs a task, they use a cognitive perspective (Fig. 4.1) to decompose the product from a holistic perspective. Then, they overlay the product reflected in mind with components and parts in a hierarchical manner (Fig. 4.2A) to perform the assigned task.

However, this explanation does not clarify how the operators adjust the relationship between changing the car body movements and their assembly actions during work or how they carry out assembly while making these adjustments. The process linking Figs. 4.1 and 4.2A is not clear. Therefore, we examine how the cognitive organisation created by combining the cognitive and action perspectives enable the act of working, along with the type of flexibility it generates.

4.9.2 Establishing the Operating Position

The most noticeable feature of action from examining UPS from a work execution perspective is how an operator's position would change relative to the body according to the work area. The change in work position would cause operators to adjust their own position and perspective accordingly. It is generally understood that this shift in perspective alters the appearance of work objects, becoming an obstacle to work stability. For example, as previously mentioned, Toyota put a 'fixed-position stop system' to maintain operators' posture within a specific range as much as possible while fixing the positional relationship with the line and body.

So, how did UPS address such changes in work positions? First, while UPS did not use lines, the operator's perspective of the workplace concerned the whole product. They coped with shifting the work position through the 'cognitive perspective' and 'perspective in action' detailed above. These two cognitive categories were specifically utilised in the assembly process by defining the exact locations of assembly work. The 'work module' directed the process and the specific part of the work process, and the work module further narrowed the cognitive focus for the actual work during the assembly process.

In terms of cognition, the critical aspect to note is that as the smallest unit of the work process, each work module were labelled, i.e. the 'engine room', 'coupe' and 'luggage room'. These labels indicate the assembly position, the front, middle and rear of the car body, and the work was divided into four assembly processes. Ultimately, standing positions for the tasks of each operators in a group were categorised into these 'coupe', 'engine room' and 'luggage room' areas. With this expression of marks, the operators understood the general location and position of the tasks they were responsible for while facing the body. However, the implications go further, as they specified the work perspective.

For example, 'coupe left' indicates that the assembly was to be performed on the left side when standing on both sides of the car; a 'luggage room' means that the work was to be performed from above after opening the top lid at the rear of the car (Engström et al., 2001). Essentially, these expressions not only guided the work position of the body but also the direct orientation of viewing the object corresponding to each work position. Therefore, the work position and viewpoint were assigned to all work modules. A change in perspective for each act occurring during work execution at UPS was determined not by experience but by the act of work and cognition. UPS's work execution design structured and organised the cognitive characteristics of human action.[7]

The meaning of organisation lies in the mutual supplementation of cognition required for assembly through a 'cognitive perspective'. This provided an overview of the entire product as a framework for the cognitive setting to enable the 'perspective in action' required to perform each task. Furthermore, the viewpoints used in each step specified by the work modules were established. Through this series of cognitive settings, the cognition required for task execution at UPS was structured. This allowed for a unified cognitive framework to be systematically utilised during any work process. Furthermore, the viewpoints were determined by the work viewpoints of each assembly process, as stipulated by each work module. The cognition for work execution at UPS was structured through this series of cognitive settings. This means that a unified cognitive organisation could be systematically organised, and used by any operator for any work process.

4.9.3 Moving Object and Human View Point

Let us consider how humans perceive objects during the process of action based on studies from cognitive theory (Rasmussen, 1986, pp. 108–110).

[7] When comparing UPS with Cell system, UPS has generally been explained as one type of Cell, because both systems have similarities with a line flow. However, there is a crucial difference between Cell and UPS: the size of the object being assembled. In UPS, the object is a car, while in the Cell system, it is mainly consumer electronics and ICT devices. In the Cell system, the assembler does not need to change viewpoints by moving assembly positions because the products are small. Thus, from cognitive point of view, cell is done by one person system. This means that Cell has a strong connection with craftsmanship type, as it does not require cognitive formation for organisational standard in UPS. These differences classify Cell and UPS as distinct systems.

For instance, when something previously in place has moved, and the arrangement has changed, or when the actor has moved something, how does the actor respond to these changes and modify their actions? Cognitive research indicates that when the movement of the actor's body leads to a shift in perspective, if the actor does not establish a fixed viewpoint and position, they cannot acquire accurate cognition, and the action cannot be completed. It is thought to be more challenging to maintain a fixed perspective while moving oneself than when in a static situation (Miyazaki & Ueno, 1985, pp. 77–79). In UPS's work, which assumes assembly tasks while moving around the body, the viewpoint, as the position of eyes to observe, is always moving, and the situation is constantly changing. Therefore, it is necessary to construct information to smoothly replace the outside world information with the operator's viewpoint.

From the perspective of 'action perspective', in terms of vision and cognition, humans must first extract meaningful information from the external world, and this extraction is called the front-end function (Rasmussen, 1986, p. 78). To capture a moving object, two types of information are described as necessary: 'External-specific information' and 'Perspective-specific information' (Ueno, 1985, p. 81 pointed by Gibson, 1979). That is, the actor first needs to identify what the target of action is, and the information acquired by identifying the object is called 'Perspective-specific information'. Next, the information formed by 'External-specific information' to determine 'one's own viewpoint' is also called 'Perspective-specific information' (Ueno, 1985, p. 81). These two perspectives are indispensable for humans to grasp moving objects. Ueno explains the actor's cognitive formation as a combination of 'External-specific information' and 'Perspective-specific information', further pointing out that when 'seeing' the external world, one extracts the position of the object and self, including perceiving changes in the appearance of the object. Cognitive scientist Rasmussen emphasises the role of the mental model formed in these series of cognitive processes (Rasmussen, 1986, pp. 136–140).

4.9.4 Map Organisation and 'Work Modules' to Generate Viewpoint Information

Based on the viewpoints of the cognitive research mentioned above, what happens when we examine the cognitive process of the UPS actors? UPS did not impose technical constraints on the work position and posture as done in line assembly. Therefore, the points where assembly work was performed would change depending on the parts to be assembled. This means the positional relationship between the actor, the parts and the assembly target components and the body would constantly change. This resulted in a shift in how the work target 'appeared' during work. A significant shift in the operator's perspective means a substantial change in the specific external information, altering the visible scene situation, the 'view' ('mie' in Japanese) for each person. Therefore, it was necessary to develop information that smoothly translated external data into the operator's perspective.

Looking at UPS's cognitive responses, Fig. 4.1 shows the entire product (the assembly target), and Fig. 4.2A displays the parts to be assembled. Both figures illustrate 'specific external information' during the movement process; Fig. 4.1 first displays the overall information, and Fig. 4.2A enhances specificity by showing part of the overall information. However, merely displaying external information would be insufficient for individual concrete assembly actions. Further, specific actions for each particular task needed to be displayed. For this purpose, 'work modules' were created to indicate the work sequence and position for individual tasks. The information displayed on these 'work modules' provided operators with unique details related to individual assembly tasks, namely 'perspective-specific information'. Thus, the charts created at UPS (Figs. 4.1 and 4.2) became action-confirming information that enabled operators to maintain their perspective and perform their work, even when changes occurred in work position, work location and posture.

As a result of these cognitive UPS domains, we focus on the fact that UPS constructed a formal route for the operators' cognitive process. Operators could acquire the necessary 'viewpoint' of each individual operator as a method for making judgements using these cognitive routes. By developing this 'viewpoint' of individuals, this approach provided space for understanding and reflection. By these technical, organisational design, each operator could obtain the conditions for autonomous actions based on their judgements. The viewpoint of each operator was considered a factor in ensuring organisational mobility for the production behaviour at UPS.

4.9.5 *Work Visualisation and Setting a Common Recognition Format*

UPS had dramatically expanded its use of cognitive functions compared to previous production systems. So, what effects did the expansion of cognitive utilisation elicit at UPS?

First, to utilise cognition, the maps in Figs. 4.1 and 4.2A were established. Thus, a connection was made between the work target and the operator's perspective. Through both figures, the cognitive processes that occurred during the work process was made 'visible' or objectively demonstrated. This visualisation of cognition through the maps signifies the creation of a crucial operational cognitive structure at UPS.

To explain, the creation of an operational cognitive structure allowed individual operators, pairs of operators and even groups of operators to gain a foundation for a shared cognitive process. This foundation of cognition represented a shared operational cognitive structure among group members. It allowed for mutual understanding of object cognition and situational judgement among members during work, and workplace understanding and execution. The visualisation and commonalisation of cognition and action at UPS allowed operators to work within the same cognitive form.

The formation of this common cognition created a 'Format' of shared cognition. If a change in the work process occurred, it became possible for operators within a group to appropriately perform work substitution and assist each other. When switching parts, understanding the functional division and assembly methods of the parts commonly enhanced the capacity to respond to work assignments. This became the basis for flexibility in individual, in a pair of operator and group work execution and changes in work content. Such work visualisation and the establishment of a shared cognitive format are referred to as the formation of a 'common context' from the cognitive theory perspective. Since workgroups acted according to a common context, the context was situated as 'organised' (Ueno, 1999, p. 70, 76).

Summarising the relationship between cognitive organisation as a shared cognitive format and work execution included the following: (1) Holistic cognitive representation, which recognised the work object from the entire eight functional collections of a car; (2) Based on this holistic understanding, they could establish a cognitive organisation that reinterprets one's work actions within the overall car assembly task, supporting work execution. (In addition to these cognitive functions, although not mentioned in the main text here); (3) It was necessary to add an 'inner monologue' and 'inner dialogue' during the action process; (4) The formalisation of standard questions for the actor to 'reflect' on their work execution during action. These questions were divided into 'part knowledge', 'structure', 'assembly', 'inspection' and 'adjustment (countermeasures)'. They allowed operators to assess their own actions by asking themselves about 'component functions', 'how to use tools' and 'patterns of the individual, component, tools, body' during work actions (Nilsson, 1992, p. 40). The standardised formalisation of these self-questioning tools was considered to be the formalisation of the cognitive process where each operator reflected on their work execution (Engström et al., 2008; Nohara, 2007). Thus, by constructing a cognitive design that used a close complementary relationship between cognition and action and reflectivity, work had become possible at UPS even in the absence of a technological line assembly environment.

4.9.6 Acquisition Viewpoints and Workability Based on Common Cognitive Format

What can be said regarding the superior characteristic of UPS's production? Notably, it lies in the creation of autonomous cognition and autonomous actions of actors as individual operators. In automobile assembly characterised by strict standardisation of work and high dependency on equipment, UPS demonstrated the possibility for each operator to perform tasks based on their own situational judgement. This was a new challenge and a significant achievement in terms of both technical and organisational aspects of the methods of work design. Specifically, there were four places (stations) within a workgroup where the work assembly process was shared. This was based on production goals and production quantity and adjusted according to the combination of processes at each station. Each station in a group had the technical autonomy to adjust the work progress within that group, serving as a buffer. Under these technical conditions, the operator and workgroups made judgements of assembly development and daily production according to the situational workgroup themselves (Engström et al., 1998).

Such mutual support actions within groups during work execution were based on the premise that the overall task picture was understood in advance. Even if it involved support for another pair within the group, it was difficult to take on the support work without understanding the method of how to work and how to set the position at each process. The shared cognitive format provided a common basis that enabled mutual incorporation, namely, supporting and helping with operators' tasks. In other words, the formation of shared cognition played a role in systematically allowing operators to acquire perspectives to establish their task cognition. It functioned as a cognitive format that facilitated mutual understanding among operators regarding work progress and execution, assuming that each operator's viewpoint had been acquired.

4.9.7 Flexibility Through the Integrating of Technological and Social Factors in the Case of UPS

Returning to the question posed at the beginning of this section, what was UPS's improvement in flexibility as a production system? The answer can be found in the fact that UPS created a collaborative relationship between the technical and material organisation and the information structure that enabled the action judgements of each actor. This led to an expansion in the flexibility of individual operators' choices in assembly actions. It allowed them to generate and secure their perspective regarding work actions while responding appropriately.

So, where can the characteristics of flexibility demonstrated by UPS be found? They are evident in the results produced by the junction of three elements: assembly-oriented product function, cognition and action, formed around 'function' as the core. As explained, UPS was based on the organisation of work-oriented product functions with assembly process design, parts classification and parts kits. This set-up comprised holistic cognitive organisation and work modules. Examining the entire organisation, particularly the work organisation where operators grasped engineering product functions through holistic cognitive organisation, the workgroup divided tasks along assembly functional processes. It is evident that an organisational structure and a principle of action differing from traditional line-based work division was realised.

The reason this paper focuses on the cognition of individual operators is to emphasise that the cognition and actions that supported assembly activities of operators at UPS were based on their innate ability to recognise and understand product functions showed in relationship between components and parts; that innate ability was organised as a structure of cognitive processes. The work of each operator was based on a technical and cognitive organisation built around product and assembly work, which was fundamental to their action choices. From a collective aspect, the workgroup adjusted productivity at the group level. In UPS's individual actions and organisational behaviour, the axis of 'assembly-oriented work function' derived from product and product information was established, indicating a solid relationship between human behaviour and technology. The role of the cognitive domain was to link technical elements with human behaviour, including cognitive domains regarding the organisation of specific work assembly production behaviour and process design as the factors creating UPS's flexibility.[8]

4.10 Examination: Comparing TPS: 'Cognitive Framework' to Control Cognition and Behaviour

4.10 Two Types of Cognitive Forms: Cognition A and Cognition B

What characteristics does UPS's method of constructing the interrelationship between cognition and action have compared to TPS and LPS? First, creating an interrelationship between human cognition and action to smooth and control the execution of production behaviour is commonly practised, to some extent, in any production system. For example, with a Ford-style conveyor belt, operators' cognition is limited to a specific part

[8] To summarise this section, several key points should be mentioned. As previously discussed, UPS adopted a holistic educational perspective, which has allowed us to realise its significance. The organisational foundation of UPS is partly grounded in the practical accumulation of vocational training. However, it extends beyond this; UPS also depends on modern standardised industrial production methods: the division of labour and standardised product formation. These three elements are, deeply interconnected with manufacturing engineering, forming the fundamental components of UPS. Notably, in addition unique Swedish conditions are integrated into these elements, specifically the theoretical and practical accumulations of STS, the grouping theory of individuals and the 'symbiosis' of technological conditions. We recognise that these Swedish conditions cultivated the potential for collective cognitive formation within organisations, facilitating the expansion of standardised cognitive processes. Although distinct, yet sharing certain similarities with LPS and TPS, the integration of the following elements is indispensable for LPS and TPS: the intended integration of modern industrial division between machine and man power, manufacturing engineering functions and the organisational cognitive conditions intended for productivities within each company.

of the line. The cognitive content mobilised there is formed around the designated work. Therefore, the range of operators' cognition is limited to the work area operators are responsible for, which aligns with the subdivisions created by line production.

However, based on the same Ford-type conveyor line in TPS, operators' cognition is actively expanded to increase the scope of task responsibilities. The cognitive methods and knowledge mobilised are also developed through processes such as on-the-job training. So, when comparing TPS, LPS and UPS, what differences are evident regarding the methods of using cognition, the cognitive structures employed and their relation to human behaviour? The following discussion refers to the 'holistic cognition common to each action'. This represents cognitive elements commonly used across individual occurrences as 'principles' or 'values' as Cognition A (hereafter, Cog. A). In contrast, we consider 'instrumental cognition used in individual work and tasks', (e.g. '5S', '5 Whys', 'Kanban' and 'Poka-Yoke'), to represent Cognition B (hereafter, Cog. B) (e.g. Liker, 2004).

4.10.1 TPS Extends Individual Work (Cog. B) and UPS Extends Shared Work (Cog. A)

First, UPS displayed Cog. A for performing work tasks, i.e. the cognitive structure for recognising individual work processes using multiple maps. After structuring cognition in this manner, how can operators acknowledge their tasks and perform actions without mistakes? Shared attention items at UPS for all Cog. B tasks were set through the 'five whats' (Ellegård et al., 1991, pp. 42–48; Nilsson, 1995, p. 80) while preparing common tools for confirmation during task execution. In the case of UPS, there was no external technological system that regulated and controlled human behaviour from the outside, like a production line. Therefore, Cog. A, which supported overall cognition, was significant for carrying out production actions. Cog. A served as the basic information for the understanding of Cog. B, which involved performing work checks.

In contrast, in TPS and LPS that use production lines, the balance between Cog. A and Cog. B differs significantly from UPS. As such, TPS and LPS characteristics are considered an expansion of Cog. B, which is the 'instrumental cognition' during work (e.g. 5S and the five whys), rather than Cog. A as the 'overall cognition common to each action'. The reason is that in a technical structure based on assembly line work like TPS, the work cycle time is limited to about a minute. Therefore, the overall cognitive area that corresponds to an operator's Cog. A is extremely small and limited compared to the Uddevalla factory because cognition is focused and concentrated on carrying out tasks within the assigned process.

4.10.2 Cog. B to be Organised in UPS

Next, we ask if the type of instrumental cognitive element of Cog. B, which is prominent in TPS, is used in UPS? The answer is yes. Specifically, this element comprises the 'five questions' formulated by Nilsson (Nilsson, 1995). These five questions are meant for operators to ask themselves when performing their actions. It allows them to form explicit questions through their actions, common to each specific behaviour for self-reflection. Thus, the 'five questions' of Cog. B penetrate each particular action, becoming a cognitive form that governs behaviour.

In addition to this, the structure of the viewing method for operating processes as shown in B in Fig. 4.2, which is a cognitive process organised by the 'work module', should be recognised as Cog. B. Thus, the cognitive structures in the UPS, comprised of Figs. 4.1A and 4.2A, are organised mutually and formatted to support assembly work. That is to say, the relationship between Cog. A and Cog. B is more systematic than in TPS or LPS.

4.11 Position and Role of Manufacturing Engineers

4.11.1 The 'Ohno Line' and UPS

How was the development of Cog. B, which achieved significant expansion in TPS, conducted? This question is important in demonstrating the organisational characteristics of TPS. The role of manufacturing engineering plays is crucial in the creation of Cog. B. Manufacturing technology accumulated by a manufacturing engineer is a unique perspective that analyses and organises on-site work from technical and work action aspects. Thus, the role of manufacturing engineers represents a new area founded by Taiichi Ohno where technology and people overlap. This is believed to have formed during the 1940s and 1950s after World War II with the 'Ohno Line' (Wada, 2009, pp. 297–315).

Some explanation points are necessary. Ohno is known as the father of TPS. He advocated for automation with a human touch, where people are deeply involved in controlling machine systems. The TPS constructed by Ohno emphasised the importance of machine control (Ohno, 1980) while creating a technological device that allowed operators to deal with when machine trouble occurs. This was accomplished by displaying the discrepancy between production quantity and planned quantity through the Andon system and installing stop cords parallel to the line. These management techniques notify discrepancies between machine technology and human motion as an 'unusual situation'. The technical management tools used in TPS are effective means for humans to regulate machines. They are also used as devices that create managerial efficiency. It is said that these have been practically conceptualised and formalised as Cog. B. Thus, TPS thoroughly uses various regulate methods from the human side against the machine system, formalising them and organisationally utilising their effectiveness to regulate technology. The development of these control technologies was the result of close attention to on-site work.

4.11.2 Position of Manufacturing Engineering and Japanese Generation

The establishment of 'the Ohno Line' as the technical approach to work process reflects a historical progression that depends on organisational composition inherent in the Japanese corporate organisational structure. Why is it considered Japanese? The organisational position and role of manufacturing technology are unique. In other words, in Western technology design, the role of the production engineer responsible for production design extends to the design of manufacturing processes and work. In contrast, the role of engineers in Japanese companies is divided into two types: production engineers who establish production processes and manufacturing engineers who also bear responsibility for work execution.

The latter, manufacturing engineers responsible for work execution, are positioned between the production engineer and work performers, playing a unique role in connecting the two. This organisational structure is noted to have resulted from the post-war economic recovery period during the late 1940s and early 1950s; there was rapid recovery in demand to cope with the increase in product quantity and to realise 'multi-skilled' work by connecting engineers with the production site (Fujimoto, 1999, p. 64). However, the position of production engineers and manufacturing engineers and the role-sharing with team leaders differ depending on the regional characteristics of the plant. This suggests that it also varies according to the organisational structure of each company (Tamura, 2006, 2023).

4.11.3 Elements of Manufacturing Engineering Knowledge Used at UPS

Notably, the manufacturing knowledge created by engineers who observed production on shop floor, and the use of this knowledge, formed the basis of UPS production design. As indicated by UPS's maps and part kits, the cognitive structure of operators was incorporated into the organisation of production concepts. Incorporating the operator's cognition implies that the operability of manufacturing actions from a manufacturing engineering perspective could not be integrated into production concepts without analysis. The position and role of manufacturing engineering were also utilised in UPS, leading to production and work design. In other words, the realisation of UPS would have been impossible without a manufacturing technology perspective.

UPS was successful in connecting the domains of work execution processes and technological organisation from a new perspective. This led to new methods of production and control systems. Manufacturing engineering is characterised by designing work processes from both work execution and operability perspectives. UPS had also designed layouts from a technological perspective while simultaneously considering operability and organising control of work through maps created from the operators' perspectives by using cognitive domain.

Thus, forming cognition from the operators' perspective created a new method of work control. Moreover, the organisation of cognition from the operators' side expanded the operators' cognitive breadth. For example, operators discovered technical issues like buffering at the installation location of a fuel engine's pipe during assembly tasks (Blomgren & Karlson, 1995, pp. 127–128). The acquisition of the operators' perspective and the expansion of work cognition are influential in enhancing product and production process understanding, the quality of the work object and the response to unusual situations or operations.

In other words, TPS and UPS share a commonality: They integrated manufacturing technology domains and elements, leading to new production concepts. The unique position and role of the manufacturing engineering domain, linking technology and human behaviour, continues to be emphasised because it connects human behaviour to productivity, particularly with the advancement of IT technology. Manufacturing engineering is expected to remain a significant domain in production reform, closely related to cognitive engineering.

4.12 Cognitive Frameworks in Work Design in TPS and LPS

4.12.1 Prominent Expansion of Cognition B in TPS and the Role of Cognition A

Comparing the 'five questions' corresponding to Cog. B used in UPS's process of action with TPS, it is evident that the existence and thorough expansion of Cog. B in TPS is significant. It presents the prominent position of Cog. B highlighting the characteristics of TPS. In other words, TPS has shaped instrumental cognitions such as the 5S and five whys in the realm of Cog. B into organisational behaviour norms and has sought to establish them organisationally. However, it is crucial to remember that this thorough method is deeply connected with Cog. A in terms of function.

The methods and means of production described in Cog. B requires value judgements in the process of work execution that relies on those methods. Judgements include the understanding and interpretation of the methods on the operator side when they use them with the agreement on using these methods and means. Therefore, Cog. B sets an understanding regarding the process as an organisationally shared philosophy. Cog. B becomes a shared organisational philosophy and concept linked with the individual value cognition of Cog. A, operating as a consistent cognitive system to enhance competitiveness as a whole. By connecting with Cog. A (which represents the individual operator's value recognition), Cog. B operates as a consistent cognitive system that enhances overall competitiveness.

TPS realises corporate success by linking Cog. B (which represents the cognitive understanding of methods) with Cog. A (the organisationally shared value recognition). This connection transforms the use of tools into a method of self-realisation that embodies organisational philosophy (Tanaka, 2005).

Therefore, enhancing Cog. A while systematising Cog. B is the foundation for improving production efficiency in TPS and LPS. The coordination between Cog. A and Cog. B and the systematisation of Cog. B take the form of a unique cognitive style and a combination of cognitions formed within an organisation over time. Consequently, this interplay of Cog. A and Cog. B, as intrinsic content, becomes a unique competitive factor that is challenging to transfer across organisations. Since Cog. B is an individual method or tool used in specific job responsibilities, it is easier to introduce. In contrast, Cog. A, which represents the value aspect, requires the 'installing thinking pattern and behaviour pattern' and on top of that, demanding responsibility (Ihara, 2017, p. 40). As a result, implement Cog. A is time-consuming because it cannot be easily localised.

4.12.2 Factors Preventing the Introduction of TPS and LPS

Therefore, integrating TPS and LPS actualises the collaboration between Cog. A and Cog. B and often encounters difficulties. Namely, adopting Cog. B, a method-oriented action tool for individual tasks, poses no issue for Japanese companies that readily accept Cog. A as a common corporate value. On the other hand, in Western organisations where operations are primarily based on job performance within specific roles integrating Cog. B with Cog. A is often viewed as a management practice that builds upon individual operators' behaviour patterns. For example, 'Kaizen' as a thinking pattern, which belongs to Cog. A, requires organisational consensus on the meaning and value of Kaizen activities or behaviours. Therefore, organisational cognitive adjustments are required. Consequently, as Womack recognised, the term 'creative tension' in thinking (Womack et al., 1990, p.102) indicates that organisational response is required to implement LPS.

The characteristic use of organisational cognition in TPS is not easily captured from the traditional perspective of the organisational division of labour. It is thought that Cog. A and Cog. B form a cognitive framework that operates the production system together. Therefore, understanding TPS and LPS is insufficient based on just a technical aspect or division of labour perspective. However, the position of the cognitive framework that integrates the cognitive domain, which mobilises technology and division of labour, should be grasped at this point. If this point is not understood, it is difficult to understand the advantages of TPS and LPS. The complexity of TPS and LPS lies in the challenge of concretely grasping this shared organisational consciousness and action (means) cognition. The integration of Cog. A and Cog. B and how to construct a cognitive framework is a managerial challenge and the most critical element for transferring LPS.

In this paper, the construction format linking Cog. A and Cog. B becomes an organisational issue that should be examined for the development of LPS. UPS has also incorporated the cognitive factors of Cog. A and Cog. B uniquely in its production execution while integrating it into the production design.

4.12.3 Final Remarks

4.13 UPS Shows Implications of Building a Technology-Cognition Interface

This paper has reviewed the historical progression of production reforms in Sweden, symbolised by UPS. From an STS theoretical perspective, an essential result of the UPS reforms is the realisation of a production method that leverages human cognitive abilities without relying on external technological devices such as production lines. The central feature lies in the integrated organisation of technological, cognitive and organisational arrangements centred assembly-oriented work design. For instance, several STS practices in Sweden have aimed to achieve work autonomy in organisations based on adjustment revisions in the line layout or by introducing buffers. The reform perspective demonstrated at the Kalmar plant can be considered a restructuring of the division of labour of production and organisations.

In contrast, UPS restructured work division based on traditional production line technology. They moved towards incorporating human cognition into production. This led to the realisation of a highly autonomous production organisation that relied on human cognition and work collectivity as technical and organisation design integrally. UPS succeeded in restoring autonomy by focusing on the inner structure of products while organising cognitive and technological domain in conjunction with each other. They effectively elucidated the relationship between these inner structures and cognitive domains, thus controlling human action on individuals and organisation both. The meaning of this restoration refers to the establishment of the actors' discretion and making room for action based on their own judgement by individuals. Autonomy here refers to the capacity for individual judgement and choice of activity depending on technological and organisational factor in production. This autonomy formed the foundation of UPS's organisational flexibility. UPS is a significant example for the realisation of new relationships between technological, organisational and human cognitive capacity in Swedish conditions.

As discussed in this chapter, the analysis of technology and cognition at UPS and the construction of their interrelationship serve as an important practical example for deepening the understanding of TPS and LPS. However, this insight is not limited to the production field. Cognitive matters provide a valuable practical example for designing and implementing human-technology interfaces as the use of IT expands across various fields of business and society. This is because of an environment where the separation and integration of technology and cognition become increasingly complex due to IT development.

Under these conditions, the significant challenge for organisations will be how to use human resources and advance organisational behaviour through shared understanding and cognitive formation rather than through direct technical management control. UPS's practices offer many insights into the organisational development and utilisation of cognitive function. UPS provides a basis for understanding the importance of a new relationship between cognition and motivation, as well as the relationship between cognition, human behaviour and technology. We suggest that the innovative contents of UPS must be re-evaluated to enhance for future applicability.

References

Agurén, S. (1984). *Volvo Kalmar revisited: Ten years in experience*, Development Council, SAF/LO/PTK.

Akaoka, I. (1989). *Sagyou Soshiki no Shin Riron* [The new theory on industrial organisation]. Chikura Shobo.

Berggren, C. (1992). *Alternative to lean production: Work organization in the Swedish auto industry*. ILR Press.

Blomgren, H., & Karlson, K. (1995). Assembly skills, process engineering and engineering design: An example from the Volvo Uddevalla plant. In Å. Sandberg (Ed.) *Enriching production*. Avebury.

Ehn, P. (1993). Scandinavian design: On participation and Skills. In D. Schulter, A. Namioka (Ed.), *Participatory design: Principle and practices*. CRC Press.

Ellegård, K. (1989). *Dokumentation av metalls medverkan i projekteringen av volvos uddevallafabrik; underperioden mellan våren 1985 och vintern 1987/88*, Metall.

Ellegård, K., Engström, T., & Nilsson, L. (1991). *Reforming industrial work: Principles and realities: In the planning of Volvo's car assembly plant in Uddevalla*. Arbetsmiljöfonden.

Ellegård, K., Engström, T., Johansson D., Nilsson L., & Medbo L. (1992). Reflective Production in the Finall Assmebly of Motor Vehicles-An emerging Swedish challenge. *International Journal of Operations & Production Management, 12*(7/8)).

Ellegård, K. (1995). The creation of a new production system at the Volvo automobile assembly plant in Uddevalla, Sweden. In Å. Sandberg, *Enriching production*. Avebury.

Engström,T., Medbo, L. (1992). Precondition for long cycle time assembly and its management-some findings. *International Journal of Operations & Production Management, 12*(7/8).

Engström, T., Jonsson, D., & Medbo L. (1998, September). The Volvo Uddevalla plant and interpretations of industrial design process. *Integrated Manufacturing Systems, 9*(5).

Engström, T., Jonsson D., & Medbo, L. (2001). Some cognitive engineering aspects applied to product and process Information within the Swedish Automotive Industry, presented CoCKEAS.

Engström T., Medbo, L., & Jonson, D. (2008). *Alternativ Montering*. IF Metall.

Fujimoto, T. (1999). *The evolution of a manufacturing system at Toyota*. Oxford University Press.

Gibson, J. J. (1997). *The ecological approach to visual perception*. Houghton Mifflin.

Granath, J. Å. (1991). *Architecture technology and human factors - Design in a socio-technical context*. Chalmars University.

Gyllenhammar, P. G. (1977). *People at work*. Addison-Wesley.

Ihara, R. (2017). *Muda no Kaizen, Kaizen no muda*, [Kaizen becomes Muda and Wasteful Kaizen. Sakurai Shoten.

Ishihara, S. (2003). 'SAF gijutsubu' [SAF technical department], *Rikkkyo Diagaku kenky*u, *57*, 1, Rikkyo University.

Karlsson, J. (1995). The Swedish sociotechnical approach: Strengths and weakness. In Benders et al. (Eds.), *The symbiosis of work and technology*. Taylor & Francis.

Liker, J. (2004). *The Toyota way*. McGraw-Hill.
Medbo, L. (2003). Assembly work execution and materials kit functionality in parallel flow assembly systems, *International Journal of Industrial Ergonomics*, 31.
Miyazaki, K., & Ueno, N. (1985). *Shiten* [View point]. Tokyo University Press.
Monden, Y. (2006). *Toyota production system* [The Toyota production system]. Daiyamondo.
Nilsson, L. (1995). The Uddevalla Plant: Why did it succeed with a holistic approach and why did it come to an end? In Sandberg, Å. (Ed.), *Enriching Production*. Avebury.
Nohara, H. (2007). *Gendai no bungyo to hyoujunka* [Modern division of labour and standard work at shop floor]. Takagake shuppan.
Nonaka, I. (1990). *Chishiki sozo no keiei* [Knowledge-Creating Company], Nihokeizaisinbunsha.
Ohno, T. (1980). *The Toyota production system* [Toyota production system]. Daiyamondo.
Rasmussen, J. (1986). *Information processing and human-machine interaction: An approach to cognitive engineering*. Elsevier Science Publishing Company.
Sandberg, T. (1982). *Work organization and autonomous groups*. ORD&FORM.
Sandberg, Å. (1992). *Technological change and codetermination in Sweden*. Temple University.
Sandberg, Å. (1995). *Enriching production: Perspectives on volvo's uddevalla plant as an alternative to lean production*.
Suchman, A. L. (1987). *Plans and situated actions: The problem of human-machine communication*. Cambridge University.
Tamura, Y. (2006). Japanese production management and improvements in standard operation: Taylorism, corrected Taylorism, or others? *Asian Business Management, 2006*, 5.
Tamura, Y. (2022). Seisan Kodo ni okeru bungyo to roudo [Division of labor and cognitive functions in production behavior: Reconsideration about Uddevalla Plant]. *Roumu Riron Gakkaishi, 30/31.*
Tamura, Y. (2023). International comparison about management organizations for lean production—A trial comparison study between Sweden and Japan—. *Touho-Gakushi*.
Tanaka, M. (2005). *Kangaeru Toyota* [Thinking Toyota). Business sha.
Ueno, N. (1999). *Shigoto no Nakano Gakushu* [Leaning during Work) Tokyo University Press.
Wada, K. (2009). *Mono Zukuri no Guwa*, [A parable of manufacturing). Nagoya University Press.
Womack, J. P., Jones, D. T., & Roos, D. (1990). *The machine that changed the world*. Rawson Association.

Yamauchi, Y. (2012). Sankagata Design to sono Atarashii Tenkai [New development in participation oriented design) *System/Seigyo/Joho*, Vol. 56.
Yuasa, Y. (1996). *Gendai Roudo Katei no Kenkyu* [A study in modern labour process). Ehime University Press.

CHAPTER 5

Operators' Work Situation in Volvo Uddevalla

Lars Medbo and Carl Wänström

5.1 Introduction

In the last two decades, much research has focused on lean production. Notably, research has examined different types of waste, tools and methods on a conceptual level with a few empirically grounded studies (e.g. Jasti & Kodali, 2015). This is unsurprising since most lean initiatives in Europe and the United States surround improving operational excellence, often achieved through event-driven projects by experts focusing on methods and tools (Jasti & Kodali, 2015). In contrast, Toyota's success is rooted in its culture of changing and improving social (i.e. human behaviour) and technical systems (i.e. organisation and artefacts) as a basis for long-term sustainability. This includes significant investment in employee development and product value streams (Liker &

L. Medbo (✉) · C. Wänström
Department of Technology Management and Economics, Chalmers University of Technology, Gothenburg, Sweden
e-mail: lars.medbo@chalmers.se

C. Wänström
e-mail: carl.wanstrom@chalmers.se

H. Nohara and L. Medbo (eds.), *Technology Convergence and System Divergence*, https://doi.org/10.1007/978-981-96-1910-8_5

Hoseus, 2008). The Volvo Uddevalla factory represents a change in social and technical system development based on experiences from several production systems designed within the Swedish automotive industry. For instance, the Uddevalla factory can be seen as the future of the automotive industry in achieving sustainable industrial work. In Uddevalla, to effectively satisfy customer needs, significant effort is placed on focusing shop floor design and operators, where value is created. Thus, it is essential to understand the operators' work environment and content to develop essential prerequisites for successful production.

As a theory of human motivation, self-determination theory (SDT) can be used to understand operators' work environment and how it impacts operational performance and health. Deci et al. (2017) developed a basic SDT workplace model comprising two independent variables—work context and individual differences. The model also includes two dependent variables—work performance and health and two mediating variables—basic psychological needs (BPN) and motivation. The mediating effect of BPN and motivation has been illustrated by several authors (e.g. De Cooman et al., 2013; Trepanier et al., 2015). Consequently, operators' BPN (autonomy, competence and relatedness) satisfaction is essential to achieve operational performance and health. In the SDT basic model, work situation focuses on the social system. However, to capture operators' complete work situations, technical system dimensions must also be considered. Therefore, this study explored operators' social and technical work situation at the Volvo Uddevalla plant and how the organisation considers operators' need for autonomy, competence and relatedness. The study scope surrounds operators' work situation. Therefore, the individual differences variable, such as studying employees' general causality orientations (Ryan & Deci, 2017), is beyond the scope of this paper. The study findings are based on operators' perceptions of their work, differing from Chapters 2 and 3. The cross-disciplinary theoretical framework is based on SDT, job design theory (JDT) and socio-technical systems theory.

5.2 Theoretical Framework

Over the past hundred years, extensive research has been conducted on work design, job design, job characteristics and their impact on performance (Parker et al., 2017). This interdisciplinary research approach (organisational psychology, industrial engineering, human factors and socio-technical aspects; see Campion, 1988; Edwards et al., 2000) and SDT have been helpful to conceptualise operators' work situation (Fig. 5.1).

5.2.1 The Technical System

This study focused on capturing the facility layout and type of production flow, as well as the equipment used to manage the work. The operating strategy (e.g. the company's production system), the organisational structure (e.g. type of organisation, hierarchy, span of control, chain of command, communication channels) and the management strategy (e.g. leadership ideals and leaders' decision authority) combined with the human resource strategy (e.g. pay contingency, introducing processes, formal education and training) all impact the work situation. The teams' structure and strategy (e.g. size, composition, roles, decision authority, purpose/goals) and support functions (e.g. industrial engineering, maintenance, etc.) also impact the work situation.

Task characteristics focus on 'how the work itself is accomplished and the range and nature of tasks associated with a particular job' (Morgeson & Humphrey, 2006). Hackman and Oldman (1976) employed JDT to identify task variety (a wide range of tasks), task identity (an identifiable piece of a task), task significance (impact on others), job feedback (direct and clear information) and autonomy as important factors affecting the work environment. The conceptualisation of autonomy has expanded to include freedom in work scheduling, decision-making and work methods (Morgeson & Humphrey, 2006). Cognitive demand has also been incorporated into JDT. This includes job complexity, i.e. the extent to which a job is complex and challenging to perform (Campion, 1988), skill variety (the job requires different skills and talents, e.g. Hackman & Oldman, 1976) and specialisation (deep knowledge in a specific area, e.g. Edwards et al., 1999). Wall et al. (1995) identified two aspects related to cognitive demand—active information processing (monitoring) and problem solving (job requires unique ideas/solutions and using the term innovation). These can be the physical demands of a job, such as physical ergonomics (appropriate posture and movement) and work environment (physical health hazards) (e.g. Edwards et al., 1999).

5.2.2 *The Social System*

Leaders can shape their subordinates' perceptions and impact their motivation (Fernet et al., 2015). Several studies indicate that transformational leadership promotes subordinates' BPN (e.g. Hetland et al., 2011). Transformational leaders exhibit four types of behaviours: idealised influence (e.g. role model, trust, respect), inspirational motivation (e.g. vision, high expectations), intellectual stimulation (e.g. encouragement, reframing problems, finding new ways, being creative) and individual consideration (e.g. coaching, personal attention) (Bass, 1985; Judge & Piccolo, 2004; Tavelin, 2013). This contrasts with transactional leadership (contingent reward, management by exception—active or passive) and a laissez-faire attitude (absence of leadership) (Bass, 1985).

Jungert et al. (2018) found that team members play an important role in increasing autonomous motivation. In a team, there is interdependence, meaning the dependence of work output from others (received interdependence) and the extent that others are affected by work output (initiated interdependence) (Kiggundu, 1981). Mullen and Cooper (1994) conducted a meta-analysis of group cohesiveness and found that task cohesion was critical to team performance. Group collective efficacy suggests that group members' shared beliefs about general effectiveness regarding multiple tasks impact group performance (e.g. Stajkovic et al., 2009). Psychological safety also affects the work of a group. This can be defined as 'a shared belief that the team is safe for interpersonal risk-taking' (Edmondson, 1999). Finally, teams can provide social support, meaning the *'extent to which a job provides opportunities for getting assistance and advice from either supervisors or co-workers'* (Humphrey et al., 2007).

An organisational culture can be analysed on three levels (Schein & Schein, 2016). The surface level includes artefacts and behaviour (Schein & Schein, 2016); however, both are handled at the organisational design level, leadership and team culture. The second level is norms and values (Schein & Schein, 2016), e.g. openness to ideas and valuing differences. The third level is underlying assumptions, such as being 'taken for granted' and a slight variation within the social unit (Schein & Schein, 2016).

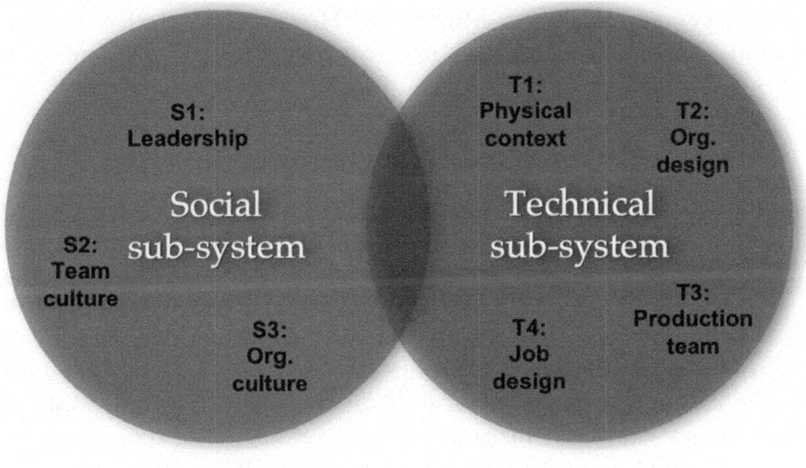

Fig. 5.1 A socio-technical model to understand operators' work environment based on the defined dimensions

5.2.3 BPN Mediate Work Performance and Health

Humans have three BPNs that must be fulfilled for an individual to flourish. A failure to satisfy these needs will result in reduced growth, integrity and wellness (Ryan & Deci, 2017). However, needs frustration differs from a lack of needs satisfaction, which means that '*not feeling volitional, competent and related*' differs from '*feeling oppressed, incapable and isolated*' (Trepanier et al., 2015). The following definitions of needs are used in this study: '*autonomy or the need to self-regulate one's experience and actions*', '*competence refers to our basic need to feel effectance and mastery*', '*relatedness concerns feeling socially connected*' (Ryan & Deci, 2017).

5.2.4 Work Performance and Health

Regarding the workplace, Deci and Ryan (2000) suggest that where psychological needs are supported and autonomous motivation is facilitated, psychological and physical well-being and improved work performance will occur, particularly in heuristic activities. Trepanier et al. (2015) suggest that autonomous motivation is related to positive outcomes, e.g.

work engagement and job performance. However, controlled motivation limits positive and promotes negative outcomes, e.g. psychological distress.

5.3 Methodology and Empirical Data

Regarding coding, Table 5.1 presents data from seven of the transcribed interviews with 17 assemblers at the Volvo Uddevalla factory that were conducted in the spring of 1993. The factory was shut down in late spring 1993. Each interview was videorecorded and lasted one and a half to two hours. The factory included five product workshops, 1–5, with seven to eight assembly teams (groups) for each. Each team assembled complete cars at full speed, about four cars per day. Product workshops had a product workshop manager and a coordinator as the manager's assistant. Product workshops 1–3 started first (1988–1989) with eight teams per workshop. Each team consisted of seven people with a work cycle of approximately 100 minutes. Cars were moved at one point during assembly between two workplaces. Product workshops 4 and 5, which started later (1989), had seven teams per workshop. Each team consisted of nine people with a work cycle of approximately 80 minutes.

The interviewed assemblers came from seven different teams—two teams from workshops 2 and 3 and one team from each of the other product workshops. From a team in workshop 2, five male assemblers were interviewed simultaneously during one interview. In the other interviews, pairs of assemblers were interviewed. Three pairs consisted of women, two pairs of men and one pair was mixed. One pair, two women, formed a two-person team in product workshop 5—the two assembled entire cars themselves with a work cycle of just over five hours.

Table 5.1 Coded interviews of the main groups based on the theory section

Technical sub-system	*Technical sub-system*
T1: Physical context	S1: Leadership
T2: Organisational design	S2: Team culture
T3: Production team	S3: Organisational culture
T4: Job design	

5.4 Operators' Work Environment and Performance (First-Order Analysis)

T1: Physical context

Physical work environment: The physical working environment was considered good. It was essential to adjust the product to an adequate position and to use the proper technique when assembling, e.g. using hand machines. High work tempo (without recovery) and repetitiveness (e.g. non-work rotation) can lead to musculoskeletal disorders. Assemblers indicated that they had become stronger over time and thus handled the load better.

> *'You have yourself to blame if you don't adjust the car (working position) so that you get good ergonomics. We know exactly when to lower, raise,* etc. *It is immediately felt in the body if we do not adjust the assembly position as we should. Some ignore it and then they become worn out'.*

Tools, machines and transport equipment: It is essential to have tools organised and the pairs of assemblers agreeing on where the tools should be placed on the tool carts. Consequently, tool carts among production teams differed. Material kit deliveries were delayed sometimes due to the Automated Guided Vehicle system being at a stop or when parts were missing from the material kits. This caused disruptions that spoiled the assembly flow with consequent productivity and quality issues.

Kit design: The material kits were generally considered good with a combination of shelves, boxes and bags. Material kits meant that even during an interruption (e.g. a meeting), it was easy to return to assembling. However, material kits were organised based on a compromise between two assembly setups, one in product workshops 1–3 and another in product workshops 4 and 5. This meant that some parts ended up in the 'wrong' place. On kit rack shelves, parts were placed in a standardised order corresponding to the assembly sequence and always in the exact shelf location. From an ergonomic perspective, heavy parts should not be at the bottom. However, this made the rack unstable. Brake pipes and mouldings could have been batched instead of having them on the racks since it became crowded and parts became tangled.

The kit design effectively identified boxes and bags with numbers and names that stipulated the assembly order. Boxes were also affected by the compromise between the two assembly systems—some parts had to be handed over between assemblers. The kit bags provided cognitive support for the assemblers to remember assembly instructions because they received instant feedback if they had missed something. Notably, materials left in bags would make it difficult and time-consuming for assemblers to pick the material themselves.

> *'It's good with the material kits - it makes a better job - if you open a bag, you get rid of it, then you know that you assembled everything and put all the parts. You have the sequence. The worst thing is if there is too much in the bag.'*

Product and design: There was frustration regarding product design and despite several proposals from operators, nothing changed. The product was designed in such a way that many different bits and sockets were required to tighten screws. To avoid constant changes, the assemblers tried to partially change the assembly sequence; they partially inserted screws as much as possible by hand and later tightened them with a machine. A change in product design was requested.

T2: Organisational design

Production planning: There were fixed start times for each car. The time was set to handle the most complicated product variants. When there were simpler product variants, operators could rest between assemblies. Consequently, teams that assembled complicated product variants did not complete fewer cars than teams that assembled simpler product variants. There was no deliberately levelled production plan for the team. From the operators' perspective, the lack of levelled production plans was problematic because sometimes there were only complicated product variants. Based on the Kaizen method for team improvement, before shutting down the factory, they reached a takt time of two hours.

Visual management board: Assemblers displayed their start times and could view weekly schedules. Team leaders updated the board with the team's production plan, noted outcomes, borrowed assemblers and reported it in the computer system.

The production engineer role: Production engineers were not part of the workshops in the beginning nor collaborating with assemblers. This led to a gap in trust and work improvement, where learning stayed within the team. However, when production engineers moved to the workshops (during the last year of production), the situation improved. The perceptions of production engineers shifted. Much criticism was evident when they were jointly located in a separate office building and were responsible for 'one screw each'. It was challenging to know which engineer to contact and it was difficult to get a hold of them. Once they were contacted, production engineers had to consult with head office colleagues, which took time. This led to negative experiences for assemblers with problems requiring immediate solutions. Over time, the production engineers' skills improved and the reorganisation of engineers being placed in different product workshops was positive.

Payment: The bonus system meant that the team received bonuses for good quality and productivity. However, the amounts were so small that it was not perceived to affect the assemblers. Salary was also affected by how many work cycles an assembler could perform and the number of specialist roles an assembler held. However, a lack of training was problematic. For example, some team leaders lacked the training to become team leaders and, thus, the related compensation. This led to operators dropping out of specialist roles.

Organisational learning: Institutionalisation of gained improvements in assembly was poor. 'If someone comes from another team and works with us, they give tips on how they do it. And when you work with someone else, you notice that they work in a different way that might be better'. Thus, the assembly methods of teams and product workshops, along with how tools were arranged, differed significantly. Production engineers wrote the assembly instructions without the involvement of the assemblers. These were used only by beginners. There was no formal organisation to share common, improved working methods, only short-term attempts at quality meetings.

T3: Production team

Relationships between the production team and the organisation: An increased exchange between different organisational factory parts would have improved a shared understanding of each other's work and how they influenced each other. Special mention was made of the relationship between assemblers and material handlers who picked the material kits. Thus, there were issues with the picking quality because assemblers and material handlers were not organisationally linked. It would have been beneficial if the material handlers belonged to the assembly team.

Team size: The assemblers generally thought that the most common team size in the factory of seven people was adequate, preferring fewer people. 'It will be easiest to get good team spirit'. The more people there are, the more difficult it would become to rotate between balances (i.e. work tasks assigned to one operator to be performed during one work cycle).

Team composition: Upon employment, assemblers started in a training workshop or training team and were placed in production teams. Criticism was directed at how this was done in the beginning. People who did not collaborate well together sometimes formed pairs of assemblers and teams. Gradually, however, team and division into pairs of operators became more efficient through reorganising and relocating operators.

Team competence: Competence was sufficient among teams. This was considered the most critical aspect of work production. In many teams, some assemblers knew practically the entire work cycles of the car. Some teams, however, demanded team members to have additional skills. A resulting dilemma was that each assembly pair typically worked hard on their two balances. This meant that they would lose the competency to work with the other balances that they had learned, especially when working at full speed. The assumption was that within a week's work, assemblers could regain competence and pace.

Team collaboration: Collaboration worked well in many of the teams. It was believed that 'you have to be a bit of a team player to work here'. Collaboration was seen as an effective way of 'putting pressure on people'. A pair of assemblers helped each other if necessary during the work cycle. If one finished their assembly before the other, indirect work was done, such as administration, sorting and feeding materials, etc. However, help with assembly if necessary would mean that both assemblers finished at the same time. Even pairs of assemblers within teams helped each other, primarily through indirect work, to maintain a good team flow. Between assemblers in a pair and between pairs of assemblers, there were many dependencies for assembly. If any disruption occurred for an assembler, e.g. in the event of missing material, replacement material needed to be ordered. These were often temporarily borrowed from another car's material kit and then handled for the delivery of the missing material. These disruptions affected the entire team's workflow and led to an uneven workload within and between pairs of assemblers. It was believed that the team leaders could have had a role in reducing such disruptions. To mitigate the consequences of disruption, it was essential for assemblers to have the skills to conduct work upstream and downstream of their own balance. However, these skills were lacking among many teams since assemblers became specialists in one or two balances. It was also desirable to consider that the assemblers' physical capacity was evenly distributed among teams so that young, physically strong male workers assembled the second part of the car while physically weaker workers did not assemble the first part (physically heavier work), which meant friction within the team. However, assemblers had relatively good opportunities to change teams to achieve well-functioning teams.

Team absence: Team leaders for different product workshop teams agreed on staffing and loaning assemblers between teams when needed. The last year of operation, a common team pool was introduced in the factory, which is expected for the five product workshops. This can reduce the need for loaning in and borrowing assemblers across teams. Introducing the team pool was intended to help teams work at full speed across all workshops, even in the absence of assemblers within a team. Experienced assemblers with significant competence would work in the team pool. Initially, some workers did not feel comfortable in their teams when the team pool was applied. However, after running-in problems, the assemblers thought that the team pool worked well. This reduced team stress in the case of absences, where teams did not have to lend assemblers, thereby providing stability and security. Remaining issues, however, were that the pool of team assemblers sometimes lacked motivation and their assembly sequence did not always correspond to the assembler they replaced. This meant the ingrained coordination in regular assembler pairs was lost and the work speed initially dropped. As such, the level of competence among the pool of team assemblers was essential. While these assemblers had no formal role in disseminating factory knowledge, they disseminated some examples of effective work task development. In the event of a planned short-term absence within a team, extra work was accomplished by starting earlier in the morning or working longer in the afternoon so that the 'lone' assembler in a pair of assemblers would not have to work independently for too long.

Specialist roles in the team: In addition to assembling cars, there were several specialist skills that operators could acquire. For example, tools, production engineering, quality responsibility, data/IT, human relations, team leadership. These skills were not assembly work related but necessary for the team's production to function satisfactorily. In traditional organisations, it is common for supervisors or other white-collar workers to perform these tasks. It was common for an operator to have one specialist skill. Often, more than one operator was a team leader—they changed once every one to two months. The team leader was in contact with the product workshop manager. They planned and formed the team and coordinated with other teams. This included much of what was perceived as burdensome administrative work. It was considered that the system of specialist competencies worked well. While this was effective for team leaders, the work was challenging because team leaders were also simultaneously assemblers on a balance. This means they had a high workload and were held accountable to the product workshop manager. If a team consisted of eight instead of seven members, this would have worked better as the team leader would have been able to take care of many indirect tasks to support assemblers. One would like the team to influence planning more, e.g. for holidays. Team leaders were expected to attend a three-day course; however, many team leaders lacked this course.

Team performance: A team typically assembled about four cars per day, where a pair of assemblers assembled four cars per day. In the beginning, many thought it was impossible to accomplish such an achievement. If annoyances such as interference were mitigated, even higher performance could be achieved. It is impossible for two individuals to work at the same speed or even for one to work at the same speed when a work cycle is repeated. However, it worked well in pairs of assemblers based on the flexibility of supporting each other and distributing work tasks. When teams perform well, members would like to benefit from this later, e.g. being able to slow down or be trained on new balances or products.

Training time in the team: 'If we had been educated more, it would have worked to help each other. For example, to level production'. Since separate training time was requested in teams, it would have made it easier to rotate between balances and address disturbances while becoming more flexible. There should also be a minimum skill level for each team.

T4: Job design

Standardised work: The assemblers largely distrusted the standardised work and allocated time to carry out their work. In the agreed-upon standardised work, the respective balance of assembler pairs should take the same amount of time. In reality, the assemblers perceived that it took different amounts of time. They often viewed the production engineers' work design as not useful. The difference between theory and practice is significant. Pairs of assemblers distributed work between themselves while adjusting work tasks depending on the product variant. There was a relatively good balance in the amount of work between various assembler pairs. However, the nature of work differed at the beginning and end of car assembly. The work in the beginning mainly involved attaching parts with screw joints. The end involved more fitting of parts. The more experienced the assembler became, the faster it was possible to fit parts at the end. In contrast, machines used to draw screw joints do not become faster over time. Levelling of production was attempted during specific periods. For example, every other car had a complicated product variant requiring more work and every other a simpler product variant requiring less work. However, batches of similar products were often assembled in terms of the amount of work.

Assembly instructions: Only beginners students used the assembly instructions. Experienced assemblers did not access the assembly instructions on the computer. It would have been easier to have a booklet with pictures for each material kit box. The assembly instructions, printed and via computer, were largely unused during production runs. Workers would become tired of many loose papers. Product changes were presented with separate documents by production engineers. Instructions with or in the form of, pictures were requested. Furthermore, when ordering parts missing from material kits, the assembly instructions did not help the assemblers to determine the right parts to order.

Work pace: Work experiences differed among assemblers. For the work to flow as smoothly as possible, many assemblers tried alternative assembly methods, such as reducing the need to use other tools. As previously mentioned, full speed indicated that each team with seven assemblers assembled four cars per day. To maintain the planned pace, it was important that the pair of assemblers coordinated well with each other and that no disturbances, e.g. incorrect or missing material or product design uncertainties, occurred. Notably, assemblers experienced differences from being a beginner to an experienced assembler. Many assemblers said they got up to speed by trying different work approaches—they gradually found more efficient ways to carry out the assembly, individually and in pairs.

Job variation: Even though there was a work cycle of hours compared to minutes on an assembly line, many still experienced the work as monotonous, especially if they only assembled one balance. Yet no one wanted to work on a traditional assembly line either. To maintain job variety while finding better assembly methods, some assemblers changed the assembly sequence—others rotated with co-workers and many would have liked to rotate to more balances. However, demands for high speed, additional skills and personal convenience limited the rotation between balances.

Job rotation to maintain skills or specialisation: In many teams, pairs of assemblers became specialised in just two balances. To maintain competence on several balances, it was considered necessary to rotate balances once a week, which was done in some teams. In a few teams, a pair of assemblers assembled an entire car at full speed and according to quality standards.

Job complexity: Assemblers often considered the importance of planning the next steps. Some balances were perceived as more difficult than others. Still, the nature of the work differed, such as remembering which part to assemble or being able to assemble so that parts fit well, which may require adjustment (e.g. interior design). The latter was perceived as the most challenging to learn. With a more developed product design, these difficulties could have been significantly reduced.

Seeing the whole: Assemblers compared their previous experiences of working on an assembly line where minimal work was performed. In the Uddevalla factory, they felt motivated by seeing the whole picture, knowing that there would be finished products. Knowing what parts are called and their car function was also helpful. The fact that some cars were sold to individual customers was particularly encouraging.

Introduction to the workplace: The first employees received an introduction and tour of the entire factory, which was good. Those who were hired later lacked this overview and understanding. The training workshop, with access to supervisors, provided safety during the work introduction. This initial period with a separate training workshop to learn the basics was appreciated. It was essential that the teaching supervisor had a high level of competence and gradually allowed the mentees to perform the assembly themselves. To get up to speed, an assembler first needed a basis of competence. After that, they could be introduced to teams that worked at full speed. Assemblers typically had a supervisor for support in the first period. The first employees spent significant time in separate training workshops with low or no work rate demands. This was perceived as devastating regarding the ability to later reach full speed within the production team. If an assembler did not fit a balance for a month, the ability to assemble the balance at full speed would be lost. After starting to assemble the balance again, it would take about a week to reach full speed.

Competence: Most assemblers had formal competence corresponding to half a car or more, i.e. three to four balances. However, in the production run, they assembled one to two balances, depending on whether they switched with a co-worker. The perception was that it was possible to learn more. Many considered it likely that a pair of assemblers could assemble entire cars themselves. However, the rate at which it was possible was uncertain. An additional balance was thought to be learned in a week. However, it took a month to reach full speed.

S1: Leadership

Leadership types: Some managers' leadership styles were perceived as good, while others were poor. Good was indicated by being easy to contact, delegating responsibilities and rights to the team, not getting involved in everything, being available, informing the team about their whereabouts, being understanding and supportive when something went wrong and being trained or providing coaching to solve problems within the team. Poor leadership was indicated as being involved in everything, not completing or getting things done, not having knowledge of assembly work, not being familiar with the team's work, having victims of bullying, not being present on the workshop floor, not giving help or coaching, i.e. having a 'do as you want' attitude.

First line manager (product workshop manager): The role of the product workshop managers primarily comprised administration. Managers lacked knowledge of the assembly work itself. They also lacked an understanding of what full-time assembly entails and the need for team training. They could not change the standardised work themselves—this was decided by the technical organisation and its production engineers (for a description of production engineers' role in the company, see Chapter 4). The focus on increasing production was evident based on the relationship with the manager. Some managers actively supported teams when problems arose, such as conflicts, through coaching and making creative suggestions. Assemblers related to previous professional experience in traditional production where issues for the team or from individual work (e.g. quality defects such as incorrect assembly) came from the manager to the employee. In Uddevalla, it was instead the case that the feedback from quality inspection 'hit' the assemblers first, who, in turn, had to talk to the manager.

Supervisors were essential for new employees learning assembly work. In the beginning, there was a lack of supervisors with sufficient competence. Notably, the ability to teach beginners to reach full speed was lacking and demands for development were not set. Even transitioning from having a supervisor to working independently in production teams was perceived as challenging.

Workshop managers' group: There was too much pressure and competition. The managers sat in Tuesday meetings and competed over which workshop made the most cars. Thus, the product workshop managers competed 'and we became the objects that would do it for them'. Continuous production and full-pace tests were perceived to be a competition between product workshops. Product workshop managers frequently criticised teams in their workshops.

Information transparency: Criticism was made regarding the distribution of information from workshop managers. Assemblers would like to know more about plans, why various tests were done, how the bonus system worked, etc. Knowledge of what the workshop managers do was lacking.

S2: Team culture

The team leaders' role: Team leaders planned the staffing, borrowing and loaning of assemblers. Within some teams, team leaders decided who would work on which balances. On other teams, the other team members took over part of the team leader's duties, so they would not become overburdened. Generally, team leaders worked well in the product workshops. Since workshop managers were often unavailable in the product workshop, the supervisors (workshop managers' assistants) performed management tasks. Supervisors were typically former assemblers and, therefore, had an easier time understanding the various issues.

Team climate: The assemblers thought this worked well within the team. All assemblers emphasised the importance of appreciating co-workers. It was 'motivating to come to work and meet them here–we have a lot of fun' and 'here, the best of all is the workmates'. However, some assemblers had previous experiences of disharmony and conflicts within the team. Team spirit and how problems and conflicts were handled varied among teams. In most teams, there was a good atmosphere with dialogue between assemblers. When issues or conflicts arose, they solved them themselves or with the help of the product workshop manager. In some teams, however, there was friction. This happened that the team split into two parts, e.g. where physically strong younger male workers worked as fast as they could, often with negative consequences on quality. In some mergers of less well-functioning teams into larger ones, there were conflicts and problems with team cohesion—even those who previously worked well together had issues, with quality suffering additionally.

Pairs of assemblers and team pools: To make the work function well, pairs of assemblers needed to cooperate reasonably. When someone was lent to a team (e.g. from the team pool), it became complicated. It would take a few cars before a pair could cooperate reasonably well. The assemblers had well-established routines for how they worked. For example, routines for when they would start assembling a new car, how the tool carts should be arranged and where racks and material kits should be placed. Over time, pairs of assemblers would establish such a routine that 'I know almost exactly what the other person is doing and will do, you hear what the other person is doing'—the assembly becomes very synchronised between them. 'Because the two of us have worked together for a long time, everything works automatically, e.g. passing machines to each other. We could build an entire car without saying a word'.

Relationships between assemblers and within the team: Since no assemblers can continually work at the same speed, it does not help if balances are evenly distributed—a pair of assemblers will not complete their work simultaneously. First and foremost, assemblers always helped their co-workers to complete work simultaneously. Even during specific work steps during a balance, the two needed to be in sync, e.g. when the car's position is to be changed. After that, the team's other pair of assemblers could receive help. Assembly teams had a sufficient grasp of how they performed during assembly in terms of time in relation to the production plan. Pairs of assemblers were usually in sync. However, as little as a minor disturbance such as a missing detail in the material kit was enough to get out of sync. Teams required a high level of competence, preferably overlapping between balances. Many teams taught each other common initiatives from assemblers, which would persuade the product workshop manager.

In teams with lower competence, assemblers did not dare switch to other balances as substitutes in case of illness.

S3: Organisational Culture

Atmosphere: Some assemblers were irritated that what was communicated before they were employed in the factory did not align with reality. Several assemblers felt that trouble they did not cause, e.g. lack of material, affected the team negatively. They fell behind schedule and had to work hard to try to catch up but ended up in a vicious circle, which led to a negative atmosphere in product workshops.

5.5 Discussion and Conclusion

5.5.1 *The Socio-technical System at Volvo Uddevalla*

The production team organisation, with seven operators per team, provided good opportunities for operators to have their BPN fulfilled. However, workshop managers were responsible for eight teams, meaning their span of control included 56 operators. Consequently, there were few examples of transformative leadership, as shown in the analysis and tables. This confirms the findings of earlier studies that indicate a variation in the psychosocial work environment of different workshops (e.g. Engström et al., 1995). In workshops with transformative leadership, there was high quality, high productivity and a positive psychosocial work environment (Engström et al., 1996a). These positive examples were not spread to the rest of the organisation because there was no structured way for organisational learning and best practices. Consequently, the workshops were like islands, functioning as a factory within the factory.

The layout and parallelisation of production made it possible to achieve high productivity and quality compared to traditional factory lines (Engström et al., 1996b). However, based on how production planning and control were conducted, the performance measures did not illustrate this potential. Furthermore, complex product design resulted in lengthy assembly time. The planning and the takt were set based on the most complex product variants, leading to several wait times for operators. Thus, many had a long wait time before they could start assembling the next car based on the production schedule. This would have been necessary in an assembly line but was not required in this concept. Consequently, management planned Volvo Uddevalla as if it were an assembly line. However, studies indicate that working in pairs with group work patterns is efficient because there are no system losses (not waiting for each other). Therefore, assembly time per car was 11 hours in Uddevalla compared to 16 hours per car in line factories (Engström et al., 1996b).

The kit concept used to manage materials supply was considered positive. It served as a cognitive support for assemblers to remember their assembly instructions and provided a product and work overview. However, the kit system made the assembly system sensitive to disturbances due to material shortages. This was probably due to the design and management of the kit concept and not the concept itself. For instance, today's automotive industry, such as Toyota, where kits have been introduced in line factories, shows that it is possible to achieve disruption-free material feeding.

Generally, the assemblers enjoyed the work and stated that they found the work to be generally good. The friendships were positive and delays were predictable (e.g. stability regarding who you work with, work hours). Making cars sold to customers (many were that) was perceived as motivating because they could relate to the customers.

The managers played an important role. However, they lacked coaching skills and deep knowledge regarding the process in many cases. They were focused on achieving goals but not on how objectives could be achieved. Consequently, there was a limited focus on learning and development. For example, operators' skills to develop standardised work were not utilised. This subsequently led to operators being largely distrusting of standardised work.

Pairs of assemblers discussed everything possible regarding their work. It was essential to get along, be on an even playing field and enjoy each other's company. Cognitively speaking, pairs influenced each other. They found it boring if they had to work alone. There would also be at risk of being stressed to finish. Many assemblers also socialised outside of work hours. Pairs of assemblers developed their own way of communicating. For example, they created their own names for parts. This could lead to problems when replacement parts are ordered.

5.5.2 *The Socio-technical Impact on operators' Basic Psychological Needs*

The Volvo Uddevalla system positively impacted autonomy (see Table 5.2). The organisation and physical design were both mainly positive. The production team structure and team culture were significantly positive. This indicates that members of the production teams felt that they could influence their daily work, both in performing work tasks and arranging equipment and tools. They could also affect work planning within the team. However, leadership was more orientated towards controlling which limited the autonomy.

Table 5.2 Socio-technical system impact on operators' autonomy

Autonomy		*Positive*	*Positive & Negative*	*Negative*
Technical sub-system	Organisational design	Reporting from the team, board, computer, time use The team leader role, decision rights, responsibility		Production planning
	Physical context	Tools, machines and transport equipment		Deliveries of material kits to assembly
	Production team	Assembly sequence own team compared to other teams and influence Analysed and budgeted times, balancing and levelling of production Handling of absence in the team Specialist roles in team		
	Job design		Variation in own work and the pair of assemblers	

(continued)

Table 5.2 (continued)

Autonomy		*Positive*	*Positive & Negative*	*Negative*
Social sub-system	Leadership	Communication	Leadership types	Pressure and prioritization from the manager
	Team culture	The assemblers' feelings about the job Team leaders and responsibility Relationships between assemblers and the team - The work itself, new colleagues Relationships between assemblers and in the team - Balancing, following plan The assemblers relation to the work tasks	Relationships between assemblers and in the team – Learning	

The Volvo Uddevalla system positively and negatively impacted competence (see Table 5.3). The operators were proud that they could assemble the quarter of a car or more and they became skilled car builders. This positively impacted their feelings of efficiency regarding their work. However, workshop managers pushed teams to work harder, which affected them negatively. Many managers focused on increasing production volume through specialisation—assemblers would specialise in one or two balances, which limited skill development. Thus, growth and learning opportunities were something operators felt missing after learning the work cycles as beginners. As such, they often knew more extensive work than what they later practised, which evidently meant a loss of competence. A positive change during the last year was reorganisation. The production engineers moved organisationally and physically into product workshops, which meant that the operators received greater support from production engineers.

Table 5.3 Socio-technical system impact on operators' competence

Competence		*Positive*	*Positive & Negative*	*Negative*
Technical sub-system	Organisational design	Feed back	Importance of supervisors for learning	Product workshop manager role (first line manager)
		Introduction to the workplace		The production engineer role
		Reach full speed		Organizational learning
		Learn assembly work as freshman		What is the hardest thing to learn
				Training time in the team
	Production team	Analyzed and budgeted times, balancing and leveling of production		Use of budgeted non-assembly time in the team
		Specialist roles in team		
	Job design	Work pace, how quickly the assembly is carried out	The work itself—how difficult	Variation in own work and the pair of assemblers
				Job rotation to maintain skills or specialization
	Physical context	Kit design in general	Kit rack shelves	Who does what, who will pick
		Kit boxes		
		Kit bags		
Social sub-system	Team culture	Pace	Relationships between assemblers and in the team—Learning	

(continued)

Table 5.3 (continued)

Competence	*Positive*	*Positive & Negative*	*Negative*
	Relationships between assemblers and the team—The work itself, new colleagues		
	Relationships between assembler—Know each other, communication		

The Volvo Uddevalla system also positively and negatively impacted relatedness (see Table 5.4). Operators were strongly positive of working in teams and the team culture created a sense of belonging, trust and helpfulness. Team operators helped each other. When one operator finished a balance, they continued to work on their colleague's tasks and then the team's remaining tasks. However, the relationships with workshop managers were not as positive. Operators felt like the managers lacked knowledge of their work and that they felt it was more important to compete with other workshops than build trust. Relationships with factory management were also unsatisfying for the operators. Therefore, there were poor relations between the managers and their team members, leading to a lack of relatedness. However, relatedness was high within groups.

Based on the above, the design with parallel product flow and extensive work cycles and content generates excellent potential for high performance and good working conditions. However, this potential was not fully utilised in terms of accomplished productivity. Based on the interviews with operators, the technical system with a parallel flow pattern, kits for material feeding and work in relatively small teams enabled BPN fulfilment from several aspects—autonomy through the fact that the teams essentially organised themselves, competence through kitting, overview and working in pairs and relatedness through organisation in teams. With leadership based on the production system's design and needs, well-being and performance would have improved significantly.

Table 5.4 Socio-technical system impact on operators' relatedness

Relatedness		*Positive*	*Positive & Negative*	*Negative*
Technical sub-system	Organisational design	Reporting from the team, board, computer, time use	The production engineer role Importance of supervisors for learning	Picking of material kit included in the assembly team's work Product workshop manager role (first line manager) Relationship between the team and the rest of the factory and the company Division into teams Payment
	Physical context	Physical work environment, load, work injuries, noise, lighting Kit boxes Kit bags		Who does what, who will pick kits
	Production team	Handling of absence in the team	Team size, suitable, good, bad	Team composition Specialist roles in team
	Job design	Takt - knowing how you perform See the whole	Work pace, how quickly the assembly is carried out	

(continued)

Table 5.4 (continued)

Relatedness		*Positive*	*Positive & Negative*	*Negative*
Social sub-system	Leadership	The coordinator role	Leadership types Communication Pressure and prioritization from the manager Conflict handling	The workshop manager's knowledge of the assemblers' jobs Workshop manager collective Manager as a bad role model
	Organisational culture			Bad respect and atmosphere Organizational relations
	Team culture	Relationships between assembler - Know each other, communication Relationships between assemblers and in the team - Balancing, following plan Team climate team spirit, friendship, enmity Relationships between assemblers and the team - The work itself, new colleagues Pool team impact Team leaders and responsibility	Conflicts and handling of problem in the team	Team size and team climate Composition of teams

References

Baard, P. P., Deci, E. L., & Ryan, R. M. (2004). Intrinsic need satisfaction: A motivational basis of performance and well-being in two work settings. *Journal of Applied Social Psychology, 34*(10), 2045–2068.

Bass, B. M. (1985). *Leadership and performance beyond expectations*. Free Press.

Campion, M. A. (1988). Interdisciplinary approaches to job design: A constructive replication with extensions. *Journal of Applied Psychology, 73*(3), 467–481.

De Cooman, R., Stynen, D., Van den Broeck, A., Sels, L., & De Witte, H. (2013). How job characteristics relate to need satisfaction and autonomous motivation: Implications for work effort. *Journal of Applied Social Psychology, 43*, 1342–1352.

Deci, E. L., Olafsen, A. H., & Ryan, R. M. (2017). Self-determination theory in work organizations: The state of a science. *Annual Review of Organizational Psychology and Organizational Behavior, 4*, 19–43.

Edmondson, A. (1999). Psychological safety and learning behavior in work teams. *Administrative Science Quarterly, 44*(2), 350–383.

Edwards, J. R., Scully, J. A., & Brtek, M. D. (1999). The measurement of work: Hierarchical representation of the multimethod job design questionnaire. *Personal Psychology, 52*, 305–334.

Edwards, J. R., Scully, J. A., & Brtek, M. D. (2000). The nature and outcomes of work: A replication and extension of interdisciplinary work-design research. *Journal of Applied Psychology, 85*(6), 860–868.

Engström, T., Jonsson, D., & Medbo, L. (1996a). *The Volvo Uddevalla Plant: Production principles, work organization, human resources and performance aspects: Some results from a decade's efforts towards reformation of assembly work*. Department of Transportation and Logistics, Chalmers University of Technology, Gothenburg (Report on Work Environment Found Projects Dnr 93-0217 and 94-0516).

Engström, T., Johansson, J. Å., Jonsson, D., & Medbo, L. (1995). Empirical evluation of the reformed assembly work at the Volvo Uddevalla Plant – Psychosocial effects and performance aspects. *International Journal of Industrial Ergonomics, 16*(4–6), 293–308.

Engström, T., Jonsson, D., & Medbo, L. (1996b). Production model discourse and experiences from the Swedish automotive industry. *International Journal of Operation and Production Management, 16*(2), 143–161.

Fernet, C., Trépanier, S. G., Austin, S., Gagné, M., & Forest, J. (2015). Transformational leadership and optimal functioning at work: On the mediating role of employees' perceived job characteristics and motivation. *Work & Stress, 29*(1), 11–31.

Gagné, M., & Deci, E. L. (2005). Self-determination theory and work motivation. *Journal of Organizational Behavior, 26*(4), 331–362.

Hackman, J. R., & Oldman, G. R. (1976). Motivation through the design of work: Test of a theory. *Organizational Behavior and Human Performance, 16*, 250–279.

Hetland, H., Hetland, J., Andreassen, C. S., Pallesen, S., & Notelaers, G. (2011). Leadership and fulfillment of the three basic psychological needs at work. *Career Development International, 16*(5), 507–523.

Humphrey, S. E., Nahrgang, J. D., & Morgeson, F. P. (2007). Integrating motivational, social, and contextual work design features: A meta-analytic summary and theoretical extension of the work design literature. *Journal of Applied Psychology, 92*(5), 1332–1356.

Jasti, N. V. K., & Kodali, K. (2015). Lean production: Literature review and trends. *International Journal of Production Research, 53*(3), 867–885.

Judge, T. A., & Piccolo, R. F. (2004). Transformational and transactional leadership: A meta-analytic test of their relative validity. *Journal of Applied Psychology, 89*(5), 755–768.

Jungert, T., Van den Broeck, A., Schreurs, B., & Osterman, U. (2018). How colleagues can support each other's needs and motivation: An intervention on employee work motivation. *Applied Psychology: An International Review, 67*(1), 3–29.

Kiggundu, M. N. (1981). Task interdependence and the theory of job design. *Academy of Management Review, 6*(2), 499–508.

Kovács, G. (2020). Combination of Lean value-oriented conception and facility layout design for even more significant efficiency improvement and cost reduction. *International Journal of Production Research, 58*(10), 2916–2936.

Liker, J. K., & Hoseus, M. (2008). *Toyota culture: The heart and soul of the Toyota way*. McGraw-Hill.

Morgeson, F. P., & Humphrey, S. E. (2006). The work design questionnaire (WDQ): Developing and validating a comprehensive measure for assessing job design and the nature of work. *Journal of Applied Psychology, 91*(6), 1321–1339.

Mullen, B., & Copper, C. (1994). The relation between group cohesiveness and performance: An integration. *Psychological Bulletin, 115*(2), 210–227.

Parker, S. K., Morgeson, F., & Johns, G. (2017). 100 years of work design research: Looking back and looking forward. *Journal of Applied Psychology, 102*(3), 403–420.

Punnett, L., & Wegman, D. H. (2004). Work-related musculoskeletal disorders: The epidemiologic evidence and the debate. *Journal of Electromyography and Kinesiology, 14*, 13–23.

Ranney, D., Wells, R., & Moore, A. (1995). Upper limb musculoskeletal disorders in highly repetitive industries: Precise anatomical physical findings. *Ergonomics, 38*(7), 1408–1423.

Ryan, R. M., & Deci, E. L. (2000). Self-determination theory and the facilitation of intrinsic motivation, social development, and well-being. *American Psychologist, 55*(1), 68–78.

Ryan, R. M., & Deci, E. L. (2017). *Self-determination theory*. The Guilford Press.

Schein, E. H., & Schein, P. (2016). *Organizational culture and leadership*. John Wiley & Sons.

Sisley, R., & Smollan, R. (2012). Emotional labour and self-determination theory: A continuum of extrinsic and intrinsic causes of emotional expression and control. *New Zeeland Journal of Employment Relations, 38*, 41–57.

Stajkovic, A., Lee, D., & Nyberg, A. (2009). Collective efficacy, group potency, and group performance: Meta-Analyses of their relationships, and test of a mediation model. *Journal of Applied Psychology, 94*(3), 814–828.

Tavelin, S. (2013), *The transformational leadership process antecedents, mechanisms, and outcomes in the social services* (PhD dissertation). Umeå universitet, Umeå.

Trepanier, S. G., Forest, J., Fernet, C., & Austin, S. (2015). On the psychological and motivational processes linking job characteristics to employee functioning: Insights from self-determination theory. *Work & Stress, 29*(3), 286–305.

Wall, T. D., Jackson, P. R., & Mullarke, S. (1995). Further evidence on some new measures of job control, cognitive demand and production responsibility. *Journal of Organizational Behavior, 16*, 431–455.

Wild, R. (1975). On the selection of mass production systems. *International Journal of Production Research, 13*(5), 443–461.

PART II

Toyota Production System and Its Socio-Technical Conditions

CHAPTER 6

Labour Management/Labour Relations and Toyota Production Systems

Masaki Saruta

6.1 Introduction

This paper uses the term 'Toyota System' to encompass the 'Toyota Production System' and Toyota's personnel and labour-management relations. The 'Toyota Production System' defines the entire system, which is supported by personnel, labour management and labour-management relations. The two pillars of the 'Toyota Production System' are the 'just-in-time' and the 'Ninnben no tuita Jidoka (Automation with a Human Element)' which are in turn supported locally by the 'management education' (controlled Education) provided by public education. The system prioritises technology, production and efficiency which are supported by human beings and employees.

While the 'Ford Production System', which is famous for its assembly-line approach to automobile production, is characterised by utilising simple and repetitive physical labour, Toyota's system goes one step further by utilising physical as well as mental and emotional abilities of

M. Saruta (✉)
Chukyo University, Nagoya, Japan
e-mail: masakisaruta@wing.ocn.ne.jp

H. Nohara and L. Medbo (eds.), *Technology Convergence and System Divergence*, https://doi.org/10.1007/978-981-96-1910-8_6

workers for management purposes, even in the same production line. This is achieved by the 'continuous improvement' and the 'suggestion system'. However, the result is that both family and community relationships are neglected.

What is needed to achieve this is 'human resource development' through in-house training; in other words, the cultivation of 'Toyota people (Toyota Man)' who are willing to continue working for Toyota's corporate objectives. Toyota's personnel and labour management are the means for achieving the 'ways of working' and 'ways of making people work', that is, hiring management (including nonregular workers), employment flexibility, long hours and overwork, flexible and unclear wages and disregard for safety and health issues. Essentially, labour unions are supposed to consolidate workers' demands and bargain collectively with management, representing the workers' standpoint; however, in the case of the Toyota Labour Union, they have concluded the so-called Labour-Management Declaration, promising to take responsibility for and actively cooperate in improving productivity. Thus, for example, even if there is an injury or accidental death in the workplace, the production line cannot be stopped for an extended period. Most of the time, it is not allowed to stop the production line. Stopping the production line is treated as a personal illness or lack of self-responsibility. The result of the production-first policy is not only to undermine workers' health and safety, but also to fail to protect them in the event of an industrial accident. A typical example of this is that even if a unionised worker 'dies of overwork or suicide', the Toyota labour union will not provide support.

Toyota has long succeeded in making high profits through the smooth functioning of the Toyota Production System (TPS). This success is due in part to the success of the management side in terms of human resource management and labour-management relations. Although the production system, human resource management and labour-management relations can be thought of as separate entities, in Japan, at least, Toyota has been able to achieve high profits over the past 60 years by making these three elements function as a single entity through the complete implementation of 'Hitozukuri' (human resource development).

The subject of this paper is Toyota's human resource management and labour-management relations in the Nishi-Mikawa region of Japan and 'management education', which is a form of local public education as a social environment. The aim is to clarify the reality of the 'Toyota System', which has controlled the entire personality of workers and made the TPS function effectively.

6.2 Flexible Employment Management

Three main factors have influenced Toyota's employment management since 2000.

The first is the intensification of global competition. The second is the mass retirement of baby boomers, the declining birthrate and the ageing population. The third is the collapse of Lehman Brothers, the recall problem, the major earthquake and tsunami, the nuclear power plant accident and the COVID-19 pandemic, all of which resulted in a drastic drop in production. Sharp fluctuations in production characterised this period and in response to this, in terms of personnel management, Toyota promoted its employment portfolio by making employment more flexible and further promoting diversity management.

A general feature of recruitment management in large Japanese companies is membership-type recruitment of new graduates, which is also the 'basic condition for flexible employment'. In the case of Toyota, instead of getting a job, you join a company called Toyota and become a member of Toyota. Moreover, employees who join Toyota are forced to join the Toyota labour union.

The basis of Toyota's employment management is the so-called 'human kanban system' (only the required number of people at the required workplace at the required time). Labour and management have recently called this 'flexible staffing' and are promoting it increasingly more.

Traditionally, Toyota has frequently provided and received support within the company, adjusted the number of temporary workers and requested dispatches to other companies. Then, in 2004, when dispatching to the manufacturing industry was legally permitted, many dispatched workers were immediately put into the field, mainly by subcontractors.

Even after 2000, the number of Toyota's regular employees in Japan has hardly increased. Instead, there has been a noticeable increase in the number of subcontracted small and medium-sized enterprise workers, temporary workers and other nonregular employees. Moreover, discriminatory and dividing policies such as wage disparities between regular and nonregular workers and between temporary workers and 'appointment tests' continue to be used. As a result, the relative status of Toyota's regular employees has improved and on-site workers are becoming lower-level managers and supervisors.

The first feature of recent employment management is thorough 'recruitment management'. It has successfully continued recruiting relatively docile and 'good new graduates' by taking advantage of large companies' wages and working conditions. In addition, hiring engineers has increased and hiring local 'limited regular employees' (gyoumu shoku, clerical positions) in the name of administrative jobs has started. The 'declining birthrate' is being dealt with by increasing the number of fixed-term workers (kikankou) and actively utilising the 'appointment test' when production increases.

Toyota promises 'lifetime employment' to full-time employees. It gives employees a sense of security with 'relatively high wages', including overtime, late-night, shift allowances and bonuses (summer and year-end lump-sum) and in-house welfare programmes. Toyota has raised all employees' morale by raising the status, promotions and salaries through personnel evaluations.

In addition, to make up for retiring periodic workers, the company recruits daily and hires after health checks to make up for labour shortages and increasing the number of 'appointees', which it uses to the fullest extent possible to attract applicants.

The second is the use of hierarchical labour markets and discriminatory employment management. In the past, we pointed out a group of 'floating workers' who cannot endure the work of Toyota and leave the company after a short period and a group of 'core workers' who remained in the company and continued working.

Toyota's turnover rate, excluding retirees, is declining due to the severe labour market and the number of new hires is also decreasing. The number of 'promoted workers' who monitor nonregular workers' work performance and health during the period and hire them after that is increasing.

In addition, under discriminatory employment policies, the ratio of low-wage and nonregular workers is higher in SMEs and microenterprises. An increasing number of companies use foreigners as nonregular employees or technical intern trainees. As a result, the hierarchical labour market in the Nishi-Mikawa area remains solid and inequality in society, including that of foreigners, Japanese–Brazilians, Chinese, Vietnamese, etc., is becoming more entrenched.

6.3 Flexible Work Time Management and Safety and Health Management

6.3.1 Work Time Management

Long hours and overcrowded work under the TPS remain the same, with two consecutive shifts and overtime and working on holidays according to Toyota's annual schedule, the 'Toyota Calendar', which ignores national holidays.

The 'Uchino Death by Overwork Trial' began in 2002 and was the first trial in history to be fought using real names. The trial generated a great deal of excitement, leading to Toyota's defeat in the Nagoya District Court in November 2007 and producing major results, including the introduction of small-group activities (QC circles) within work (paid work) and card readers (time cards). At the same time, however, labour and management have been reforming work time management from 'prior reporting' to 'after-the-fact reporting' of overtime work.

Furthermore, long working hours remain the same and the number of Toyota workers working more than 360 hours of overtime per year, which had been decreasing after the collapse of Lehman Brothers, has since increased, reaching 4,484 in 2014 (Asano, 2004, Saruta, 2008). Most are clerical and technical workers, especially engineers and 3,840 of them work overtime hours over 360 hours per year.

6.3.2 Safety and Health Issues

As far as the author is aware, work-related accidents, deaths from overwork and suicide continue unabated at Toyota and related companies. This is evidenced by the unceasing stream of court cases in which Toyota work-related accidents are recognised. Occupational accidents constantly threaten Toyota workers. The following are some of the Toyota work-related accidents, mainly fatal, that the author has heard about or seen.

There were 22 cases before the nationwide Karoshi Hotline was established in 1988 and if the '10 suicides' are included, the number of fatalities is 31. There have been 29 cases since then and with 29 fatalities, including temporary workers who worked at Toyota workplaces or were dispatched from other companies. The total number of Toyota fatalities listed here is 60. However, this is only part of the story; the actual number remains unknown.

Although we did not specifically investigate Toyota-related companies, we saw or heard of approximately 26 industrial accident cases at various meetings. Of these, 16 cases are believed to have been fatal accidents, but the number of fatalities, including 73 at Daido Steel, is 89.

The exact figures for each year are not entirely clear. Still, for reference, the number of fatalities in 2002, which is relatively straightforward, is as follows: the number of fatalities while working in service, as published by the company in its internal magazine, is eight and the number of fatalities while working, as reported by Katsuji Tsuji (Ritsumeikan University), based on the magazine The Economist, as published by the Toyota Workers' Union, is 68. The number of deaths, including 'overwork deaths' seen and heard about by the author is eight, including two-period workers. Thus, the exact figures are unknown.

Incidentally, the author has been directly involved in eight court cases, including four related to Toyota, three related to subcontractors and one related to both Toyota and Denso, as witnesses, written statement and chairpersons of support groups. Furthermore, each trial was a long-term trial lasting several years.

Another article reports the total number of fatalities at Toyota and related companies as 148 in all (Saruta, 2020). It is not known whether these fatalities were recognised as work-related accidents, except for a small number. It can be assumed that most have been treated as private illnesses (self-responsibility). Tsuji classifies the treatment of the deaths in Toyota into three categories: (1) treatment as a saint or saintly person, (2) ordinary death of a very average common believer and (3) treatment as a demon or evil spirit (Katsuji Tsuji, 2011, p. 565). According to this classification, all those who have appealed to the courts are employees who have been 'treated as (3) demons/evil spirits' (Tuji, 2011).

With regard to nonregular workers (period workers) and workers in related subcontracting companies, careful health checks are carried out when contracting or recontracting employment and workers with health problems are ejected from the company before they fall ill in the workplace. One subcontractor manager said, 'Our company has no deaths from overwork. We will retire them before that happens'.

The fact that the reality of 'work-related accidents, overwork deaths and suicides' at Toyota has not been made public until now is, first, largely the fault of the trade union, which treats work-related accidents as a 'personal matter' and does not recognise them as work-related accidents. Second, there is a lack of awareness of workers' rights and third, there is a lack of corporate social responsibility. As a result, both labour and management have not disclosed the reality of the situation and worker deaths have been kept secret.

The reason why the victims' families cannot give their 'real names' and demand improvements is that work-related accidents are considered a private illness and for which employees are self-responsible. If they offer their real names and confront the company, they must be prepared for harassment, bullying, blame and isolation by those around them, especially in a corporate town.

Trade unions are inherently responsible for protecting workers' 'safety, health and life'. Still, the Toyota Workers' Union, for example, does not take up the serious issue of 'death by overwork' or 'suicide by overwork' of its members in labour-management negotiations, saying it 'does not get involved in personal issues'. This is because Toyota's labour and management declared in the 'New Labour-Management Declaration' that they are committed to 'mutual trust and mutual responsibility between labour and management' and act according to 'common values'. Therefore, even if a worker dies of overwork, the union will not strike, let alone engage in collective bargaining and is reluctant to defend workers' rights or fight for shorter hours and work reform.

6.3.3 Flexibility in Wage Management

Since the 1990s, wages have been based on the principle of ability. However, since then, 'performance and performance-based pay' have been introduced, mainly for clerical and technical (white-collar) workers and now even for skilled factory workers (blue-collar), whose lump-sum pay is directly related to the company's performance.

In recent spring fights, wages have been continuously lowered; however, in the spring, fights under the Abe administration, at the request of the Prime Minister, resulted in base increases and rises in lump-sum payments after a long time, including at Toyota, partly due to high profits. In addition, in the 2014 Spring Struggle, wages were increased for temporary workers (workers with a fixed-term contract) and for the first time, a base increase differential based on length of service was introduced.

However, the increasing ratio of lump-sum payments that depend on company earnings and the violent changes in wage income due to sudden changes in lump-sum amounts and the significant decrease in overtime pay due to drastic changes in the economy have made the lives of workers with the heavy burden of mortgages and education costs more precarious. This is because the families of Toyota employees remain the typical 'male breadwinner' model, with their livelihoods dependent on their husbands' income.

A conspicuous feature of wage management around 2020 was the government-led Spring Struggle and the nondisclosure of base increase details. Under such employment management, work time management and wage management, how has Toyota continued to 'develop human resources' and motivate its employees? This is examined below.

6.4 Education and Training Management—Toyota's 'Hitozukuri' (Human Resource Development)

Toyota has succeeded in developing employees who have the skills and techniques (proficiency) to do their jobs, understand the TPS and have mastered the Toyota Way (respect for humanity and continuous improvement).

Toyota's 'system for developing motivated Toyota people' is based on formal education and 'informal activities', with workplace education as the central pillar. The three primary aims of Toyota's capability development are (1) to create people who can think, (2) to create people who have the spirit and ability to do things and (3) to foster a 'corporate mindset'.

The aim is to create obedient 'Toyota people' who have mastered the 'Toyota Way'. This is achieved by small-group activities (QC circles), a suggestion system, voluntary study groups (training for promotions, TPS study groups, etc.) and human relations activities (Toyota Club, P/T activities, in-house organisations, educational background, place of origin and position). Additionally, dormitory associations are used to build a web of human relations within the company.

The trade union's role in 'Hitozukuri' is to cooperate with 'Toyota man zukuri' under an integrated labour-management relationship. There, labour and management work together to mobilise the company's physical, mental and emotional needs without regard for the lives of employees–union members' families or the community. Examples include ideological propaganda that if they work as the company and their bosses tell them, the company will develop and this is the best way for workers to 'stabilise and improve their lives and personal growth'. Additionally, propaganda activities are used, such as when a cross-company minority union critical of Toyota (All Toyota Labour Union) was formed, sabotaging its formation and urging workers to stay away from heterogeneous groups. This includes publicity campaigns. In addition, when workplace accidents or 'overwork/suicide' occur in the workplace, trade unions are not able to help the victims' situation at the time or to assist the victims' families in their appeals for clarification of the facts and resolution of the matter. The Toyota Workers' Union, in effect, acts as a bulwark to protect Toyota from criticism by workers and society.

6.5 'Management Education' and Toyota's 'Human Resource Development'

Toyota's 'human resource development' has also been supported by the spiritual features of 'management education' in the Nishi-Mikawa region. Aichi Prefecture has never become an innovative prefectural government. It is an unusually conservative region for a prefecture with a large city and has consistently accepted 'management education'.

What is 'management education'? It is characterised not only by competition for entrance examinations under a hierarchical order at high school and university but also by the thorough guidance of students (in terms of higher education and lifestyle). Moreover, unlike in Sweden, parents bear a heavy burden of the huge education costs, including tuition fees for cram schools, high schools and universities.

In sum, the similarity between 'management education' and Toyota's 'human resource development' is that 'students and employees are treated as children'. It is characterised by not allowing democratic discussion at teacher–staff meetings or other meetings and by disregarding the autonomy and individuality of the students.

Management methods are also common, for example, school songs, the national anthem (Kimigayo) and company songs, hair and uniform inspections, the 5S (Seiri, Seiton, Seiketsu, Seiso, Shitsuke [Arrangement, tidiness, cleanliness, cleaning and discipline]), one-minute speeches and reflections, QC circles and group activities, collective responsibility, road safety campaigns, group behaviour training, textbook certification and information manipulation. The measure of management throughout is discriminatory. Selective personnel evaluations and assessments are based on the unitary values of 'academic ability' and 'competence'. As a result, children are subjected to (1) discrimination and selective 'ability-based' competition and exclusion, (2) loss of free time due to long hours of detention and (3) a wide range of schools for entrance examinations through guidance.

This 'management education' is supported by the teachers, who, due to the thorough 'teacher management' approach led by the Board of Education, are not allowed to speak or discuss openly in the workplace and meetings are limited to 'reporting, communication and consultation'.

Specific management methods of 'management education' include: (1) denial of student autonomy through rule-based guidance, (2) extreme restrictions on off-campus life, (3) cramming based on study guidelines and failing-out education, (4) manipulation of the percentage of students going on to higher education and the ratio between public and private schools and thorough career guidance based on internal examinations and (5) Increasing burdens on teachers and students by accepting 'schools promoting the creation of distinctive schools' and 'schools commissioned for research' as a means of 'human development'; (6) Deprivation of free time through long hours of detention at school, including club activities and homework; (7) A system of mutual surveillance among students; (8) Collectivist education through group training and group responsibility; (9) Disciplinary and physical punishment of teachers by applying 'penalties'; and (10) expulsion from the school setting and nonattendance at school.

Thus, 'management education' is an educational system that does not tolerate different people's existence and is prone to 'bullying' of the 'weak'. This is extremely compatible with Toyota's 'human resource development'.

6.6 Motivation Management

How has Toyota managed to motivate these workers? The following is a summary of Toyota's internal documents and voices heard from workers.

The workers nurtured by Toyota were workers who continued to work for the company, even at the expense of their body, mind and livelihood and who worked obediently despite long hours and overcrowded work within the framework of the 'motivational management' developed by Toyota. The most important factor motivating them is, firstly, 'motivation by economic and social stimulation. The aim is to climb the ladder of 'advancement, promotion and salary increase' by performing their assigned tasks without a flaw. This is so-called career advancement and money. This is why, under the second type of personnel management, i.e. 'fewer people', the third type, 'participation' and 'enjoyment of work', are emphasised by management and the workers are forced to continue to work. Thus, Toyota workers continued to improve productivity by reducing 'standard working hours' and 'lead times' through constant improvements in the manufacturing, development and production preparation departments. This also led to an increase in depression, overwork and suicide in the engineering department after 2000. Toyota, whose primary objective is to strengthen competitiveness, has continued to stimulate its workforce under a thoroughly 'Flexible Manpower Line' policy, uses various means of motivation such as QC circles, suggestion system, job enlargement and job enrichment, desire for self-actualisation and participation, with the trade unions giving their full support to this. Particularly noticeable in recent years is strengthening assessments through 'meritocracy' and performance-based management of 'advancement, promotion and salary increases'. The widening wage differentials through 'meritocracy' and the introduction of performance and performance pay have further widened the wage gap between workers, intensifying the competition between them. At the same time, violent economic fluctuations have led to wild fluctuations in wage income.

Regular workers and nonregular workers (fixed-term workers) have traditionally been completely separated and discriminated against in terms of wages and working conditions. Still, with the recent increase in the number of nonregular workers and the extension of contract periods to two years and 11 months, on the one hand, the number of 'appointees' has increased, discriminating and dividing fixed-term workers and on the other hand, fixed-term workers have been asked to participate in company events, etc. The company is trying to unify regular and nonregular workers by asking them to participate and cooperate in company events.

Traditionally, period workers have been motivated mainly by economic incentives (Expiration benefits and expiration incentive payments, Lump-sum payment on condition that the employee continues to work without a break until the expiration date of the contract) aimed at all workers, but in recent spring fights, discriminatory base increases based on length of service have been introduced among period workers.

6.7 Labour Relations Management

The Toyota Workers' Union brings people who are discriminated against and divided workers together as company people. Labour unions in many large Japanese companies have a union shop system and a check-off system, forcing them to take control of all workers. Furthermore, labour-management relations at Toyota can be described as 'management-led labour relations' and a system of 'mutual trust and mutual responsibility' has now been established under the 'New Labour-Management Declaration'. Besides, most subcontracting company workers, except Toyota Group companies, are unorganised.

The labour-management relations chart drawn by Toyota shows that 'shared values' have created 'the establishment of mutual trust and mutual responsibility', but more accurately, Toyota's management and right-wing union leaders are trying to 'establish mutual trust and mutual responsibility' and are using their power to impose 'shared values' on employees through personnel and labour management and other measures from the top. At its annual convention in fiscal 2023, the All Toyota Federation of Trade Unions, which comprises the labour unions of the Toyota Group, revised a part of its 'Platform', which is similar to a constitution, for the first time since its formation. Among the four items in the charter, 'mutual understanding between labour and management' was changed to 'mutual trust between labour and management', which is the same as in Toyota's charter.

It is customary for workers at Toyota to leave the union when they are promoted to the rank of head of a production line and being a union officer is also part of the promotion route. Furthermore, labour and management share the same values and labour-management relations at Toyota can be described as a 'subordinate–supervisor' relationship. In addition, at the heart of the trade union is a structure in which trainee workers who have graduated from the company's high school and have learned the 'Toyota Way' sit at the heart of the trade union.

Discussions between labour and management are mainly based on 'labour-management talks' rather than 'collective bargaining' and the main theme these days is the promotion of greater competitiveness by labour and management. The main feature is that class conflict has almost disappeared within Toyota. Furthermore, in labour-management talks, instead of labour and management facing each other, they are now arranged in a triangle of desks, with management, middle management and the trade union talking side by side. Meanwhile, at Toyota and its small and medium-sized subcontractors, disputes between regular and nonregular workers continue to occur frequently, including deaths and suicides by overwork, depression, work-related accidents, dismissals and nonpayment of wages.

'Toyota-style industrial relations' have now spread throughout Japan, and 'management education' and 'management to teaching staff' are becoming widespread and deep-rooted, not only depriving workplaces of democracy, as the organised trade unions of large companies become extremely distorted, but also setting back local democracy.

The management-led Spring Struggle has turned into a 'government-led Spring Struggle', and now Prime Minister Kishida of the LDP government has not only demanded a wage increase but has even invited the Prime Minister to address May Day at the national centre, RENGO (Japanese Trade Union Confederation), under which the unions at Toyota have demanded regular (annual) pay raise, base increases and lump-sum payments in the Spring Struggle, and the management responded in full. However, the background to this was a long-term decline in real wages due to a sharp rise in prices and other factors, the expansion of discriminatory wages for work of equal value by making wages performance-based, and the division of regular and nonregular workers, and the introduction of wage discrimination between nonregular workers. Furthermore, Toyota has gone so far as to declare that base increases will not be made public on the grounds of reducing disparities.

6.8 Conclusion

Under such labour-management relations, Toyota has been subdividing and simplifying the labour of individual workers into one-minute segments through personnel management and while 'completing processes by function' as an organisation (so-called kumi), has made workers multi-skilled (so-called multiprocessors) and has made them responsible for 'normal work' and rework as well as abnormal treatment.

In addition, by having workers participate in 'part of the conception' through the creation of 'standard worksheets', QC circles and suggestion systems, the company has succeeded in mobilising them to improve productivity, physically, mentally and emotionally.

Under Toyota's just-in-time system and the 'human kanban system', fluctuations in work hours are the norm. In addition, work hours have become increasingly uncertain as flexible work hour systems, such as the discretionary work system, have been introduced and expanded. A further expanded and deepened 'deworking time-aware discretionary working system' such as Free Time & Location has been introduced for administrative and technical staff since December 2017. In terms of wages, starting in 2021, there was a bifurcation between 'job standard pay' determined according to job position and 'job individual pay' based on evaluation. In addition, 'regular salary (annual) increases' were abolished and consolidated as personnel evaluation pay. As always, Toyota justifies these policies by saying that it aims to be a company where hard work is rewarded.

Toyota's personnel and labour management is moving towards greater flexibility, i.e. the use of nonregular workers, 'dework hours' and wage determination based solely on personnel evaluations. Additionally, the amount of the base increase in the spring labour offensive was kept secret under the guise of narrowing the wage gap between small and medium-sized enterprises. When the base increase amount is not disclosed, there is no room for individual workers' demands and claims. It is hard to deny that this disregard for the humanity and individuality of workers has led to a spate of occupational accidents, deaths from overwork and suicide in safety and health issues.

Toyota has continued to reign as one of Japan's leading companies, generating high profits while attracting and repelling large numbers of Toyota workers, nonregular workers and subcontract workers through a campaign to strengthen competitiveness in which management and the labour union work in unison. As a result, the following situations have arisen within Toyota and Toyota-related companies: (i) a lack of democracy; (ii) disregard for decent work; (iii) flexible employment, work hours and rigid livelihoods and disregard for the work–life balance; (iv) acceptance of wage differentials by gender, enterprise size and employment type and loss of worker solidarity; and (v) disregard for respect for the autonomy and individuality of children.

Toyota's labour and management have been avoiding and opposing the class struggle. A state of class annihilation has emerged within the company, with individual disputes being cut off. Therefore, on the one hand, many labour disputes take place at Toyota and in small and medium-sized subcontracting enterprises.

Toyota's development is unlikely to lead to an egalitarian and universalist 'welfare society' such as Sweden as it fosters competition and fixes discrimination between workers who are 'personality controlled' by the companies.

References

Asano, K. (2004). Toyota seisan hoshiki to rodo jikan [The Toyota production system and working hours], *Chingin to shakai hosho* 1382 pp. 4–33, 1383 pp. 29–44. Junposha.

Saruta, M. (Ed.). (2008). *Toyota kigyoushuudan to kakusashakai* [Toyota Group of Companies and Disparities in Society]. Minerva Shobo.

Saruta, M. (2020). Toyotashistemu to rosai,karoshi,jishi -40 nen no Toyota chousa,kennkyuu o furikaetute [The Toyota system and "Industrial accidents, death from overwork, and suicide"-A review of 40 Years of Toyota Research and Studies]. *Chukyo Business Review*, No. 42.

Tuji, K. (2011). *Toyota jinjihoushiki no sengoshi* [Postwar History of the Toyota Personnel System]. Minerva Shobo.

CHAPTER 7

Toyota Production System and Modern Society

Hikari Nohara

7.1 Structure of Chapter 7

This chapter comprises two parts, namely, the study of the Toyota Production System (TPS) from 7.2 to 7.4 and influences of the system over society in 7.5. Section 7.2 describes the transformation of the Ford Production System (FPS) into the TPS and provides comparisons between the TPS and the FPS. Section 7.3 describes Toyota's response to workers' aversion to short-cycle work on the line within the framework of the TPS. Section 7.4 explains Toyota's method of handling the division of conception and execution, which is a fundamental difficulty embedded in Taylorism, from the viewpoint of workers' aversion to short-cycle work on the line. Lastly, Sect. 7.5 provides an overview of the influencing characteristics of the TPS over human society.

H. Nohara
Hiroshima University, Higashihiroshima, Japan

H. Nohara (✉)
Nagano University, Ueda, Japan
e-mail: nohara@nagano.ac.jp

H. Nohara and L. Medbo (eds.), *Technology Convergence and System Divergence*, https://doi.org/10.1007/978-981-96-1910-8_7

The author's main concern is that the very essence of the TPS, itself, probably influences human society, i.e. the human way of thinking, to a great extent. Alternatively, as far as the production system is concerned, the TPS is the crystallisation of modern rationality. However, ambivalence is observed in modernity. That is to say, although we focus on a production system itself, the TPS is one of the best examples of modern rationality; however, once we widen our horizons beyond production systems and towards society, we find the realisation of danger embedded in modern rationality.

As the last remark of this introduction, an explanation will be given as to why the author focuses on production systems. As described in the introductory chapter, 'production system' in this book is a shop floor production system. Essentially, a production system is the organisation of activities, processes and resources used to transform inputs into outputs, i.e. finished goods.

Why then does the author focus on a production system, although it is not the real situation of Toyota's manufacturing site? The reason is that, for the purpose of getting a step higher in the level of understanding, the author really wants to find common language not only among Western and Japanese researchers and practitioners but also among admirers and critics of Japanese manufacturing systems. In this regard, the production system is a kind of abstraction that is deprived of particularities from any particular society.

However, as an abstraction, a production system could be a common language beyond the difference of real and concrete situations among countries.

The other reason is to find the transferability of the manufacturing system/organisation from one country to another. As will be described in Sects. 7.2.1 and 7.2.2 when wanting to transfer one manufacturing system, we have to analytically divide it into a production system (the mechanism of how to manufacture) at the shop floor; societal and technical constraints, which the production system faces in the latter country, and societal and technical given conditions, which make it possible to transform the original production system into the modified production system, adapt to the new circumstance of the latter country.

Now, the author would like to start the subject itself.

7.2 Toyota Production System and Its Socio-technical Conditions

Since the end of 1980, several practitioners and researchers have been interested in whether the TPS is transferable because of its high efficiency. However, when we try to transfer or reform a production system in a factory, we have to factor in societal and technical conditions to find the optimal solution. We call such an approach to reality a 'Socio-technical Approach', which differs to technology determinism. In the case of the automobile industry, this approach is considered very explicitly as in the case of the new Uddevalla production system (UPS) but implicitly in the reform of the TPS.

Japan was almost completely defeated and surrendered to the allied forces in 1945. Therefore, after World War II, Japan had faced tough societal and technical difficulties such as poverty, famine, collapse of industries and factories, capital shortage and so on. As a result, it was impossible for Japanese auto companies to transplant FPS without any modification. What elements of the FPS should have been modified? What kind of societal and technical conditions had made such modifications possible? When we solve these, we eventually understand why the FPS transformed into the TPS.

Furthermore, when you try to transplant the TPS abroad, it will be important for you to consider these questions for the purpose of your success.

7.2.1 Societal and Technical Constraints Toyota Faced After World War II.

The FPS had been the ideal model globally to manufacture an automobile among its makers. However, it was impossible for Japanese makers to directly implement this model in post-war Japan. First, the product (car) market had been very limited after the war because most Japanese citizens could not afford to buy a car. Therefore, such a limited market forced Toyota to incorporate small volume production instead of typical mass production. This meant introducing semi-specialised machines and deskilled multi-functionalist workers, which were more suitable for small volume production. As far as car price was concerned, until the end of the 1970s, because of relatively low income, Japanese had required low price cars.

Therefore, it was of highest priority for Toyota to realise the low price of an automobile during this era. To realise the lower price, Toyota had done its best to cut production costs. Because of small funds, they had to save investments for human and material resources. Such effort brought about three results. The first was the principle to solve a defect when and where it occurred (*Jidoka* in the case of a machine, in Japanese). The second was having multi-skilled workers instead of single-skilled who were engaged in single-specialised operations during their working life. The third was for the company to minimise capital in kind, staying in a production line, because of capital shortage. The necessity of this capital minimisation not only introduced the just-in-time method but also resulted in work improvement rather than equipment improvement, i.e. an equipment investment.

In retrospect, these results had provided Toyota a great advantage after the 1980s. During the 1980s, Japanese companies had rapidly expanded the export of automobiles to North America. However, the price determination power was monopolised by the US Big Three auto companies. In such a case, the first result (the principle to solve a defect when and where it occurs) and third result (a minimisation of capital) produced advantages for Toyota while competing with American auto companies.

7.2.2 Societal and Technical Conditions to Transform the FPS into TPS

As abovementioned, the FPS required modification to a great extent. However, why did such modifications succeed? Societal and technical conditions transformed the FPS into TPS. In this regard, three elements were key: workers' being free from literacy problem, workers' positive participation into a labour process and manufacturing engineers coming into being.

First, in the USA, workers on the line had been mostly supplied by southern and east European immigrants who were not fluent in English. As a result, the management faced literacy problems. By contrast, line workers in Japanese auto plants were local and mostly supplied by small farmers in rural areas. These farmers were functionally equivalent to immigrant workers in the case of FPS. In this regard, Japanese management never faced the literacy problem, and they could expect to more easily train freshmen to be multi-functional workers rather than deskilled specialists. In addition, because of this language advantage, Japanese auto companies could mobilise workers into more complicated jobs, i.e. responses against change and irregularity on the line within regular working hours, and KAIZEN activities within or after regular working hours.

Second, workers' participation in the labour process was positive, where they had stable labour-management relations under an overwhelming influential power of the management after the Toyota labour dispute in 1950. This has been the pre-condition for workers' positive participation because the management had been able to propose workplace reforms without any interruption from a union, and the management had been a single hegemony influencing workers at the workplace.

Compared with monotonous and repetitive short-cycle work on the line of the FPS, any work with conception encourages operators and their positive participation into a labour process. In this regard, the Toyota company established a very systematic way to introduce conceptual work (see 7.4.1) on the shop floor. An operator's daily routine is a short-cycle standardised work (SW) that includes no conceptual work. However, several types of conceptual works have been embedded on an operator's work outside of SW. In the case of veteran operators, they have been invited to join the team that has conducted the revision of SW from the viewpoint of a manufacturing engineer (see Chapter10). The operator has also been engaged in a work of response to change and irregularities on the line (see Chapter 10). In addition, all operators are mobilised to and joined in KAIZEN activities (see 7.4.2), the participation into which has been mandatory. In many cases, an operator's KAIZEN activities have included the improvement of an operation and an equipment from the manufacturing engineer's viewpoint, which has been a conceptual work. Needless to say, each individual operator's achievement in this regard has been strongly reflected on their performance rating.

Third, Toyota has manufacturing engineers on each shop floor, who are the mediators and so-called translators between design/production engineers and operators on the line. They have the knowledge of engineering science alongside the knowledge regarding how to do real operations, i.e. standardised short-cycle work on the line. This has been a very unique category of the production engineers almost peculiar in Japanese plants.[1] This category has made it possible for Japanese management to prefer work-oriented improvements on the shop floor rather than equipment-oriented improvements, i.e. automation-oriented improvement like in most Western plants. It has also made it possible for Toyota's management to realise workers' positive participation into a labour process as described above. That is to say, this manufacturing engineer's support has enabled workers to join the activity of revising SW, the response to the change and irregularity and KAIZEN. It means that manufacturing engineers can express operators' experiences-based wisdom in common scientific language, which is understandable among people without experience on manufacturing site. In addition, some qualified veteran operators have been selected as members of the manufacturing engineering section.

Overall, the combination of societal and technical constraints to which Toyota has faced, and societal and technical given conditions in post-war Japan, made it possible to transform the FPS into TPS.

7.2.3 *Toyota Production System as a Chain of Nexus Between Ends and Means*

What is the TPS? In this regard, it can be expressed as a chain of nexus between ends and means, based on the founder of the TPS, i.e. T. Ohno's (1978) own description (Fig. 7.1).

7.2.4 *Toyota Production System Compared with Ford Production System*

As previously mentioned, the TPS was born as a result of societal and technical conditions. However, these are not the system itself but conditions to provide characteristics peculiarity to the system. Then, next is showing you these characteristics of the TPS compared with those of the FPS (Table 7.1).

Regarding the above comparison, first, the TPS had peculiar characteristics such as low cost, high quality and short lead time, different from

[1] In the case of Uddevalla, they had a small number of de facto manufacturing engineers. In addition, some researchers working for the university, such as T. Engström, L. Nilsson, and L. Medbo, had taken on the role of manufacturing engineers.

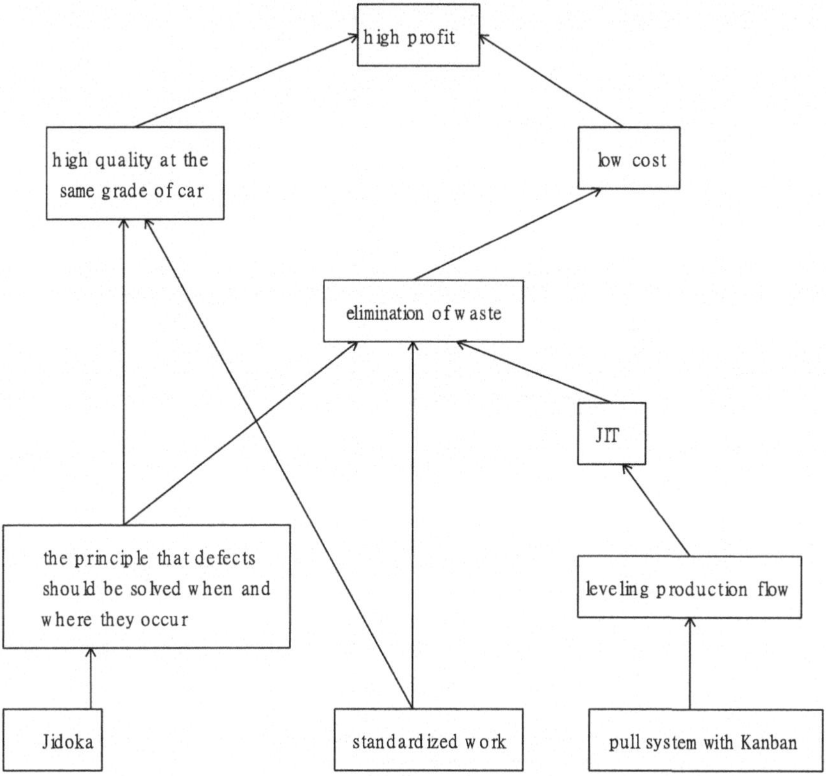

Fig. 7.1 Toyota Production system as a chain of nexus between ends and means
Note (1) An arrow means a nexus between ends and means. The beginning of an arrow is a means, and the arrow tip shows an end. (2) SW is regarded as One Best Way for assembly at the moment. Therefore, it is also regarded to contribute to realising high quality. (3) *Jidoka* includes Stop Code System and a machine using an automatic stop device. This concept requests us to pay attention to human and machine. (4) In the case of the principle that defects should be solved when and where they occur, the defects mean defects of products (complete car, component and part) and machines. Therefore, this principle asks us to pay attention to two aspects, namely, a man and machine. The principle also implies three points. That is to say, at first, they have to find not only the defects of products, and components and parts, but also the defects of machines and equipment. Second, they have to find causes of defects and find measures against their recurrence. Third, they have to repair products and equipment

Table 7.1 Comparison between the TPS and FPS

Category	*TPS*	*FPS*
Production volume	High-mix, low-volume production	Single-item mass production
Machine	Semi-specialised machine	Specialised machine
Skill on the line	Deskilled multi-functionalist	Deskilled specialist
Human buffer	Abolishment of the fixed number per team and of relief person	Narrow job classification and relief person
Material buffer	Minimisation of material buffer (JIT)	Mass volume of intermediate buffer
How to handle defect	Principle in which defect should be handled immediately when and where it happens (*JIDOKA*)	Repair section on the end of a line
Division of production engineer	Production engineer and manufacturing engineer on the manufacturing site	No manufacturing engineer
How to make a standardised work	① A production engineer makes a draft of SW, thereafter a team of a manufacturing engineer, group leader, team leader and veteran operator modify the draft based on real operation on a line ② Frequent reorganisation of SW ③ SW is the point of reference in terms of continuous improvement (see 7.4.2)	① A production engineer makes a SW and gives it to a manufacturing site ② Fixed SW
An operator's work	① SW and its rotation ② Intense concentration on SW ③ KAIZEN ④ A qualified operator handles irregularities on the line	Single-skilled worker
Production flow	Pull system Levelling production flow	Push system Fluctuating production flow

the traditional FPS. Such characteristics were birthed from the disadvantages of post-war Japanese societal and technical conditions. However, such characteristics have changed into advantages when Japanese transplants in the USA were forced to compete with Big Three auto firms in the domestic market. Namely, during the 1980s and 1990s, the market size per single model became smaller and smaller and the market fluctuation became wider and wider. As a result, it gave great advantage over the traditional FPS that Toyota's production techniques were suitable for rather smaller markets and were more flexible in terms of being able to relieve the impact of production fluctuation caused by the market and other production flow change.

Second, regarding the work content in terms of SW, no inner dialogue is present on labour process in both cases of TPS and FPS. In this regard, division of conception and execution is strictly kept on an operator's SW. In addition, no dialogue is present among team members on the line in both cases of TPS and FPS and no team work is observed.

In comparison, an inner dialogue is present in an operator's work in the case of the UPS. Therefore, conception and execution are integrated on labour process. In addition, team work is observed within a team (see Chapter 2). These points will be described later on.

7.3 New Stage of the Toyota Production System Since the Beginning of the 1990s

7.3.1 *Fundamental Characteristic of Assembly Work on the Line: Lack of Contextually Meaningful Connections Among Series of Operations*

Since the introduction of the conveyor belt at the Highland Park Ford plant at the start of the twentieth century, line balancing has been the sole criterion of how to conduct SW. This SW strictly instructs an assembly operator on the task to be performed (Meyer, 1981 and Hounshell, 1984).

In the case of short-cycle time, line balancing is only one principle used in standardisation. That is to say, first, all assembly operators along an identical conveyor belt line have the same cycle time for work. Second, every operation is divided into fragmented pieces, as many as possible, to eliminate waiting time in short-cycle work. Third, some pieces are used only to fulfil the cycle time without resulting in any waiting time. This is the way that engineers compose SW from these pieces. As a result, no contextually meaningful connection is observed among such pieces within a cycle. To summarise, line balancing divides each operation into fragmented pieces in an attempt to completely fill the cycle time without any lag time between motions.

In general, as far as a SW is concerned, operators can memorise motions and parts even in random order, but the volume and accuracy of memory are very limited in such cases.

Many decades since the introduction of the belt conveyer line at Ford factories, the automobile industry has been bothered by workers' aversion to work on the line. Such labour crisis became severe in the 1970s and 1980s in Western countries and trickled over to impact Japan by the end of 1980s. In Japan, Toyota, most notably, had faced the severest labour shortage among Japanese auto companies. For instance, applicant numbers were far less than the planned employment number in 1991 and 1992 (Nihon Keizai-Shinbun March 10, 1992).

7.3.2 Toyota's Solution: Autonomous Complete Process and Its Conditions

Against labour shortage, upper and middle managers set a goal: to make the work more comprehensible to workers. Accordingly, they thought that if operators could regain some contextually meaningful connections between series of operations, operations would become more understandable. The term 'contextually meaningful connections' indicated that operators were able to recognise where they were and what they were doing at any moment, on the way to completing a product, only when they introduced the rule that the order of assembly sequence was determined from the viewpoint of component functions and the ease of assembly.[2] It is because of human inherent cognitive capability (see Chapters 2 and 3).

To improve work, several procedures were employed. The first method involved introducing the Autonomous Complete Process (ACP), in which one component's function was completed within a single work group and each operator's series of operations had as much contextual meaning as possible (see Asao et al., 1999 and see also Chapter 10). In this case, 'autonomous' means that a certain function is completed within a single work group. Corresponding to the ACP, the second procedure was that the grouping of parts was reorganised from the viewpoint of completing

[2] When Nohara asked middle management in 1994, what was the criterion on how to organise the order of assembly sequence, Nohara was told 'component function and the ease of assembly'. As far as Nohara understands, this empirical criterion corresponds to the assembly-oriented product structure theoretically defined by T. Engström (see Chapter 2).

component functions and ease of assembly. The presence of manufacturing engineers made it possible to realise the ACP. Third, instead of being arranged along the assembly line, parts were placed in a kit box for each ACP, which was named Set Parts Supply (SPS). Because of this, an operator could easily operate on an assembly line.

The fourth procedure was introducing the buffer function. A long single conveyor belt line was divided into several short lines, each of which was termed a 'mini-divided line'. Each mini-divided line was entrusted to one workgroup, which was responsible for several numbers of the ACPs on the line. Furthermore, several automobile bodies could exist between mini-divided lines. This was termed the 'intermediate buffer'. This buffer absorbed fluctuations in product flow and the fluctuation of each worker's pace within a particular group. In addition, many assembly objects (e.g. cars) on the assembly line in relation to the number of operators improved the possibility of absorbing fluctuations in product flow and of each employee's work pace (buffer mechanism). In sum, since the mid-1990s, the production system with such procedures became very different from the traditional TPS.

In the long run, development of these work-focused technical means resulted in such improvements as increased flexibility of product flow and assembly sequences with more contextually meaningful connections within the assembly line system.

7.3.3 More Than the Solution: Standardised Assembly Sequence as the Adaptation to Plants' Globalisation

As described above, Toyota's solution against labour shortage has been the ACP. We will now discuss Standardised Assembly Sequence (SAS or *Hyojyun Kojyun* in Japanese) which was introduced from the mid-1990s (Nohara, 2006). Using the view of a component function, every vehicle was divided and classified into 4 units, which were sub-divided into 8 units, then into 54 units (what Toyota managers call 'function groups'), and finally into 108 units (called 'subgroups') for all vehicle models and types. This step-by-step classification process was common and standardised to all Toyota plants worldwide (see Niimi et al., 1997). Thus, every vehicle consisted of 108 units in any case. Furthermore, assembling order among these 108 units was also standardised to all Toyota plants worldwide. This order was termed as the SAS.

From an organisational viewpoint, one section was designed to be responsible for several of the 54 units (function groups), and a workgroup comprises around 20 members. One workgroup was designed to be responsible for several of the 108 units (subgroups), which meant that each workgroup completed more than one function of the subgroup.

The relationship among an organisation, grouping of components, SAS and ACP will be explained using the Kyusyu–Miyata plant as an example (Nohara, 2006). The Kyusyu plant was the first one to be built after the establishment of these new concepts. In this regard, the Kyusyu plant was the physical realisation of these concepts. An assembly line was divided into 11 lines, which are named as mini-divided lines. Each mini-divided line was entrusted to a certain workgroup. Therefore, 11 lines resulted in 11 workgroups. These workgroups completed 108 units altogether. This meant that each workgroup completed several of the 108 units. The process in which each subgroup of component function was completed represented the ACP. Therefore, each workgroup was responsible for several numbers of the ACPs among 108 units.

By contrast, it is commonly said that one vehicle model comprises around 30,000 individual parts. According to managers, these individual parts constituted 1500 assembly units (what we co-authors call delivery units), which were arranged along the assembly line or in kits to be assembled in a vehicle body.[3] It was common to all Toyota plants that around 30,000 individual parts constituted 1500 assembly units from the viewpoint of component function and ease of assembly. In the case of the Kyusyu plant, those 1500 assembly units constituted 108 units (subgroup), which were entrusted to 11 workgroups, each of which had around 20 group members.

An assembly sequence of automobiles is standardised according to SAS. This assembly sequence consists of assembly processes of 108 units (subgroup). All or at least most of such 108 units are aggregates of the ACPs. The assembly sequence of each operation inside the ACP is standardised according to 'contextually meaningful connection'. Therefore, assembly sequence is consistently standardised not only among ACPs of 108 units but also inside the ACP all over Toyota's plants.

Prior to initiating the ACP programme, assembly work had been standardised by the precise areas of the vehicle of a specific model at a

[3] The number, i.e. 1500, was based on the co-authors' interview with the management at the beginning of the 2000s.

particular plant, which is the traditional concept of standardisation of assembly line work. Namely, *standard* of SW varies from situation to situation. In this regard, this very standard is not the universal standard but the specific standard in a particular situation.

However, in the mid-1990s, Toyota shifted to a new level of standardisation. As stated, this meant that assembly sequence among these 108 units was standardised for all models and Toyota plants. As far as SAS was concerned, a worker always performed the same work, regardless of the vehicle model and location of the plant. This meant that the quality of assembly work became more independent from individual worker's competence and attitude to work. This was very important from the management's viewpoint because they had been worried about whether or not they could find royal, diligent and competent workers abroad.

Furthermore, the establishment of two concepts, i.e. ACP and SAS, meant that Toyota management got the 'common language' across all Toyota plants and different production sections. That is to say, standardisation was only one criterion of how to organise operations. The ACP and SAS were made based on this principle. In this regard, the two were common criterion on how to improve assembly work in any case beyond plants and production sections.

In sum, the ACP was the response against labour shortage, whereas SAS was not the direct response to it. While the management investigated answers to the problem raised by labour shortage, they had conceived the ACP and SAS based on the standardisation principle. However, SAS gave management a gift beyond a solution against labour shortage. Although a traditional standard had been the specific standard in particular situations, SAS meant a universal standard applied to any vehicle model and across all Toyota factories. Therefore, we name it as the second stage of standardisation.

Ultimately, we will define the term as the 'New Toyota Production System' (NTPS). As described above, in the 1990s and 2000s, the TPS had reached a new stage in which they introduced the ACP and SAS. The ACP means the recovery of contextually meaningful connections in assembly work to some extent, and SAS is the new level of work standardisation regarding the assembly sequence among 108 component units. In this regard, the system in the 1990s and 2000s or NTPS, differed from the TPS, which had prevailed until the end of the 1980s. Therefore, in our definition, the NTPS has three points different from the traditional TPS. Namely, the first one is the introduction of contextually meaningful connection as the criterion of how to organise assembly work (ACP). The second is the standardisation of assembly sequence among 108 component units (SAS). In addition, these two points are accompanied by the regrouping of parts and kit system (SPS), as a complimentary tool. The third one is the introduction of an intermediate buffer and new buffer mechanism.

Outside of these three points, nothing different is observed between the TPS and the NTPS. As far as comparisons go within Fig. 7.1, no difference is observed between TPS and NTPS. Nevertheless, once we focus on the work content, i.e. the ACP and SAS, we will find the difference between the TPS and NTPS.

In retrospect, let us reconsider the conclusion of the book, *The Machine that Changed the World* (Womack et al., 1990). As far as the TPS is concerned, the system had changed to a great extent, from the view of work content just after this publication. It regarded the UPS as out-of-date craftsman work. However, the UPS included the new type of standardisation of work beyond that book's imagination. Standardisation is one of the most distinguishable characteristics of mass production opposite to the craftsman work. Therefore, let us briefly explain our difference from this book.

Where inner dialogue was excluded in assembly work in the NTPS, the UPS took the opposite path, making the inner dialogue the core component of assembly work. In the case of the UPS, when an operator recognises contextually meaningful connections in the sequential order of motions and in the parts arrangement, based on the assembly-oriented product structure, the operator is able to conduct motions smoothly and accurately mediated by the inner dialogue. In addition, these motions are guided by facial communication with co-workers and other feedback mechanisms such as partly empty kitting fixtures and co-workers' ongoing work.

To reduce differences in work performance, a number of means for work process standardisation were developed, including work modules, intra-group work patterns and formulated and standardised questions for the inner dialogue (see Chapters 2 and 3), which corresponded to three different phases of production processes, i.e. knowledge of: materials and structuring, assembly, and inspection and adjustment.

That is to say, in the NTPS, the results of work-focused improvement based on contextually meaningful connections was realised as an objective means separated from individual competence, such as the ACP, grouping of parts and components and SAS. In the UPS, contextually meaningful connections between series of operations were an indispensable condition of inner dialogue. Work was performed following those connections.

To summarise, the NTPS and UPS had the same mid-term goal, i.e. reintroducing contextually meaningful connections between series of operations in the labour process, but its role differed in each company.

Finally, let us briefly explain the following points again. What elements of the TPS should have been modified, and what was the reason for modifications? In addition, what kind of societal and technical conditions had made such modifications possible?

In this regard, by the end of the 80 s, Japanese auto companies had faced a severe young labour shortage because of very monotonous, segmented and repetitive short-cycle work. This labour shortage attacked Toyota more severely than any other company because Toyota's cycle work was the most segmented work. The reason was as follows. Toyota's manufacturing engineering section had the highest capability to analyse work. Therefore, they could analyse work and segment it to a greater extent. From the viewpoint of line balancing, this effort of the manufacturing engineering section took the correct direction, but the work, as a result of this effort, was the extreme segmentation of work content. Therefore, young workers became averse to the line work.

Therefore, interestingly, Toyota's manufacturing engineering section rescued Toyota from this labour crisis again. Because now the new target, i.e. more comprehensive work different from traditional one, was given to the manufacturing engineering section. They had found answers in the ACP and SAS.

7.4 Conceptual Work in the Toyota Production System

7.4.1 *Conceptual Work and the Autonomous Complete Process*

In general, when a craftsman is ordered to make some product, first, they conceptualise the product in terms of the shape, materials, specifications and so on.[4] Second, the craftsman conceptualises arrangements;

[4] Generally speaking, when a customer orders a product, the customer's order is very general and abstract in terms of specification. Therefore, we can say that the conceptualisation of product is also entrusted to a craftsman.

deciding what kinds of tools and machines to use, the layout of machines, how to make a process, the staffing, etc. This is the conceptualisation of a production plan. Finally, each operator decides how to individually operate under the foreman's instruction and On-the-Job Training, which is a conceptualisation of operation. These three kinds of conceptualisation were controlled by a foreman in a factory in the USA. In this regard, a manufacturing shop floor was the so-called foreman's empire. In the end of the nineteenth century, one of the most difficult problems in a factory for management was to deprive a foreman of the power to control workers' operations at a plant.

The management is unable to control workers if it cannot deprive a foreman of the abovementioned three types of conceptual work. As far as the conceptualisations of a product and production plan were concerned, the management depended on the development of technology and a class of engineers as such developments were relatively independent of workers. However, the last conceptualisation was kept under a foreman's and an operator's control. Now, it was the turn of Taylorism (scientific management). It opened the way for the management to take power over how to operate, depriving the operators of their conceptualisation of operation. As a result, a worker's operation was divided into segmented elemental motions, then, each operator's operation was reorganised as a combination of these elemental motions. In the case of the Ford type conveyer line of assembly, the reorganisation was also done from the viewpoint of line balancing; therefore, contextually meaningful connections had been entirely lost in combination between elemental motions.

Eventually, the dominant characteristic of work on the line in the twentieth century had been monotonous and repetitive short-cycle work, in which elemental motion had no contextually meaningful connection with each other. Because of this characteristic, workers became averse to such work. Therefore, the management was urged to take a measure to attract workers. That measure was mainly an automated factory without workers in Western plants. However, it failed. Instead, the management took the path to improve work content itself to attract workers in the case of Toyota and Uddevalla, Volvo.

For Toyota, the assembly work on the line changed from one that an operator did not understand in terms of what they were doing into one that they were able to understand from the viewpoint of contextually meaningful connections between respective motions. What about in Uddevalla? Instead of repetitive and monotonous short-cycle work, they tried to introduce a cycle longer than 20 minutes. However, it was keenly recognised through the Kalmar experience that a cycle time longer than 20 minutes with traditional learning resulted in decreased efficiency. Such a long cycle time needed to introduce a new learning strategy, i.e. holistic learning instead of additive learning (see Chapter 3). The holistic learning strategy meant that an operator had the knowledge of overview, which is the knowledge of something perceived as a whole. Based on this knowledge of overview, through repeating the inner dialogue, the operator was able to individually think what the next operation was, which is the conceptual work in the step of operation (see Chapter11).

By contrast, what was realised in the ACP at Toyota? It was the recovery of contextually meaningful connection among respective elemental motions within a short-cycle work. As a result, an operator became able to understand what they were doing while working, from the viewpoint of contextually meaningful connections among respective motions. This could be the starting point to regain conceptual work in the step of operation, but Toyota did not take the next action towards conceptual work. Instead, for Toyota, a worker's conceptual work was pursued outside SW, namely, KAIZEN activities; the revision of SW and the response to irregularity.

That is, we could recognise the recovery of contextually meaningful connections among respective elemental motions. However, it became the tool to promote the expansion of a scope in which everything was standardised. The scope in which standardisation was not the principle to organise work was transformed into the scope in which the main principle of how to organise work was standardisation. Nevertheless, the area in which the principle of standardisation was not carried out remained. In this case, the scope was classified into sub-scopes. For instance, when irregularities (deviations) happened, they were classified either into irregularities experienced for the first time on the particular shop floor or into ones experienced repeatedly. The former irregularities were entrusted to manufacturing engineers, supervisors and maintenance persons. With the latter, they pursued standardisation on how to handle such repeated irregularities. As a result, the latter become entrusted to qualified operators on the line who are qualified to handle such repetitive irregularities (with regard to the hierarchical structures of non-standardised operations, see Chapter 10, and see also Nakaoka et al., 2005). In the case of overseas Toyota transplants, the main topic was the extent to which the latter could be entrusted to supervisors on the shop floor (see Fujimoto, 2004).

From here on, let us examine KAIZEN activities, which are the conceptual work pursued outside of SW.

7.4.2 KAIZEN Activities

7.4.2.1 Relationship Between Standardised Work and Intellectual Skills at a Toyota Plant

Well, where is the conceptual work? In the case of the TPS, only execution is included in SW. While working, an operator is required to pay attention only to the ongoing operation. However, after introducing the ACP, SW took on a new characteristic, i.e. an operator could understand why one particular operation was succeeded by another one, based on the contextually meaningful connection. Even in such a case, the operator could perform the job based on just memory. However, when surveying contextually meaningful connections between series of operations, conducting operations became easy for the worker because they generatively emerged from memory according to the contextually meaningful connection. In this regard, as a result of introducing the ACP, repetitive short-cycle SW became much easy for all operators.

Therefore, SW itself does not include any conceptual work in the TPS and NTPS, the latter of which implies the ACP. In this sense, there is no intellectual skill in the TPS and NTPS. Nevertheless, there are some intellectual skills that are included in an operator's daily work in the field outside SW in the Japanese automotive factory. Such intellectual skills are built in the process of production system development and function as means of the system rationalisation, i.e. efficiency increase. At Toyota, the management's strategy is 'the simpler, the better' as far as SW is concerned. Workers' mental potential should be put into KAIZEN activities. Therefore, they do not use a strategy to enrich SW itself.

Three kinds of intellectual skills are performed by an operator. The first one is KAIZEN (improvement) activities, which means activities after every day's SW for the purpose of improving everyday work, an equipment and environment that all operators are involved. A lengthy continuation of KAIZEN activity creates excellent and loyal operators. Second, such operators are also involved in the process in which a draft of SW is modified and confirmed. Third, operators are involved in the activity of responding to frequent irregularities in the manufacturing process (see Chapter 10).

7.4.2.2 Characteristics of KAIZEN activity

First, we will examine the characteristic of KAIZEN activity. After SW is completed, the KAIZEN activity is practised within or after working hours and conducted by a group leader and a team leader. This involves all team members. The targets of KAIZEN activities are the elimination of ***muda*** (waste) and Levelling Production (continuous flow processing), whereas its methods include Five Ss, Five Whys, and ***mieruka*** (visualisation).[5] Targets and means are given to operators beforehand, and they are driven to think about KAIZEN by their superiors and colleagues.

The objects of KAIZEN are an operator's own SW and everyday working environment; the operator should be more knowledgeable about and more familiar with their work and environment than anyone else. In addition, thinking about the improvement (KAIZEN) of their own work and environment is an intellectual activity, although such thinking activity is prohibited while engaging in SW. Because of these two points, i.e. knowledgeability and familiarity of objects and thinking activity, participating in KAIZEN activities becomes interesting for workers. It is almost like 'playing a game'.

Let us compare such a shop floor with one where an operator simply repeats SW aimlessly. It is very clear that the former shop floor is superior to the latter one in terms of quality and quantity of products. In addition, such KAIZEN activities endlessly continue, so the disparity between the former and the latter become wider.

Paying attention to such expansion of disparity, we will next examine the characteristics of an operator's 'thinking' activity within Toyota. As indicated previously, the particular targets of KAIZEN (elimination of waste and Levelling Production) are given to an operator in advance by the management, alongside the means for realising these targets. Therefore, the scope for 'thinking' is already given, so the operator has no autonomy in deciding the scope of thinking in terms of problem setting or what means should be taken to solve the problems.

Generally speaking, from the viewpoint of mental independence, it is fundamentally important that a person is able to decide what the most

[5] According to Medbo, in Western literature on KAIZEN and problem solving, KAIZEN means methods such as A3, Toyota Business Practices (8 steps), Practical problem solving (7 steps) and Kata, all of which include tools such as PDCA, 5 'Whys', ***mieruka*** and quality control tools. Therefore, KAIZEN is not just tools but entire methodologies for institutionalising work improvement.

significant problem is. As previously mentioned, a worker was encouraged to highlight deviations and to join problem solving such as making suggestion for improvement in the case of Toyota. Nevertheless, the worker is not allowed to establish problem setting. When a worker is shut down from problem setting, where do they go? In this situation, the person is forced to do their best in problem solving without thinking whether the problem was worth thinking or not.

As targets and means are given beforehand, it becomes inevitable that most operators eventually think identically about what to think and how to think. Effectively, almost all operators start behaving as if they were single person. This is the strong point of Toyota's shop floor, which other companies find difficult to follow. In this sense, KAIZEN activities are not only the practical activities in the production process, but they are also powerful means that mobilises workers to pursue the same target as management. Therefore, we can say that KAIZEN activities are one kind of HOSHIN KANRI just opposite to Management by Object.

Let us take the five 'Ss' and the five'Whys' as examples of means of KAIZEN. The five 'Ss' are principles of an operator's attitude and behaviour required by the management in everyday work on a shop floor. The five 'Whys' relates to asking yourself at least five times why a trouble happens when it occurs. Such procedures are very simple but excellent means to achieve a standardised way of thinking about KAIZEN, by which anyone can easily handle these means without any special competence. These standardised methods are provided to all parties and allow any operator to reach high performance in KAIZEN activities.

In addition, it means that all employees identically think about and pursue KAIZEN. Other means are prohibited for an operator to use. This unilateral decision-making about a worker's way of thinking leads to a uniformity of thought. All team members ultimately operate as one single person, which is the characteristic of totalitarian society.

If you are given five 'Ss' and 'Whys' in advance as a method/tool, and you are asked to think about KAIZEN based on these means, it becomes much easy for you to find better solutions in KAIZEN activities than in situations in which you have to determine a method on your own. However, it also signifies that determining a method is omitted from an operator's thinking activity.

Of course, standardisation is the main principle of how to manage the modern factory. However, it is epoch-making that the principle of managing the factory has reached a new stage whereby workers' very way of thinking is also managed based on the principle of standardisation. Although it may mean the realisation of ultimate efficiency in the modern factory, workers' way of thinking may reach the stage of uniformity, which is the opposite of diversity/plurality.

Generally speaking, it is a characteristic of modern mass production factories to repeatedly manufacture the same products in the same way. Therefore, they also pursue the standardisation of the product and the way to produce it, which results in a uniformity. In this sense, it is inevitable that the modern factory has the tendency to obliterate human plurality and change it into the uniformity of thought. Toyota's KAIZEN activity is the result of management consciously and systematically pursuing this tendency. Therefore, it becomes a really strong weapon in the modern factory in terms of efficiency. Uniformity means that everyone thinks in the same way, and a situation in which everyone thinks in the same way means a situation in which the receptor towards totalitarianism is now prepared (Arendt, 1951). The effort towards the systematic pursuit of efficiency in the modern factory now contributes to the birth of a large quantity of citizens who internalise a uniform way of thinking. What will be the future of our society?

Medbo's explanation of the Swedish Lean (See 12.3.1) is appropriate in terms of the Western situation. However, when KAIZEN activity is characterised in the case of Toyota, we described it based on our interviews and documents written by direct parties of Toyota. As a result, the question of how to relate particularities of Toyota's KAIZEN activity to rather generalised acceptance of KAIZEN in Western literatures still remains.

7.5 Excellence and Threat of the Toyota Production System

7.5.1 Excellence of the Toyota Production System as a Societal Sub-system

From the view point of a manufacturing shop floor, an excellent production system is the system which manufactures a product cheaper, qualitatively better and quicker than any other system. Let us characterise the TPS, considering this comparison.

Five points are key, namely: the physical mechanism independent of workers' competence, ways of working, criterion of KAIZEN activities, the special category of engineer and the relationship between manufacturing section and other ones.

First, let us consider the physical mechanism. From the viewpoint of production flow, many parts and components are assembled one by one to a chassis and a car gradually emerges on the line. However, in the TPS the starting point of production flow is set at the lowest part of the flow, i.e. a completed car. Then, they take the JIT method, which means assembling some amounts of parts at a particular point on the line, just the same amounts of parts are supplied to the point from just one step ahead of the workstation on the line and again such same amounts of parts are made at the workstation. This production information is carried by the KANBAN to all branches of the upper part of the production flow. This change of the starting point of the production flow is called the change from the Push System to the Pull System. This is the paradigm change to realise simultaneous production and reduce overproduction to a great extent. In addition, it means that production flow becomes continuous and its fluctuation becomes a little, which are necessary pre-conditions for JIT's functioning. Therefore, as a result of introducing Levelling Production, they succeed in minimising the amount of parts inventory.

In this way, the whole production flow is connected by KANBAN and runs simultaneously. It is also started by the pull system and regulates Levelling Production. Furthermore, to better help managers and supervisors grasp production flow, irregularity and trouble at a glance, various devises for visual management are set along the line.

The above mentioned are physical mechanisms of the TPS. Now, let us examine ways of working, i.e. how managers and supervisors behave while the mechanism is working. First, managers and operators carefully look at these systematised whole production flow and check whether any kind of wastes are found. In addition, they struggle with how to eliminate these wastes mainly outside working hours. This is so-called KAIZEN activities. When they find a defect or irregularity on the line, they try to correct them when and where they happen even if the line stops. By contrast, the line never stops in the case of the FPS, even when facing troubles. Defects are gathered and repaired at the repair shop at the end of the line. The new way of handling defects (TPS) is entirely opposite to the traditional way (FPS). When taking the new way, it becomes radically easy to identify reasons why defects occur. Furthermore, this can avoid facing huge amounts of defects, effectively reducing costs. In addition, outside the daily routine of working hours, managers try to find fundamental reasons for trouble instead of the mere treatment of symptoms. In this very case, five 'Whys' and 'Ss', etc. are taken for problem solving.

Now, let us think about the concept, i.e. standard, which is the point of reference whether managers, supervisors and operators work appropriately. In the case of assembly work, SW is defined for the first time, and then it is strictly kept. Any wisdom and idea found on the manufacturing site are thrown into the process of making SW and reflected on its new version. It is not allowed that such wisdom and idea are kept as individual belonging even if they are found by an individual. Therefore, the best wisdom and idea are expressed on the next version of SW in any case.[6]

Under such pre-condition, when trouble occurs on a manufacturing site, is it because of the operator's deviation from the SW or does it occur in spite of worker's obedience to SW? When it is the former, corresponding workers are re-trained to follow SW; in the latter, revising SW to a new one becomes necessary. In any case, the characteristic of trouble becomes clear because there exists SW on a shop floor as the effective criterion.

In retrospect, it is necessary for them to have very high capability to analyse and put together again real operations on a shop floor for the purpose of making more effective and stricter SW. Having a staff office of manufacturing engineers just next door to the manufacturing site of the same floor makes it possible. Only the existence of such manufacturing engineers makes it possible to realise the product with high quality, low cost and short lead time.

On the process of the revision (KAIZEN) of SW, the tool to improve (KAIZEN), such as five 'Whys' and 'Ss', are given beforehand to workers, so they are guided to a certain direction to some extent. That is to say, the pattern of thinking is also standardised. Therefore, how to KAIZEN becomes very efficient.

Regarding the relationship between manufacturing shop floor and its upper section of production flow, one distinguishable characteristic is the one room system (see Liker, 2004), which means that each responsible person of a product design, process design, production planning and manufacturing is located in the same room. It is a kind of Concurrent

[6] It was based on the interview with management in around 2004.

Engineering. Because of it, for instance, it becomes much easy for the product designer to make a design and concurrently consider, beforehand, how to assemble parts to a body.

In addition, sometimes the proposal to re-design a product from the viewpoint of more easily assembling it is given to and accepted by a product design department, even after product design is completed and entrusted to the manufacturing site for the purpose of making a product identical to its blueprint. One manager of a manufacturing section assured that 30% of such proposals are accepted (see Tanaka, 2005 p. 59). In the case of part assembly, a different colour or figure is taken among similar parts to distinguish an individual part easily (see Tanaka, 2005, pp. 98–99). Such a feedback system functions well.

As highlighted in various parts in this book, manufacturing engineers' presence and authority function very well in keeping good communication between manufacturing shop floor and an upper section of production flow, such as product design and production planning. They sometime find defects that occur in manufacturing sites, but the origin of which come from inappropriate product design, etc. In such cases, they translate a problem on a manufacturing site into technical terms which product designers are able to understand. Such one kind of translation works well for the mutual understanding between manufacturing site and upper parts of production flow.

So far, from examining characteristics of the TPS, we can say that the TPS is a very rational production system from the viewpoint of how to make products with low cost, high quality and short lead time.

7.5.2 Toyota Production System as a Threat to Human Society

7.5.2.1 Pre-condition of standardisation

Although the TPS is excellent as a production system, it also threatens human nature and human society. As you already know, Toyota management believes that there is One Best Way on how an operator works. This is the strong pre-condition when they look for KAIZEN anytime and anywhere. Once they accept this pre-condition, it is the best way to find this One Best Way. They declare that it is now the SW and all operators follow the new SW, from the viewpoint of efficiency. This results in huge cumulative effect of efficiency for a long time.

In addition, if they find a defect on the line it means that this SW includes some problem; therefore, such an SW needs to be revised based on KAIZEN activity. On the second thought, you will find the relationship of a spiral staircase (virtuous circle) between standardisation and KAIZEN. The fact that they find this relationship is epoch-making from the viewpoint of efficiency.

In retrospect, let us reconsider the concept, i.e. standard. The standardisation of a product, part, tool and machine, and the way to operate is, of course, the consistent way to pursue high efficiency in the case of modern mass production factory. Because of standardisation, we can enjoy material wealth in modern society.

However, please look at the concept, i.e. SW on the line, carefully. In the case of Toyota, SW is officially acknowledged as the One Best Way to work as far as all available knowledges of work are mobilised at the particular moment. However, why is it the best way? What is the point of reference for the judgement? When we judge it is the best, it is not so clear whether or not the criterion is appropriate. Nevertheless, we do not examine mostly whether or not, such a hidden criterion is really appropriate in the case concerned.

Let us use two examples. The first one is the arrangement of letters on a keyboard. It seems to us to be a strange order, but why is it accepted among people? It is not the best way to type quickly and easily. This order is the heritage of the age of the mechanical typewriter in which they needed to avoid the type bars getting entangled under the keyboards. It is the QWERTY keyboard. In the age of computers, keeping the QWERTY keyboard is not necessary for us. However, we have gotten used to it globally (see Hashimoto, 2002). Namely, the QWERTY keyboard has become a 'de facto' standard. In this regard, a prevailed standard is not necessarily the One Best Way from the viewpoint of rationality, i.e. quick and easy typing.

Second, let us think about assembly work on the line. Whether it is the traditional FPS line or the TPS line before introducing the ACP, the criterion over how to organise SW is only line balancing. The concept, i.e. standard itself has no ability to judge whether or not the criterion, i.e. line balancing, is right or wrong. While they can employ necessary number of employees on the assembly line, SW based on the line balancing as the criterion has no problem from the viewpoint of a labour market condition. However, once they face the labour shortage, line balancing will lose the status as the criterion how to make a SW. Therefore, we had better think about three points in this regard. First, standard itself does not necessarily mean One Best Way. Second, any standard has its own hidden criterion how to organize standard way. Third, is the presence of One Best Way really the truth in any case?

At the beginning of the 1990s, Toyota could not employ the necessary number of workers because, in those days, there were lot of opportunities for employment and it became famous that working on the line, particularly at Toyota, was hard for workers. It meant that their car sales were very smooth; however, they could not obtain the necessary number of employees. This was a severe dilemma from the management point of view.

As a result, they added another criterion for line balancing in organising assembly work. It was to have contextually meaningful connections among series of operations. Because of the combination of the two criteria, i.e. line balancing and contextually meaningful connection among series of operations, they introduced the new concept, i.e. the ACP. Far more than that, taking human cognitive capability, management reorganised, more radically, assembly work in the case of Uddevalla, Volvo. That is to say, workers learned the outline of the whole architecture of an automobile and the whole assembly process before they were put into assembly work. Then, operation order was set, following the order by which an automobile was assembled. This operation order was organised for each operation to emerge one by one on their mind based on human cognitive characteristic. Moreover, parts were arranged corresponding to the operation order. It made it possible for an operator to avoid assembling the wrong parts to the body. A standard operation order was instructed to an operator; however, it was changeable while working within long cycle time. According to Engström, the term, 'standard' that signifies a suggestion to operators, is not mandatory.

Therefore, it is not a pure technical problem how to decide the 'standard' in the case of SW. Namely, when we see a certain type of SW, we can find the corresponding judgement of the management in terms of labour market and a worker's capability as a background. Alternatively, a certain social relation is implicitly embedded in a particular 'standard'. Therefore, it is correct only under the certain specified societal relation that a certain SW is One Best Way to work.

In addition, let us think about the role of 'standard' from the viewpoint of training. The goal of training is to fashion an individual able to execute SW. In any case of a vocational training for technical qualifications even beyond the factory, a standard level of skill is the goal of training. As far as vocational training is concerned, we take it for granted because we are now talking about how to get the necessary labour force. However, vocational training is entirely different from expected school education, which should support people to grow up by themselves. The former forces people into a certain pattern required by the labour market, whereas the latter supports people to develop their potentialities towards citizens with plurality.

In conclusion, even if something is standard within a certain organisation or society, this standard could not be automatically justified. We have to examine further whether we take this standard as our own point of reference.

7.5.2.2 *Standardisation How to Operate vs. Unification How to Think*

As described in 7.4.2. KAIZEN has the following characteristics. First, the ends of KAIZEN (elimination of waste and Levelling Production) and the means for the ends (five 'Whys' and 'Ss', etc.) are given to workers beforehand. As a result, a worker's concern is concentrated only on problem solving. In addition, even in problem solving, the means over how to solve are given in advance.

In the case of ends and means being given in advance, what to think is the same because ends are already given before the start of the thinking process. Furthermore, how to think becomes spontaneously similar among workers because tools regarding how to think are also given in advance. As a result, all participants think as if they are a single worker. This is efficient for standardising solution finding; however, it also means that participants omit their process of finding proper tools in terms of individual thinking process. Alternatively, such proper tools are already given standard tools. Namely, we can say that *standardisation* is now introduced into the very way of thinking for humans. It is the ultimate realisation of efficiency. By contrast, it is left without cautious consideration whether the criterion of how to standardise is appropriate.

Let us consider the relationship between the so-called five 'Whys' and one 'How' from the viewpoint of the relationship between problem setting and problem solving. As you know, five 'Whys' mean that when facing trouble on the line one should ask themselves five times the reason of the trouble. This process to find a fundamental reason for the trouble appears to be the point of problem setting, and one 'How' afterwards seems to be the problem solving.

However, please remember the KAIZEN activities (see 7.4.2). As you know, KAIZEN aims to eliminate waste and Levelling Production, which are already given by the management. In this sense, the problem setting is given as target setting to operators by the management.[7]

Well, let us return to the topic about the introduction of standardisation into the way of thinking. It means that the management systematically penetrate into and manipulate human inner thoughts based on standardisation. Such management that consciously manages operators even in their own inner world may not be found in the field outside of Japanese companies. It may violate the de facto principle of modernity, because the principle regarding how to organise modern civic society is that no one is allowed to interfere in an other's inner world without equal partnership and mutual approval. Rules and laws should only control behaviours of humans outside of their inner world. Only the bilateral dialogue among human beings with equal partnership is allowed to interfere in or refer to the other's inner world.

The introduction of standardisation into the way of thinking also means that the ways of thinking become the same among workers. Once the principle of standardisation becomes the principal way of thinking in a person's inner world, standardisation transforms into uniformity regarding the way of thinking. It means that everyone pursues the same direction. Therefore, the KAIZEN activities result in the highest efficiency in a factory because of the management's very conscious and systematic pursuit of standardisation. Concurrently, however, the introduction of standardisation into the way of thinking means that uniformity in way of thinking becomes superior in a factory.[8] If this tendency also becomes superior in society we have to say that the receptacle for totalitarianism is now prepared, in the sense that everyone thinks in the same way. Such

[7] Masatomo Tanaka is the former manager of the section responsible for the development of Toyota's TPS. He refers to target setting (see Tanaka p.53). According to him, the section manager makes a KAIZEN plan and breaks it down for each shop floor.

As you notice, the problem setting is given from the upper management to a team on the manufacturing site. After this setting, the process of five 'Whys' starts. In this regard, it is similar to the description in Sect. 7.4.2.

[8] Various problem solving take place in a team through the KAIZEN activities. Different individuals with different views contribute differently to solve problems. However, the stronger the influential power of standardisation becomes within a team, the narrower the scope of the way of thinking in terms of problem solving becomes.

uniformity is not equal to totalitarianism itself, but it means that we are now prepared to welcome totalitarianism.

Examining the point further, workers are driven to join in the KAIZEN activity by the management on the shop floor at Toyota. In the case of Toyota, the way to improve (KAIZEN) is also given to workers in advance. That is to say, management provides problem setting and problem solving in advance.[9]

In retrospect, let us think about the human condition, not about the slave condition. What is an essentially necessary condition for you to be independent of anyone else as a human? In the case of thinking activity, the most important point is for a person to decide, by themselves, based on and supported by dialogue among peers, what is the problem to be worth solving, but not to decide, by themselves, how to solve. That is to say, the human condition is the problem setting but not the problem solving. When you neglect this point, you will drop into the world where you only find petty bureaucrats, who are capable only to handle with clerkly jobs given by their supervisors, and who are indifferent to what is the purpose of their own clerkly jobs.

It is reminiscent of Adolf Eichmann as an example of petty bureaucrats (see Arendt, 1963). He successfully handled transportation difficulty regarding how to gather Jewish people together, who were scattered all over Europe, and how to transport them efficiently without any resistance to Auschwitz.

In this regard, he was a very capable official. However, he never thought the purpose of his transportation matters. Although he was not a member of the board of decision-making of the Final Solution (the Holocaust), he knew the purpose very well because he attended the meetings as an observer and was responsible for transportation matters. One day, he looked back at his work and said, 'I was good at desk work'. Arendt characterised the essence of this way of thinking as 'thoughtlessness', and she has warned that the accumulation of innumerable such 'thoughtlessness' invite totalitarianism into our world.

[9] The main text is referring to the KAIZEN activities in Japan, particularly at Toyota. However, there exists differences in the target of the KAIZEN activities between Toyota and the West. According to Medbo, regarding KAIZEN, in the West, the goal/result is the primary focus, whereas in Japan, the primary focus is on the way of getting to the goal.

Returning to petty bureaucratises, in our modern society, most of us work for some kind of bureaucratic organisation, essentially belonging to a certain sub-system of a whole system. Therefore, we are locked in the work within the sub-system, the end of which is given by the sub-system, which is one step superior to our own sub-system. In our sub-system, mostly, the target of the sub-system is already given by a step superior to our system. Namely, when constituent members of an organisation are given a problem that their superior sets in advance, it is not regarded to be questionable whether it is really worth solving for them. However, only when constituent members are allowed to reconsider by themselves whether it really has a priority as the problem to be set, they will find a possibility for them to be equally independent decision-makers within the organisation. Not only Toyota but also any bureaucratic organisation is not open for its members to raise this fundamental question. Alternatively, workers on the manufacturing site are excluded from the decision-making process.

However, the very characteristic of Toyota's KAIZEN is on improving the already given standardised way of building a car from the viewpoint of how to eliminate waste and realise Levelling Production (stability in processes). Therefore, regarding the KAIZEN activities at Toyota, no room exists for workers to think about the problem setting, although it is a fundamental matter for human independence. Furthermore, with Toyota, team members are driven, under their group leader (GL) and team leader (TL)'s initiative, to think about KAIZEN, within the scope of a given problem setting. In addition, for avoiding diffusion of directions of thinking activities, tools on how to think are given in advance. Therefore, all GL, TL and team members have the same target and tools as if they were a single person, which is hard to be realised in any other organisation, except for Toyota.

From here, let us think again about tools that drive workers to think in a particular way. The five 'Ss' and 'Whys', visualisation, etc. are excellent tools that are thought out. They are also good standardised ones, which anyone can handle. These standardised tools are handed over to workers, allowing them to efficiently struggle with KAIZEN, without having to look for tools by themselves. It also means to patternise their way of thinking and investigating processes into only one standard. As a result, everyone on the shop floor thinks in the same way.

Standardising the way of thinking is uniformising thinking. In the case of Toyota, this is the reason why any middle management speaks confidently with their own language, even though this 'own language' is almost the same when we interview them in any case. In the case of the KAIZEN activities at Toyota, tools are already given beforehand and they are ordered to 'think' based on these tools. When compared with the case in which workers have to find the most appropriate tool by themselves, as if in a fog, Toyota's KAIZEN activities are very thinking-saving, therefore very efficient. This is efficiency from the viewpoint of a production system, but not effective from the viewpoint of how to grow a thoughtful human being with his/her own way of thinking. When tools of KAIZEN are already standardised beforehand, 'thinking' activity becomes easy for everyone. Nevertheless, this 'thinking' activity lacks the activity where one should find the focusing point by themselves. In this regard, this thinking activity lacks the thinking of how to find tools for KAIZEN, which means that it is a thinking activity without an essential point of thought.

Such routinisation and standardisation of the way of thinking will result in uniformity, having all team members thinking as if they were the same single individual. This is not a particular case of Toyota, because it is common in a modern factory in which the same product is produced in the same way. However, it is the very characteristic at Toyota that the management consciously mobilises operators to participate in the KAIZEN activity in the same pattern of thinking. This method has, now in Japan, penetrated into offices, hospitals, schools and other environments beyond a factory. Furthermore, it is also learned in factories in the Western world.

In retrospect, totalitarianism is realised when a certain ideology seizes the hearts of people who are in situations where their way of thinking become uniform, so that everyone thinks in the same way[10] (see Arendt, 1951). However, people are originally different. Everyone has different qualities, experiences, ways of feeling and ways of thinking. These beings bring each other the image of reality which they barely grasp,

[10] Even if operators take the same standardised tool, such as the five 'Whys' for example, to solve a problem, it does not necessarily mean that all take the same trajectory of thought to find the same answer. In this regard, we have to distinguish between thinking in the same uniform way and taking the standard and same method for problem solving as the starting point to think. However, please remember the three pre-conditions. First, the targets are given in advance. Second, tools except for tools already given (such as the five 'Ss' and 'Whys') are excluded beforehand from operators' consideration. Third, only a group leader decides which operator's answer is the fundamental reason of the final why. These three pre-conditions aligns everyone's thinking to almost be the same.

then compare them with each other. As a result, they constitute their common reality and collectively agree on a common and acceptable way of achieving solutions. Only such a way of doing things will be the breakwater against totalitarianism (see Arendt, 1958). These contexts are condensed into the concept, i.e. 'plurality'.

Now, let us come back again to actuality at a plant. It is the characteristic of a modern factory where they produce the same product repeatedly. Therefore, as a result of pursuing standardisation, it is unavoidable that their ways of thinking are made uniform. That is to say, we cannot help saying that the force to grind human plurality works and prompts uniformity of human way of thinking in a modern factory. Namely, this uniformity of the human way of thinking is a chronic disease in modernity. Among other things, the KAIZEN activity at Toyota has the tendency in which the above disease works most consciously and systematically. This is a very strong weapon in a modern factory from the viewpoint of production management. Concurrently, when we think about modernity in the case of such ultimate rationalisation, such uniformity of the human way of thinking may seem horrible to us.

This comparison is reminiscent of the post-Fordist debate (during the 1990s) regarding the evaluation of the Japanese production system between Western researchers and Japanese ones. Western researchers who were critical of the Western production system evaluated the Japanese production system highly. As a result, their admiration for the Japanese system complemented the self-recognition of Japanese researchers who were very positive of the Japanese system. Therefore, Japanese researchers who were critical of the Japanese system were embarrassed at that time.

On the second thought, this book emphasises the problem solving-oriented characteristic of Toyota's KAIZEN, but Medbo emphasises the problem setting-oriented characteristic of KAIZEN in the West. It is unclear whether Medbo's description is right. However, it is very probable that Japanese (Toyota's) KAIZEN, i.e. activities with problem solving-oriented characteristic, has transformed into Westernised KAIZEN, i.e. one with problem setting-oriented characteristic. Otherwise, the introduction of KAIZEN into the Western plants ought to have faced severe refusal from labour unions because, generally, Westerners are more individualistic than the Japanese.

7.5.2.3 Concentration on a Production System But Indifference to a World Beyond a Production System

As you know, the five 'Whys' is one of the famous tools of KAIZEN activities. According to Ohno, we should ask 'why' five times when a trouble occurs. Otherwise, we cannot reach the essential reason as to why something occurs (see Ohno, p. 35).

Let us pay attention to characteristics of this five 'Whys'. First, it can be used anytime and anywhere when trouble occurs, without having to think over the context in which the trouble occurs. This way, it is transformed into a pure technical skill which anyone can use without any professional competence.

Second, it seems that Ohno emphasises for operators to have to find a fundamental origin, but not to find a secondary phenomenon when facing trouble. Therefore, asking five times symbolises such a thorough inquiry; it can even be four, six or more times. Most texts about the TPS refer to this point as follows: namely, although workers repeat why, many times, they can find fundamental origin.

However, on this thought, two questions come to my mind. The first question is, what is the fundamental origin? The second one is, who judges it as the fundamental one? If these two questions are overlooked, we cannot understand when questioning switches from *why to how* to solve a problem.

Therefore, what is the fundamental origin? For instance, there are three candidates for the fundamental origin, all of which are found by team members, A, B and C respectively (see a different example in Liker et al., 2008). Member A's one can be solved within their shop floor. Member B's one needs the revision of the product design, which cannot be solved within the shop floor. Finally, Member C's one needs the revision of the component design of a supplier. In these cases, A's answer to 'why' is most easily accepted as a fundamental origin by team members, TL and GL. For B, it is not easy to respond to whether the answer is accepted as a fundamental one or not among team members and its supervisor because it is beyond TL and GL's authority. It is up to GL's judgement and/or that of the product design department. What about C's one? This will be regarded as the answer over the fundamental origin at least at the present moment. Whether it is the fundamental origin is strongly related to whether or not the solution to the origin is given within the boundary of GL's authority.

In any case, once the answer is regarded as a fundamental origin, everyone is ordered to quit to pursue more fundamental origins. What does that mean? Workers try to pursue the fundamental origin until they cannot find any appropriate measure corresponding to the origin within the boundary of their GL's authority. In other words, when they reach the boundary, they quit to repeat 'why'.

That is to say, regarding when workers quit repeating 'why' is settled in advance; the boundary within a shop floor or a plant, which is controllable by team members and their GL.

However, is it possible that each individual team member judges whether or not each answer to 'why' is the fundamental origin while being engaged in their KAIZEN activity? If the judgement is fully entrusted to each individual regarding whether or not a certain answer to 'why' at some step is the fundamental origin, then each judgement will be different to some extent. It only becomes clear after the practice of the answer to the last 'why', whether the judgement of the fundamental origin is correct. In addition, it is hard for team members to understand clearly what the boundary of their shop floor or plant is because of a lack of knowledge of overview regarding their shop floor or plant. This knowledge is the necessary condition to judge what the fundamental origin is.

Therefore, the only person who is able to judge what is the fundamental origin is the GL. This is because individual team members cannot stop to ask 'why' by their own judgement but only with the authority of the GL. Of course, it is not guaranteed whether the GL can really judge the fundamental origin, but a given official authority allows a GL to judge whether the tentative answer to 'why' at the present moment is really a fundamental origin. In addition, as far as the knowledge of overview in terms of a shop floor or a plant is concerned, the GL is generally superior to any team member. Because of Japan's internal promotion system, a GL has a lot of experience as an operator and gets their promotion grounded in that experience. In addition, the GL has the power within their own jurisdiction. Therefore, it is a GL who judges when they stop the endless process of questioning 'why'. Once a GL decides, each individual team member engaged in the KAIZEN activity must quit asking 'why', even if they think that it is still necessary to continue asking 'why' to better reach the fundamental origin.

In sum, individual team members must continue asking 'why'. If they do not, they are enforced by a supervisor to continue asking. It is a

GL that judges when any member concerned should stop the process of asking 'why'. What is the boundary for this continuation? It is extendable to the area beyond the border of their shop floor, and it is possible to extend beyond all shop floors to even a product design department in small cases. However, in most cases, the boundary is limited within the GL's jurisdiction. Now, when the thinking is stopped within the jurisdiction it means that an individual worker's thinking activity is locked and closed within the shop floor. Namely, his/her horizon is limited within his/her daily routine as a small sub-system.

One's daily routine is a worker's everyday life, and it includes thinking activities. Alternatively, workers offer not only their physical capabilities but also their mental capabilities to the company. The way to offer one's thinking activity is very characteristic; one simply locks their thinking activity within a range of daily routine. This endless imprisonment of thinking activity results in particular 'habits of the heart' (see Beller et al., 2007) in which a person has the tendency to lock their thinking activity in a sub-system in which they directly belong to. It is a certain internalisation of particular 'habits of the heart'.

In retrospect, it is the very characteristic of A. Eichmann who was not interested in the problem beyond the sub-system in which he directly belongs to, although the problem beyond his system is terribly serious, i.e. genocide. Such 'habits of the heart' will promote in society the growth of Eichmann-like personality types.

Such types of people are akin to hamsters in a hamster wheel, moving the wheel with joy, although they are not interested at all in the reason why they move the wheel. In this regard, Tanaka refers to *the stretch target* (see Tanaka, 2005, p. 46). According to him, operators are given, from above, a hard target to realise, i.e. the stretch target which they have to stretch their competence to realise and finally reach to. Once they have reached the target, they are acknowledged and admired among their own group.

That is to say, the target is set by management and is realised only when operators make efforts at full stretch. In addition, rules of a game such as five 'Whys' and 'Ss' and so on are given in advance to participants of a game, although these rules are not particularly mentioned in the management's remarks. As described above, the target is difficult to realise but is still possible. In such a game-like circumstance, which is set on purpose, team members repeat to play a game so many times. As a result, what happens then? In such a case, they lose the sense of playing a game as it is transformed into their own reality. They are just locked in a closed sub-system in their real world and are enforced to move the hamster wheel without reflection over the meaning of moving the wheel. However, they now fall into the illusion in which they are able to grow up by themselves. In this regard, they are deluded to think that they are alive and free in an open and big system, although they are locked in a small and closed sub-system. Everyone is convinced that they are living in a wide and open system and pursuing total optimisation. However, everyone is actually pursuing nothing but their own partial optimisation as they like.

In sum, people now live in finely sub-divided systems, and they pay attention only to this small world, pursuing its partial optimisation. As a result, all are very busy in this small world without the concern of total optimisation. It seems that while people devote themselves to move their hamster wheels without their concern in whole, it may happen that a demagogue stands out and comes into power because of people's vague fears and expectations. If it happens, we will be thrown into a steep downhill towards totalitarianism.

References

Arendt, H. (1951). *The origin of Totalitarianism*. Secker & Warburg.

Arendt, H. (1958). *The human condition*. The University of Chicago Press.

Arendt, H. (1963). *Eichmann in Jerusalem*. The Viking Penguin.

Asao, U., Fujita, E. Nohara, H, Saruta, M., & Yamashita, T. (1999). *Syakaikankyo no Henka to Jidosya Seisansisutemu*. Houritsubunkasya.

Beller, R., et al. (2007). *Habits of the heart*. The University of California Press.

Fujimoto, T. (2004). *Nihon no Monozukuri Tetugaku*. Nihon Keizai Shinbunsya.

Hashimoto, T. (2002). *Hyojyun no Tetsugaku*. Kodansya.

Hounshell, D. (1984). *From the American system to mass production, 1800–1932*. The Johns Hopkins University Press.

Ihara, R. (2017) *Muda no Kaizen, Kaizen no Muda*. Tokyo: Kobushisyobo.

Liker, J. (2004). *The Toyota way*. McGraw-Hill.

Liker, J. (2008). *Toyota culture*. McGraw-Hill.

Meyer, S. (1981). *The five dollar day*. State University of New York Press.

Nakaoka, T., et al. (2005). Syokuba no Bungyo to "Henka to Ijyo heno Tiou." *Journal of Humanities and Social Science (Nagoya City University), 18*, 1–51.

Niimi, A., et al. (1997). Development of the new human-conscious automobile assembly plant. *Annals of CIRP, 46*(1), 381–384.

Nohara, H. (2006). *Gendai no Bungyo to Hyojunnka*. Takasugasyuppann.

Ohono, T. (1978). *Toyota Seisan Housiki*. Daiyamondosya.
Tanaka, M. (2005). *Kangaeru Toyota no Genba*. Kabushikigaisya Bijinesusya.
Womack, J. P., Jones, D. T., & Roos, D. (1990). *The machine that changed the world*. Palgrave Macmillan.

CHAPTER 8

Cell Production System and Its Conditions

Uichi Asao

8.1 Introduction

Since the early 1990s, Japan has had innovations in production systems, particularly in the assembly processes of electrical, precision machinery and other items. The conversion refers to the transition from traditional production methods involving assembly lines with conveyor equipment to the cell production system (CPS) involving assembly system without conveyor equipment. In 1997, 23.2% of factories in Japan utilised CPS for assembly processes in the primary machinery industry, whereas 25% were 'planning to use CPS in the future' (Tsuru, 2001, p. 56).

Since the early twentieth century, the mainstream assembly system for mass-produced goods in Japan has been a long assembly line equipped with conveyor systems. Automobile production serves as a standard

U. Asao (✉)
Nagoya, Japan
e-mail: uasao_2902_1972@yahoo.co.jp

H. Nohara and L. Medbo (eds.), *Technology Convergence and System Divergence*, https://doi.org/10.1007/978-981-96-1910-8_8

illustration of this process.[1] Our Japanese research group[2] was keenly interested in the introduction of CPS into the assembly processes of Japan's prominent machinery industry. The reason for this was our awareness of the implementation of a novel assembly system at Volvo's Uddevalla plant in Sweden around 1990. This novel assembly system was recognised for its ability to enhance efficiency and humanise the work process, as detailed in Chapter 2. We were primarily interested in determining whether Japanese assembly plants for electrical, precision machinery and other items implemented a system comparable to that of Uddevalla plant. We also want to understand the true nature of the CPS, the reasons behind its implementation and its potential for further promotion.

This chapter begins with an overview of the CPS of Japanese Company N, which manufactures electrical and precision machinery items and can be considered one of the most typical examples of CPS implementation. Since the end of the 1990s, the C enterprises group, which includes Company N, has used CPS in their assembly workshops, both in Japan and at their overseas plants. In particular, Company N was the first and most successful enterprise in the C enterprises group to implement CPS.

Our research group visited Company N five times between November 2000 and December 2003, conducting interviews mostly with the manufacturing department manager and section managers, excluding leaders and operators of workgroups. These interviews provided the information used to describe Company N's manufacturing system.

This study then examines the characteristics of CPS in comparison with traditional production methods, as well as why the transition from traditional to CPS has advanced.

[1] In the mid-1970s, there were examples of assembly lines without conveyor equipment or attempts to restore humanity in Japanese electronics and vehicle manufacturers' mass production assembly workshops (Masuda, 1976; Nakano, 1976). However, these attempts did not become widespread.

[2] At that time, this research group consisted of Eishi Fujita, Hikari Nohara, Masaki Saruta, Motohiko Yamashita and Uichi Asao. Yutaka Tamura has joined later. Yamashita was a researcher on labour issues in the field and is now deceased. The other five are all co-authors of this book.

8.2 Cell Production System of Company N

Company N was the leading manufacturer of certain computer peripherals (Product X). Other companies belonging and not belonging to the C Enterprises Group produced Product X. Company N employed 1,691 people, and the annual production of Product X was approximately 1.25 million units (as of 2002). Product X had two assembly lines: one for the main unit and one for its subunits. This section focuses on the assembly line for the main unit (referred to as Product X hereafter).

8.2.1 Conversion from Traditional Production System to Cell Production System

Product X was assembled using the traditional production system for approximately ten years, beginning in 1989. There were six assembly lines with conveyor equipment, each 180–210 m long and with approximately 50–100 operators. They assembled 1,000–2,000 products per shift, with work cycles lasting 20–30 s.

Between November 1998 and June 1999, Company N eliminated all six assembly lines for Product X and switched to CPS, a production system with small work areas known as assembly cells. The assembly cells had about one to ten operators and no conveyor equipment. The number of assembly cells increased and decreased with varying production volumes, expanding to 40 cells when production was high but decreasing to 14 cells in March 2003 and 15 cells in December 2003. The work cycle time ranged from approximately 40 s to 45 min (see Table 8.1).

8.2.2 Types of Assembly Cell

The assembly cell's main equipment consisted of workbenches (which also served as component shelves), small inspection machines and assembly tools, which could be easily moved. As a result, multiple types of assembly cells are used in an assembly plant to assemble the same product type. There are two basic types, namely mini-assembly line (Type 1) and assembly by a single operator (Type 2) , with subtypes rabbit chasing method (Type 2a) and stationary method (Type 2b). As shown in Table 8.2, this is the case for Company N.

In Company N's Type 1, approximately 10 operators shared the entire process (mounting, inspection and packing) in the assembly cell, and the

Table 8.1 Comparison of production systems (Company N's assembly plant)

	Traditional production system (1989-June 1999)	*Cell production system (CPS) (Nov 1998–Dec 2003)*
Assembly line and cell	6 assembly lines with conveyor equipment Line lengths range from 180 m to 210 m	14–40 of assembly cells Main equipment in the assembly cell: workbenches (also serves as components shelves), voltage and image inspection machines, and assembly tools
Number of compo-nents	40–60 components	About 80 for large numbers of components
Number of operators	Approx. 50–100 persons per line	1 person-approximately 10 persons per assembly cell
Work cycle time	20–30 s	Many assembly cells: approximately 40–170 s Few assembly cells: approximately 10–45 min

product was assembled in a work cycle time of 40–170 s. The reason for keeping the number of operators to 10 is that it is easier to exercise the group's power with this number of people. Type 1 cells comprised a large proportion of the total number of assembly cells, with 10 of 15 cells being Type 1 (as of December 2003).

In Type 2, a single operator oversaw all processes (or mounting and inspection processes) in the assembly cell, and the product was assembled in approximately 10–45 min. Type 2 is classified into two types: Type 2a, in which several operators (about two to four) assemble the product while moving between processes; and Type 2b, in which one operator assembles the product with minimal movement between processes. Type 2b was not widely adopted because it increased the number of jigs and tools and the man-hours required to supply components.

As previously stated, the various types of assembly cells fall into three categories: Types 1, 2a and 2b. Compared to traditional production systems, the external features that all of them share can be summarised

Table 8.2 Typology of assembly cells (Company N)

	Mini-assembly line (Type 1)	*Assembly by a single operator (Type 2)*	
		Rabbit chasing method (Type2a)	*Stationary method (Type2b)*
Number of operators	Approx. 10 persons	Approx. 2–4 persons	Approx. 1–10 persons
Work content	Each operator assembles the parts of the products	Each operator assembles the entire product while moving between processes	Each operator assembles the entire product in a fixed position without moving between processes
Work cycle time	Approx. 40–170 s	Approx. 10–45 min	Approx. 10–45 min
Product characteristics	Relatively large products with a large number of components	Relatively small products with a small number of components	
Duration of use and reason for use	Mainly at the start of production of new product models and during periods increasing production Short production start-up periods	Mainly during periods of stable production and reduced production Long start-up period of production (around six months)	
Percentage of all assembly cells	60–80%	20–30%	about 10%

as follows. First, long assembly lines with conveyor equipment are eliminated; second, many small work areas called assembly cells are located within the plant. Third, one or a few operators (approximately 10) assemble products in an assembly cell.

8.2.3 Results of the Cell Production System

The main results obtained by Company N from the conversion from traditional production systems to CPS were increased productivity, improved quality, increased flexibility, reduced floor space and improved operator morale.[3]

8.2.3.1 Increased Productivity

As shown in Table 8.3, an assembly shop for one model of Product X converted one long assembly line into six assembly cells, resulting in a 17.5% increase in productivity. This was due to a decrease in the number of employees from 60 to 51.

Table 8.3 Comparison of production systems (assembly shop for one model of Product X)

	Traditional production system	*Cell production system (CPS)*
Number of assembly lines and cells	1 assembly line of 210 m	6 assembly cells
Production volumes	1,200 products per shift per day Scheduled working hours per shift per day: 8 h	1,200 products per shift per day (200 products per assembly cell) Scheduled working hours per shift per day: 8 h
Work cycle time	21 s	140 s to approximately 20 min
Number of employees	Subsection managers: 2 Support personnel: 8 (Relief persons: 2 Troubleshooters: 2 Repairpersons: 2 Material handlers: 2) Operators: 50 Total: 60 persons	Subsection managers: 1 Support personnel: 2 (Relief persons: 0 Troubleshooters: 0 Repairpersons: 0 Material handlers: 2) Operators: 48 (including 6 cell leaders*) Total: 51 persons
Productivity	20 products per person per day	23.5 products per person per day

* The cell leader is the name used by Company N, usually called the group leader

[3] For information on the effects of implementing CPS in Japanese companies, see, for example, Ito (1997), Okukubo (1999), Tsuru (2001), Iwamuro (2002) and Sakamaki (2006).

8.2.3.2 *Improved Quality*

In the traditional production system, products were assembled and inspected together at the end of the assembly process, making it difficult to determine which process the defect was found. In the CPS assembly process, back-end operators checked the work done in the front-end process. For example, they checked for missing components and misaligned assembly positions. If a defective product was found, it was removed from the assembly process and reworked by the workgroup leader (cell leader) or technical staff before being returned to the assembly process.

These in-process checks determined which process produced the defective product, thus lowering the number of defective products. As a result, the percentage of products that did not require rework before shipment rose to nearly 98–100%. This was 3–5% higher than in the past.

8.2.3.3 *Increased Flexibility*

CPS made it easier to switch models, introduce new products and deal with fluctuations in production volumes. In the traditional production system, changes in models were made on a monthly basis and production had to be levelled. CPS allows for the production of multiple models in each assembly cell, eliminating the need for production levelling. Switching to new products is now possible by reorganising existing assembly cells based on product size and production volume.

In terms of production volume fluctuations, under the traditional production system, for example, if the production of a specific model on an assembly line decreased from 1,000 units to 300 units, the operator assignment for each process had to be changed. In CPS, the same model is often assembled in multiple assembly cells, so when production volume decreases, the number of assembly cells can be reduced without changing the operator allocation.

8.2.3.4 *Reduced Floor Space*

CPS, which removed large equipment such as conveyors, reduced the area of the production floor by eliminating unnecessary space between work processes and integrating inspection and packaging processes into the assembly cell. For example, converting one assembly line into six assembly cells saved 50% of floor space.

8.2.3.5 Improved Work Morale

There are no direct indications that the conversion to CPS has improved operators' work morale. However, the following summary of interviews with the manufacturing department manager and the assistant section manager suggests increased work morale.

· In the traditional production system, if a product was defective, the operator would quietly let it slide unless pointed out to them. This was because someone outside the assembly line was in charge of it. Operators in assembly cells had to work with an awareness of production volume and quality targets. In other words, the collective sense of responsibility shifted. For example, to avoid working overtime, workgroups were required to meet production volume targets within the scheduled working hours. In addition, if a quality issue required rework, they had to pull the product from the warehouse and fix it themselves (September 2001).

· Operators in the assembly cell began devising ways to increase work efficiency. The traditional production system limited the work rate to seconds, so operators were likely aware that they only had to work within that range. The assembly cell had an hourly production volume target, but there was no second-by-second constraint, so operators became aware of the importance of working as quickly as possible (March 2003).

· In Type 2a (rabbit chasing method), operators were less wasteful. For example, in Type 2a, there were only two to four operators, and unlike the mini-assembly line (Type 1), operators worked silently without talking to each other. They became more aware of each other's work as they witnessed the assembly process (March 2003).

8.3 Basic Characteristics of the Cell Production System

An overview of the Company N's production system clarifies the basic characteristics of CPS compared to traditional production systems as follows: quick response to market fluctuation, significant elimination of 'Muda',[4] strict efficiency control, and autonomy and completeness of work. These are explained below.

[4] 'Muda' is one of the key terms in the Toyota Production System. It refers to various factors of production that only increase costs without adding value at the production site.

8.3.1 Quick Response to Market Fluctuation

The CPS is an assembly system that has significantly increased production technology's ability to respond to severe market fluctuations (product diversification, shorter product life and lower prices).

Here is an example of market changes for Product X produced by the C enterprise group. The number of produced Product X models increased from two in 1985 to nine in 1998. Furthermore, the production period for the same model was two to six years in the 1980s, but it had been reduced to around one year by the end of the 1990s. In contrast, 30% of similar product production was transferred to China in 2001 (with 50% expected to be transferred in 2002 and 70% in 2003), and product prices fell by about 15–20% per year between 1999 and 2001.

In CPS, the large and fixed equipment (conveyor belts) was removed from the assembly line, and instead assembly cells are formed by equipment (e.g. workbenches, small inspection machines and assembly tools) whose layout can be changed quickly. As a result, by implementing CPS, plant management could now easily change the layout of the assembly workshop in response to severe fluctuations in the product market.

8.3.2 Significant Elimination of 'Muda'

As with the Toyota Production System, cost reduction is achieved by eliminating several types of Muda from the production process. Muda was produced in various ways under the traditional system. Company N identified the following Muda in the traditional production system: balance loss, handling loss, wasteful movements, spreading loss from one process to the entire one, double-checking, work-in-process, product inventory and extra space, among others. In CPS, these Mudas were reduced by removing conveyor equipment, improving assembly cell layouts and handling multiple process by a single operator.

For example, in assembly cells, the number of processes for which one assembly operator is responsible has increased, resulting in a longer work cycle. As work cycle time increases, time losses[5] in the assembly process

[5] As explained in Chapter 2, time loss is divided into three categories: balance loss (including variant loss), handling loss and system loss, with system loss typically being the most important. The concept of system loss does not exist in Company N's Muda or the Toyota production system's Muda, nor is it measured. However, it is officially recognised. System loss is defined as loss caused by variations in operators' actual work time, which

decrease. According to Company N, the balance loss was set at approximately 5% in the traditional production system, but as a result of the conversion to CPS, it is now 2–3% on average.

8.3.3 *Strict Efficiency Control*

A third characteristic is the strict control of assembly cell efficiency. There is no conveyor equipment in CPS to regulate work speed. Therefore, plant management cannot directly regulate work speed and must instead rely on workgroups or individual operators. In this situation, plant management implemented strict efficiency controls in each assembly cell to meet the production plan. For example, production control boards and electronic noticeboards were installed in each assembly cell to clearly display production goals and results on a daily and hourly basis. Furthermore, management targets (income, expenditure, profit, etc.) for each assembly cell were established monthly, and these targets were compared with actual results.

8.3.4 *Autonomy and Completeness of Work*

In CPS, individual operators' or workgroups' assembly work has an autonomous and complete characteristic. In large assembly plants that have introduced CPS, dozens of assembly cells are located in parallel instead of a few long assembly lines.

Changing the layout of an assembly plant from a single line to multiple cells results in a dramatic shift from serial product flow to parallel product flow.[6]

In contrast, when six assembly lines are converted into approximately 15–40 assembly cells, as in Company N's assembly plant for product X, the change in product flow is not as dramatic. However, even in this case, the products that used to be assembled on a single assembly line are now divided into multiple assembly cells. Therefore, if we look at individual assembly lines, the product flow has shifted from a serial product flow to a parallel product flow.

appears as waiting time, extra personnel allocation and so on. Among the Muda identified by Company N, 'double-checking' and 'spreading loss from one process to the entire process' are associated with this system loss.

[6] For a description of how the products flow, see Chapter 5 in Ellegård et al. (1992).

In contrast to traditional production systems, where several workgroups were typically located on a single assembly line, CPS employs a small number of workers or a single workgroup in a single assembly cell. First, the work is functionally complete in an assembly cell because the product is fully assembled. Second, within the scope of the production plan, individual assembly cells work autonomously,[7] unaffected by the work of other assembly cells.

However, as seen in the case of Company N, the proportion of mini-assembly lines (Type 1) is higher than assembly by a single operator (Type 2). Individual operators in Type2, particularly the stationary method (Type2b), perform autonomous and complete work, whereas this is not the case in Type1. Although the flow of products assembled in the assembly plant is generally parallel, the flow of products assembled in Type1 is serial. Therefore, in Type 1, individual operators' work is fragmented and lacks autonomy and completeness.

8.4 Social and Technical Conditions

8.4.1 Social and Technical Conditions That Required the Cell Production System

Prior to the widespread adoption of the CPS, the traditional production system was considered the most efficient assembly system for mass-produced goods. Why was it necessary to renew? Specifically, what social and technological conditions necessitated the use of CPS? Two major factors are the emergence of problems in traditional production systems and the labour-intensive nature of assembly processes.

8.4.1.1 Emergence of Problems Contained in Traditional Production Systems

First, the most important condition for converting to CPS was that the traditional production system's latent problems became apparent due to changes in the market environment at home and abroad (product diversification, shorter product life and intensified price competition). There were numerous underlying issues with the traditional production system, the most significant of which Company N identified as follows:

[7] This autonomy is the definition of technical autonomy (see Chapter 2).

- the occurrence of various types of Muda;
- large, expensive equipment with high capacity, such as conveyor belts;
- expenses and time spent on model changes or product upgrades;
- high number of indirect/support personnel (e.g. subsection manager, relief person, troubleshooter, repairperson and material handler) add little value;
- operators' intellectual capacities are not fully utilised (they perform simple and ordered tasks).

These issues were not apparent in a market with few product variants, relatively stable growth in sales volumes and less intense price competition. However, as shown in Product X, market conditions shifted dramatically in the 1980s and 1990s as products became more diverse and shorter-lived, production was rapidly transferred overseas (particularly to China), and price competition heated up. The rapid development of digital technology also contributed to the shift in the product market. This has increased competition in the development of digitally enabled products. Changes in the product markets for electrical and precision machinery outpaced changes in the product market for automobiles, and as a result, the traditional production system was no longer an efficient assembly system.

8.4.1.2 Labour-Intensive Nature of Assembly Process

The second necessary condition is the labour-intensive nature of the assembly process. Currently, automated assembly systems have not been developed to respond quickly to severe fluctuations in the product market, and most of the assembly process is dependent on human manual labour. Under reliance on human manual labour, the CPS is an assembly system that attempts to better use the operator's skills to respond quickly to changes in the product market.

By the way, in the case of automobile assembly systems, which are also labour-intensive, a change in the labour market in the form of a shortage of young male labour was one of the primary factors that led to the renewal of that production system (see Chapter 7). However, in the case of electrical and precision machinery products, the labour market was not a prerequisite for transforming the production system.

In the assembly processes of electrical and precision machinery products, the female workforce, including middle-aged and older part-timers, was relatively high compared to the automotive assembly process. According to Company N's 2002 projections, there were 1,208 regular workers, 30%–40% of whom were female. Female made up an even higher proportion of the assembly workshop for Product X. For example, in an assembly cell (Type 2), all 12 workers were female regular employees, and in four mini-assembly lines (Type 1), approximately 18 of the 40 operators were female.

Many of them worked for the same company for shorter periods than male regular employees, and company management saw them as a labour force that subsidised the household economy. They had fewer job opportunities because it was difficult for them to relocate away from their current area of residence. Hence, assembly plants, such as those for electrical and precision machinery products, could easily hire workers from the local labour market where they were located.

8.4.2 Social and Technical Conditions that Enabled the Cell Production System

Even if the conditions necessary for conversion to CPS as described above exist, this does not guarantee that new production systems will be easily introduced and spread. This is because various factors that have kept traditional production systems in place for a long time can impede the introduction of new production systems. What social and technological conditions have enabled or facilitated the transition to CPS, thereby removing these obstacles? Three major factors are management's broad discretionary power, manufacturing engineers who are well versed in production processes and product characteristics.

8.4.2.1 Broad Discretionary Power of the Management

First, in Japan, plant management has extensive discretion over job restructuring and workforce utilisation, implying that regulations imposed by laws or trade unions are extremely weak (see Chapter 6).

In assembly workshops that have implemented CPS, each operator's work content has changed significantly compared to assembly workshops with traditional production systems. An increase in work cycle time represents the increase in work content of one operator. In the case of Company N, the work cycle time increased from 20–30 seconds on a traditional assembly line to approximately 40 s to 45 min in an assembly cell. In particular, Type 2 (assembly by a single operator), in which one operator is in charge of mounting the entire product and inspection and packaging. Company N had an employee organisation that organised social events for its employees, but no trade union. When transitioning to CPS, plant management did not consult with this employee organisation.

This increase in work content was carried out without opposition from trade unions or workers because plant management wielded significant power over job restructuring, implying that trade unions had little influence over job regulation.

Furthermore, plant management had the authority to employ regular workers and nonregular workers, such as many part-time and contractor's employees, as assembly workers. Contracting companies employed the majority of Company N's nonregular workers. They numbered 471 and constituted approximately 28% of the total workforce (2002 estimates).

One of the advantages of the CPS is its ability to respond quickly to fluctuations in production volumes. This assembly system adjusts production by increasing or decreasing overtime hours, adopting or not adopting a shift system, and increasing or decreasing the number of assembly cells in operation. Adjusting production by using shifts and increasing or decreasing the number of assembly cells typically entails increasing or decreasing the number of operators, which is made easier by using nonregular workers such as part-time and contractor's employees.

Unlike regular workers, they typically have fixed contract terms, allowing plant management to easily increase or decrease the number of operators by hiring, renewing or terminating the contracts of nonregular workers, rather than hiring and firing regular workers. At the time, these nonregular workers were not subject to much legal or trade union regulation in Japan. This facilitated the transition to the CPS.

8.4.2.2 Manufacturing Engineers Who Are Well Versed in Production Processes

The second condition is that Japanese manufacturing workshops typically employ manufacturing engineers who are well versed in production processes. One of the characteristics of the division of labour in the Japanese manufacturing workshop is that manufacturing engineers who are well versed in production processes collaborate with production engineers or group leaders and experienced (veteran) operators, among other things, to solve various problems (see Chapter 10.4). These characteristics aided the transition to CPS by making it easier to develop low-cost equipment in-house, improve assembly cell layouts and workbenches and so on.

Because CPS requires more equipment than traditional assembly lines, developing low-cost equipment in-house rather than purchasing expensive equipment from outside companies is challenging. At Company N, for example, the image inspection machine for product X used to cost approximately 6 million yen (60,000 euros[8]) per unit, but due to ingenuity, the cost was reduced to approximately 200,000 yen (2,000 euros). Furthermore, the cost of packing equipment for shipping products was approximately 3.5 million yen (35,000 euros) per unit, but manufacturing the equipment in-house reduced the cost to around 300,000 yen (3,000 euros).

8.4.2.3 Product Characteristics

Third, CPS assembled various products, including photocopiers, printers, laptop computers, digital cameras and mobile phones. Many of these products are smaller, contain fewer components and require less assembly time than automobiles.[9] These product characteristics help operators recognise the product's overall structure. Therefore, operators could pick up assembly operations quickly.

In fact, many of Company N's assembly cells were mini-assembly lines (Type 1), where operators learned their own processes in the assembly cell and then systematically learned the processes that came before and after them, sequentially expanding their learning process. Then, workers chosen from among the operators who had learned all of the processes in Type 1 were assigned to Type 2 (assembly by a single operator). In this way, Type 2 is an extension of Type 1. Many Japanese plants have long employed additive learning methods. In short, CPS was introduced without reforming learning methods, based on the product's characteristics.

[8] Converted at the average exchange rate in 2000 (1 euro = 99.5385 yen).

[9] Some high-end photocopiers took 13 h to assemble a single unit, which included the main body and several subunits. However, high-end photocopiers were typically assembled on a mini-assembly line (Type 1), with the exterior process taking approximately 25 min to complete. (This information is based on an interview with Manager N conducted by our Japanese research group during a visit to Company C's T factory in March 2006).

8.5 Conclusion

This chapter analysed the case of Company N as a representative case of the CPS implemented in Japanese assembly plants for electrical and precision machinery products since the early 1990s. It thereby clarified the fundamental attributes of CPS and the societal and technological conditions for transitioning from traditional production systems to CPS.

When comparing CPS to traditional production systems, CPS stands out due to its four fundamental characteristics: quick response to changes in the market, significant elimination of Muda, rigorous control of efficiency, and the autonomy and completeness of work. These basic characteristics have improved efficiency by boosting productivity, enhancing quality, augmenting flexibility, reducing floor space and elevating the work morale of operators. Conversely, CPS has also implemented specific enhancements to make work more humane. This is because it has granted workgroups or individual operators the autonomy and completeness of work.

There are three types of assembly cells. The basic types are mini-assembly line (Type1) and assembly by a single operator (Type2), with Type2 further divided into the rabbit chasing method (Type 2a) and the stationary method (Type 2b). When comparing these assembly cells based on their level of autonomy and completeness of work, Type 2b is characterised by a high level, Type 2a falls in the intermediate range and Type1 is considered low.

The quantity of assembly cells and combinations of types fluctuated due to modifications in products and production volumes. Many assembly cells were classified as Type1, which were short assembly lines. However, to minimise the additional distance between assembly processes the conveyor equipment was eliminated from the conventional assembly line. Therefore, the workgroup on this new assembly line experienced greater autonomy and completeness than the traditional assembly line. However, the individual operators' work was fragmented and lacked autonomy or completeness. The CPS successfully achieved simultaneous improvements in work efficiency and humanisation. However, it should be noted that there were more Type1s, which prioritise improving efficiency rather than humanising work. This was probably influenced by factors, such as the extensive discretionary power of plant management and the absence of pressure in the local labour market.

Furthermore, five specific social and technical factors were necessary and facilitated the transition from traditional production systems to CPS: the emergence of issues inherent in traditional production systems, labour-intensive nature of the assembly process, extensive decision-making authority of management, expertise of manufacturing engineers in production processes and the unique characteristics of the product. Among these factors, the extensive discretionary authority of management and the presence of manufacturing engineers are the most prominent features of large Japanese manufacturing companies. It is uncertain whether the trends in product markets, such as product diversification, shorter product life spans and increased price competition, will persist in assembly industries like electrical and precision machinery. Additionally, whether the advancement of artificial intelligence will alter the labour-intensive assembly process remains uncertain. Despite numerous uncertainties, it can be anticipated that the transition from traditional production systems to CPS will progress in situations where the five social and technical conditions or comparable conditions are present.

References

Ellegård, K., Engström,T., Johansson, B., Nilsson, L., & Medbo, L. (1992). *Reflektiv produktion: Industriell verksamhet i förändring (Reflective production: Industrial activity in transition)*. AB Volvo (in Swedish).

Itō, A., ed. (1997). *Seru Seisan Hōshiki Dōnyū Jitsurei Shiryō (Examples of introduced cell production system)*. Āban Purodyūsu Shuppan Bu (in Japanese).

Iwamuro, H. (2002). *Seru Seisan Hōshiki (Cell production system)*. Nikkan Kōgyō Shinbun Sha (in Japanese).

Masuda, S. (1976). Shokuba no Nōdōsei Kaihatsu wo Mezashita Jyūnan na Datsu Konbea to Jishu Kanri Hōshiki (Flexible breakout from conveyor and self-management approach aiming at development of activeness at workplace). *Kōjyō Kanri (Plant Management)*, 22(1), 10–16 (in Japanese).

Nakano, T. (1976). Jidōsha Kumitate Kōjyō ni okeru Ningensei Kaifuku eno Hairyo (Consideration of the recovery of humanity at vehicle assembly plant). *IE Rebyū (IE Review)*, (102), 133–137(in Japanese).

Okukubo, M. (1999). Sayōnara, Beruto Konbeyā Seisan (Farewell to belt conveyor system). *Kōjyō Kanri (Plant Management)*, 45(7), 53–84(in Japanese).

Sakamaki, H. (2006). *Kyanon Hōshiki no Seru Seisan de Ishiki ga Kawaru, Kaisha ga Kawaru (Awareness and company will change with Canon 's cell production system)*. Nihon Nōritsu Kyōkai Manejimento Sentā (in Japanese).
Tsuru, T., ed. (2001). *Seisan Shisutemu no Kakushin to Shinka (Innovation and evolution of production system)*. Nippon Hyōron Sha (in Japanese).

PART III

Analytical Comparison of Renewals of Production Systems Between Sweden and Japan

CHAPTER 9

An Analytical Comparison of Learning Strategy

Lennart Nilsson and Hikari Nohara

9.1 Traditional Learning

As previously described (see Chaps. 3), there are three levels of knowledge: details, relationships and overview. Even if knowledge of the relationship between details is expanded and accumulated, it does not automatically reach the level of overview. Achieving an overview requires explicit focus on the strategy designed to reach this level, which is known as a holistic learning strategy.

In a traditional Ford-type assembly line, an operator's time was entirely occupied by a single segmented task learned through repetitive training

L. Nilsson
Department of Pedagogics, Gothenburg University, Gothenburg, Sweden
e-mail: Prof.Lennart.H.Nilsson@gmail.com

H. Nohara
Hiroshima University, Higashihiroshima, Japan

H. Nohara (✉)
Nagano University, Ueda, Japan
e-mail: nohara@nagano.ac.jp

H. Nohara and L. Medbo (eds.), *Technology Convergence and System Divergence*, https://doi.org/10.1007/978-981-96-1910-8_9

and memory. This represented the traditional learning strategy, which focused solely on the knowledge of details.

9.2 Toyota's Learning Strategy

Toyota's learning strategy falls somewhere in between the traditional and new learning strategy. Once a Toyota assembly operator becomes used to a certain operation, they shift from the current operation to one of the next-door operations as either a pre-process or post-process task. Such shifting occurs many times during a worker's work life, otherwise known as job rotation. In this case, operators learn work based on an 'additive learning strategy'. This work utilises the knowledge of detail and of the relation between details; however, it lacks knowledge of overview.

Through this job rotation, the worker becomes an unskilled multi-functionalist. This differs from an unskilled operator who is locked into a single task on the assembly line. However, the worker is only able to handle a number of unskilled tasks, each of which requires no expertise.

Japanese management has no concern with knowledge of overview, because they do not expect ordinary assemblers to become all-round assemblers who can complete all tasks independently. Regarding 'additive learning strategies' with no concern for ways to reach the knowledge of overview, the Traditional and New Toyota Production Systems are identical.

In retrospect, the Japanese are not adept at achieving overview, so most of them are not interested in the whole, but rather in the detail. Therefore, most Japanese are good at incremental improvements, but not good at breakthroughs. In other words, they are good at tactics but not at strategy. This seems to be a lengthy social and cultural tradition, which may be reflected in Toyota's learning strategy.

9.3 New Learning Strategy

Let us briefly examine the significance of the new learning strategy and work characteristics of the Uddevalla Production System (UPS)(see Ellegård et al., 1991; Nilsson, 1995). As previously described (see Chaps. 3, 7 and 11), inner dialogue facilitated integrated work in the UPS. For this inner dialogue process to function smoothly, two conditions were required:

The first condition was that operators must maintain all three levels of knowledge. In other words, operators could effectively manage a task depending on the knowledge of detail only when they could clearly comprehend the overall target (the architecture of the complete automobile) through knowledge of overview, and also comprehend the connections between series of operations as a result of their knowledge of relations.

Based on this viewpoint, the holistic learning strategy was adopted in the UPS. A beginner receives overview training as the first step in understanding the total structure and entire operation process of assembling an automobile. The learning process begins with learning approximately one-eighth of the assembly process, and so on. The UPS believed that operators will not have access to holistic knowledge even if they spontaneously accumulate knowledge of several detailed operations. Instead, the worker must be explicitly instructed in the overview (see Chaps 2 & 3).

In a traditional Ford-type plant, operators were only required to have detailed knowledge of their specific tasks. Conversely, job rotation along the assembly line is a serious method implemented in Japanese plants, enabling operators to gradually become 'multi-skilled', or more precisely, unskilled multi-functional. However, with job rotation, operators are not provided with overview knowledge in advance and lack an understanding of the entire structure and assembly process of a complete automobile. Therefore, it becomes challenging for a single operator to handle the entire assembly process, though this is typically not expected of them, except in special cases. Conversely, every operator in the UPS was expected to eventually become an all-round operator with regard to assembly work.

The second condition required that parts and components be systematically arranged in alignment with each assembly operation. This requires classifying and grouping numerous parts for a complete automobile, depending upon the assembly-oriented product structure. Essentially, smooth inner dialogue demands a shift from the principle of parts grouping and arrangement; i.e. chronological sequence should be visualised as parts arrangement in a spatial dimension.

The Uddevalla plant had six separate assembly work shops, each of which had seven or eight workgroups that independently assembled a complete automobile from a bare body. These small parallel workgroups comprised seven to nine operators. Their work cycle times ranged from

100 to 80 minutes, respectively. An operator's work on an automobile body was divided into large clusters of work ('work module') (see Chaps. 2).

These clusters thereby formed various work patterns ('intra-group work patterns'), which were based on 'the assembly-oriented product structure' designed for automobile assembly. These work modules were modified according to the workers' competence, their competence overlap, personal preferences, etc. (see Chaps. 2).

Such work modules and intra-group work patterns established a standardised 'situation' that operators could familiarise themselves with in advance. In certain situations, one operation may lead to a new situation, and the subsequent operation could lead to another, and so forth. That is to say, intra-group work patterns based on work modules are situational changes that operators notice in advance. Such situational changes are introduced to operators beforehand as standardised patterns (see Rasmussen, 1986; Suchman, 1987). Therefore, the responding action to situational changes becomes standardised compared to handling unpredictable situational changes. In addition, operators encounter these patterns several times.

These techniques simplify the forecasting of situational changes for operators, making their inner dialogue much easier, quicker and more accurate. In summary, long-cycle assembly work is the accumulation of operations that respond to situational change, functioning as a situated action. However, since the change is adjusted in advance, it makes it more convenient for operators to respond.

To summarise the above points with regard to learning strategy:

1. The UPS introduced the first significant reform since the introduction of the Ford Production System that paid attention to operations and the operator in terms of cognitive capability.
2. The UPS' operators used a holistic learning strategy, gaining overview of an automobile's total architecture and assembling process. This approach allows operators to engage in inner dialogue and take initiative independently without managerial intervention.

References

Ellegård, K., Engström, T., & Nilsson, L. (1991) *Reforming industrial work: Principles and realities: In the planning of Volvo's car assembly plant in Uddevalla*. Arbetsmiljöfonden.

Nilsson, L (1995). The Uddevalla plant: Why did it succeed with a holistic approach and why did it come to an end? In Å. Sandberg (Ed.), *Enriching production*. Avebury.

Rasmussen, J. (1986). *Information processing and human-machine interaction: An approach to cognitive engineering*. Elsevier Science Publishing Company.

Suchman, A. L. (1987). *Plans and situated actions: The problem of human machine communication*. Cambridge University Press.

CHAPTER 10

Comparative Analysis of the Division of Labour in the Assembly Workshop

Uichi Asao, Yutaka Tamura, and Eishi Fujita

10.1 Introduction

In this book, we primarily discussed the assembly systems of the Volvo Cars Uddevalla plant in Sweden (hereinafter referred to as the Uddevalla plant), Toyota Motor Corporation in Japan (hereinafter referred to as Toyota) and the Japanese electrical and precision equipment Company N. This chapter[1] examines how individual operators or workgroups carry

[1] This chapter is a shortened and revised version of Chapter 2 (co-written by Asao, Tamura and Fujita) in Nakaoka et al. (2007).

U. Asao (✉) · Y. Tamura
Nagoya, Japan
e-mail: uasao_2902_1972@yahoo.co.jp

Y. Tamura
e-mail: tamura.yutaka@aichi-toho.ac.jp

E. Fujita
Graduate School of Humanities and Social Sciences, Nagoya City University, Nagoya, Japan
e-mail: fujitaeishi1950@gmail.com

H. Nohara and L. Medbo (eds.), *Technology Convergence and System Divergence*, https://doi.org/10.1007/978-981-96-1910-8_10

out their operations based on the division of labour in these three distinct assembly systems, and how changes in the assembly systems may alter the nature of the division of labour in the workshop. However, the analysis may not be adequate because the information gathered through surveys and other means is only partial.

To analyse the structure of the division of labour in the workshop, we must identify which operations and how the workgroups in the assembly workshop are responsible. The production process of products can be divided into three broad categories: product design, production preparation and mass production process. Table 10.1 highlights the qualitative differences between operations involving assembly workgroups in the production preparation and mass production processes.

First, operations involving assembly workgroups are classified as standard ('routine operations') and non-standard ('dealing with changes and problems'[2]). The standard operation sheets describe standard operations.[3] Meanwhile, non-standard operations involve dealing with various work-related issues, such as quality or equipment defects, changes in operator composition or production volume, changing production methods or new product preparations.

There are two types of non-standard operations. The first is an operation for which the workshop has determined how to deal with previous experiences ('non-routine operation'). These operations have defined procedures to varying degrees. The other is the 'unprecedented operation', which the workshop is experiencing for the first time and for which no solution can be determined in advance.

In the case of non-routine operations, having a defined work procedure is important because when a similar problem arises, it will not be handled quickly without a defined procedure, thus reducing efficiency. A workshop that constantly strives for efficiency will have a clear approach to dealing with various work-related issues to avoid inefficiencies.

Non-routine operations are classified as those that must be performed by qualified operators ('operations requiring qualifications') or those that unqualified operators can perform ('operations requiring no qualifications'). Operations requiring qualifications are those performed by

[2] The term 'dealing with changes and problems' comes from Koike (1996). Nakaoka et al. (2007) provide a critical view of this term.

[3] The term 'standard operations' is synonymous with 'standardised work' as used throughout this book, including Chapter 7.

Table 10.1 The classification of operations involving assembly workgroups in the production preparation and mass production process

Standard operations (described in standard operation sheets, that is, routine operations)			
Non-standard operations = 'dealing with changes and problems'	*Dealing with Problems* Detection of quality defects Deduction of causes of quality defects and implementation of relevant countermeasures = improvement Return to original position in case of equipment malfunction Deduction of causes of equipment malfunction and implementation of relevant countermeasures = improve-ment *Dealing with Changes* Changes in operator composition (replacement of absentees, training of inexperienced operators) Changes in production volume (change in work cycle time, reallocation of operation) Preparation for the production of new products and changes in production methods	Non-routine operations (Dealing with events experienced in the past on shop floors: Countermeasure methods are determined.)	Operations requiring no qualifications Operations requiring qualifications
		Unprecedented operations (Dealing with events experienced for the first time on shop floors: Countermeasure methods are not determined.)	

Note 1. The contents under 'dealing with changes and problems' were derived from Chapter 2 of Koike et al. (2001).
2. An example of 'operation requiring qualifications' is one handled by a worker who possesses a specific qualification, such as a 'qualified troubleshooter for specialised equipment' (qualified troubleshooter).

operators capable of troubleshooting equipment, such as those qualified to handle abnormalities. Meanwhile, an example of operations requiring no qualifications is when the operator presses a button on the control panel to return the equipment to its original position in case of equipment malfunction.

When a workshop encounters an unprecedented operation for the first time, the source of the problem is identified and addressed to prevent similar problems from occurring. However, if the occurrence of similar problems cannot be eliminated, some procedures are typically established to prepare for the re-occurrence of similar problems. In other words, the unprecedented operation may evolve into a non-routine operation, although the method of dealing with the problem may vary in severity.

The workshop's division of labour structure is examined in terms of which types of workers are in charge of the operations categorised above.

10.2 Swedish Automotive Assembly Workshop: Parallel Production at the Uddevalla Plant

Between 1988 and 1993, the Uddevalla plant produced passenger cars with an innovative assembly system that replaced the traditional assembly line. It was a system of multiple workstations with small parallel workgroups working together to complete cars in very long-cycle times (parallel production system; see Chapter 2 for more information. In this subchapter, the description of the assembly system and division of labour at the Uddevalla plant is as understood by the authors.). This new assembly system also significantly changed the operator's standard and non-standard operations. However, the system's potential was not fully realised because the Uddevalla plant was closed after a few years. Therefore, the division of labour presented below must be viewed as a developing one.

10.2.1 Division of Labour in Standard Operations and Its Characteristics

10.2.1.1 Overview of Assembly Operations

The Uddevalla plant had six parallel assembly workshops and approximately 30 workgroups, each of which assembled an entire car. The following is an overview of the contents and procedures of assembly operations (standard operations) in workgroups, using three parallel assembly workshops (Nos. 1–3) (For details, see Chapters 2 and 5, Engström et al. (2001) and Chapter 5 of Tamura (2003)).

Each assembly workshop had several subassembly workstations, such as doors and instrument panels, near the car body workstation and several workgroups of seven operators were arranged in parallel. Each workgroup was given a car body and many components stored in six material kitting fixtures (three in pairs) at the edge of the designated area.

Before beginning the assembly of the entire car, each workgroup completed several preliminary tasks. Preparatory operations primarily involve placing the material kitting fixtures next to the car body and keeping the necessary components and tools within easy reach.

The assembly operation at the car body workstation was divided into two parts: the first and second halves of the vehicle. The first half was spent assembling two main component groups ('leads for electrics, air and water' and 'drivelines' component groups) to the car body, and the second half was spent assembling two more main component groups ('sealing and decor' and 'interior' component groups) to the car body.

After two pairs of operators (four operators) finished the first half of the car at the workstation equipped with a tilting device, the car body was transferred to the workstation equipped with a lifting table. The remaining three operators (one on the left side, one on the right side and one on the front and back of the car) completed the second half of the work to finish the car.

The operators double-checked their work at the end of each pair's operations. The two pairs (four people) who assembled the first half of the car also performed subassembly operations, such as doors, and all seven people did the same amount of work. In addition, the operators assisted each other by carrying components, ordering tools, resetting them and so on.

Each operator's work cycle time ranged from 100 to 500 min, with a normal cycle time of 110 min. In assembly workshops Nos. 4 and 5, the minimum work cycle time was 80 min. Several workgroups of nine operators built a complete car in those workshops.

10.2.1.2 Features of the Division of Labour

The main characteristics of the division of labour are based on an overview of assembly operations.

First, small workgroups assembled the entire car, with pairs of operators assisting one another.

Second, individual operators' assembly operations differed significantly from traditional assembly lines' subdivided and fragmented work.

The work cycle time was extremely long, and the assembly operations performed at the car body assembly workstation corresponded to the main component groups described above, resulting in many interconnected work elements. The operators also checked each other's operations and worked on subassemblies such as doors.

In addition, each operator was trained to perform at least one-fourth of a car's assembly work (including subassembly work). Before the closure of the Uddevalla plant, more than one-third of all assembly operators had acquired the ability to assemble more than half of a car, and about 25 operators could assemble the entire car (Ellegård, 1995).

Third, in addition to assembling the cars, the operators prepared the necessary components and tools. Material handlers were typically in charge of performing these preparatory operations on assembly lines.

Fourth, each workgroup assembly operation took place with the car body nearly stationary at the workstation. Thus, the operation of each workgroup was unaffected by the progress of the other workgroup's operation, ensuring the workgroup's technical autonomy.

10.2.2 *Division of Labour in Non-standard Operations and Its Characteristics*

Because there is very little relevant information on the content of non-standard operations involving operators and the characteristics of the division of labour in non-standard operations at the Uddevalla plant, we will focus on dealing with quality defects and absenteeism and competence development for indirect work.

10.2.2.1 Dealing with Quality Defects

Operators were responsible for performing quality checks during assembly operations, such as adjusting various defects, correcting incorrect assembly and performing final checks (Engström et al., 1999, pp. 194, 195). In 1991, while assembling a Volvo 740, an operator noticed that one of the fuel pipes had come into contact with the fuel tank. He suspected that if the car ran, the fuel tank would be damaged by friction with the fuel pipe and reported this to plant management. In response, the plant's process engineers analysed the problem, took temporary action and suggested a solution to the design department at the headquarters plant (Torslanda) in Göteborg. However, at the headquarters, not only the assemblers but also the engineers did not immediately recognise the importance of the issues raised (Blomgren & Karlson, 1995, pp. 127, 128). The issue was solved in 1992, when the fuel pipes were redesigned during a model changeover.

10.2.2.2 Dealing with Absenteeism

When the plant first began operations, if a workgroup was understaffed due to an absentee, the team leader (i.e. group leader) would coordinate the number of operators among the groups at a meeting prior to the start of work. Later, a common pool team was formed in the assembly shop, and its members began to take over for absent operators. The pool team consisted of operators with extensive work experience and competence (see Chapter 5).

10.2.2.3 Competence Building for Indirect Work

In addition to assembly competence, many operators received training in areas such as teaching, maintenance, production engineering and quality control. Typically, the operators possessed at least one special competence. In traditional plants, white-collar employees performed all of the indirect work.

Operators were also given the opportunity to receive training (3-day course) to become team leaders. Team leaders were appointed by the workshop managers from among operators who possessed the necessary competencies and wished to become team leaders. They were frequently rotated among group members every 1 to 2 months.

Furthermore, as operators developed broader competences, the workgroup members were able to redistribute the work within the group among themselves. For example, in an eight-member workgroup, two operators capable of assembling a complete car could form a small subgroup of two members, while the remaining six members could form one to three small subgroups (see Ellegård, 1995 and Chapter 5).

10.2.2.4 Characteristics of Division of Labour and Competence Building

The following are the characteristics of the division of labour and competence building for non-standard operations involving operators at the Uddevalla plant.

First, if the operator notices a quality defect in the assembly work, such as an incorrect assembly, he or she will address it oneself if he or she is capable.

However, for defects such as the one described above (design failure of fuel pipes), which necessitate a change in the component's design, engineers at the plant or headquarters handle the problem, while the operator only finds the problem. Nonetheless, this case shows that the operator's ability to detect quality defects is dependent on the difference in the assignment of the assembly operations under consideration.

At the Uddevalla plant, one operator assembled fuel pipes and fuel tanks to the car body; he was in charge of a group of related work elements. As a result, he was able to easily recognise his work and pay attention to the interrelationships of the assembled components, allowing him to identify problems.

In contrast, the Torslanda plant divided the same operation into three operators. The division of labour was as follows: one operator would attach the fuel pipe to the fuel tank, the next would assemble the pipe to the car body, and the last would assemble the tank to the car body and connect the pipe to the tank. Furthermore, these operations were carried out at various work stations along the line, and individual operators were not trained to check the operation performed at other work stations. These factors made it difficult for operators to understand the overall relationship between fuel pipes and fuel tanks, preventing them from detecting problems (Blomgren & Karlson, 1995, p. 128).

Second, the traditional division of labour is being reconsidered, and job integration is being pursued so that operators can learn specialised skills such as teaching, maintenance, production engineering and quality control. The following are the main reasons for pursuing this integration of duties.

At the Uddevalla plant, a single workgroup assembled an entire car, requiring additional work during the assembly process. For example, if each pair responsible for a long-cycle task does not thoroughly check their work, quality issues will arise, necessitating rework in the subsequent process. Improving the operators' quality-related skills was an important aspect of ensuring quality.

Furthermore, factory layout issues and other factors resulted in a variety of work-related issues, including missing components and delays in receiving components kits. To improve the ability to respond to such problems, operators were encouraged to acquire special competencies.

Third, the problem of absenteeism was initially coordinated among workgroups before being replaced by a pool team of experienced operators.

Fourth, workgroups were given some administrative autonomy, as evidenced by their ability to redistribute operations within the group and participate in the selection of team leaders approved by the workshop manager. This administrative autonomy was made possible by ensuring technical autonomy through the assembly of complete cars in small parallel workgroups (see Chapter 2).

10.3 Japanese Automobile Assembly Workshop: Toyota's 'Autonomous Complete Process'

As revealed in Chapter 8, in the early 1990s, Toyota introduced the 'autonomous complete process' (hereinafter referred to as ACP) to its complete car assembly lines. The introduction of ACP caused some changes in the division of labour on the assembly line.

10.3.1 Division of Labour in Standard Operations and Its Characteristics

10.3.1.1 Changes in Work Assignment

Table 10.2 shows how the division of labour for installing rear power seats (electrically positionable rear seats) changed before and after the implementation of ACP.

The rear power seat assembly operation (standard operation) consisted of five coherent work elements (1–5). Prior to the implementation of ACP, these work elements were assigned to five operators (A, B, C, D and E) across three workgroups (X, Y and Z). Following the implementation of ACP, this assembly operation gained two new work elements (+3 and + 5), bringing the total number of work elements to seven. Furthermore, the previous five work elements (1–5) were assigned to two operators (F and H) in workgroup Y.

Furthermore, prior to the implementation of ACP, the work contents of the five operators included operations other than the work elements (1–5). Following the implementation of ACP, the work contents of three operators (F, G and I) included operations other than the work elements, while the remaining one (H) only contained the work elements (4 and 5).

Table 10.2 Assignment of assembly operations (rear power seat)

Work elements	*Assembly line before introduction of ACP*		*Assembly line after the introduction of ACP*	
	Assignment of work elements	*Assignment of nonwork elements*	*Assignment of work elements*	*Assignment of nonwork elements*
1. Loading of a seat frame into the car body from the line side	Operator A in workgroup X	Yes,	Operator F in the workgroup Y	Yes,
2. Attach the frame to the car body with bolts	Operator B in workgroup X	Yes,		
3. Connect the frame connector to the wiring harness attached to the floor	Operator C in the workgroup Y	Yes,		
+ 3. Double check of elemental operation 3	—	—	Operator G in the workgroup Y	Yes,
4. Place a sheet over the frame	Operator D in workgroup Z	Yes,	Operator H in the workgroup Y	No,
5. Connect the seat cushion connector to the frame connector	Operator E in workgroup Z	Yes,		
+ 5. Double check of elemental operation 5	—	—	Operator I in the workgroup Y	Yes,

Note 1. This table is based on Asao et al. (1999, pp. 140, 141).

2. If 'Yes' is entered in the 'Assignment of nonwork elements' section, it indicates, for example, that operator A is also responsible for work elements other than rear power seat assembly operation.

10.3.1.2 Features of the Division of Labour

Two characteristics define this shift in labour division. First, following the implementation of ACP, the operation of the rear power seat assembly was consolidated into a single workgroup (Y). This assembly operation consists of functionally coherent work elements. In other words, this division of labour reduced operational fragmentation by grouping functionally coherent work elements into a single workgroup.

Prior to the introduction of ACP, in order to thoroughly pursue work allocation without balance loss within a short cycle time of one to several minutes (see Chapter 2 for the definition of balance loss), functionally coherent work elements were fragmented and assigned to multiple workgroups (X, Y and Z). As a result, each workgroup frequently completed work elements without functional cohesion. It was an operation that was unrelated to 'the product's functional logic' (Boyer & Freyssenet, 2002), which clarifies the assembly operation sequence.

However, even before the introduction of ACP, the percentage of functionally coherent operation within a workgroup ranged between 30% and 50% on Toyota's final assembly line; after the introduction of ACP, this percentage increased to 70% to 90% (Asao et al., 1999, p. 117).

Second, following the implementation of ACP, certain related work elements ('operation series') were assigned to specific operators (F and H). Toyota defines a 'operation series' as 'defining the smallest unit that subdivides operation'. For example, it is the process of taking one part, temporarily fastening, tightening and connecting it to the wire. In particular, operator (H)'s work content consists only of related work elements (4 and 5), and its operation is more coherent than the other operators (F, G and I).

Prior to the introduction of ACP, in order to thoroughly pursue work allocation that minimises balance loss, assembly operations were sometimes subdivided into multiple work elements, and these related work elements ('operation series') were also assigned to multiple operators (A, B, C, D and E). An 'operation series' limited the work's segmentation.

10.3.2 Division of Labour in Non-standard Operations and Its Characteristics

The characteristics of the labour division for non-standard operations in assembly workshops with short work cycle times (typically 60 s) will be clarified.[4]

This workshop was the final stage of the final assembly line for complete cars, with a workgroup of 20 workers. The main operations included installing the front, rear, left and right doors, installing luggage (rear trunk), injecting fluid into the radiator and others.

10.3.2.1 Dealing with Quality Defects

Gaps, wrong or missing components, scratches on the car body and poor assembly are the common quality defects in this workshop. When operators found such quality flaws, they would repair them within the work cycle time if possible, and if they could not, they would post a paper stating that the flaws could not be repaired in the car. Cars marked with this paper are later removed from the assembly line and reworked before moving on to the next process.

For example, there were two reasons for the gap between the installed door and the fender. The first was that the operator was unfamiliar with the work, and the second was that the operator found that the bolt for installing the fender was not properly tightened and re-tightened it, leaving no time to check the gap. In cases where the operator was unable to check the gap, a paper was displayed to indicate so.

Meanwhile, in response to quality defects, there was a separate rework operation (known as 'analysis') from the main-line operations. Prior to the cars being inspected, these reworkers dealt with any quality issues that were found on the assembly line. Concretely, they repaired defects that could be fixed on the spot, such as incorrectly installed or missing components, and if they could not, they investigated the cause of the defect and

[4] This description of the assembly workshop was based on Chapter 2 of Koike et al. (2001). Chapter 2 describes two assembly lines at a large automotive company in 1998: one with a short work cycle time and another with a slightly longer work cycle time. Although the name of the large automobile company is not revealed, the description suggests that this is a case study of Toyota's assembly line for complete cars. Since 1998, the survey has been conducted at an assembly workshop following the implementation of the ACP.

returned the defective product to the workshop where it originated. Only a few operators (two per shift) could handle the rework operation.

10.3.2.2 Dealing with Equipment Malfunction

Many operators used multiple pieces of equipment on the assembly line. For door installation, two pieces of equipment were used: a conveyor lifter that lifted the door from above and an auxiliary device that installed the door.

If that does not resolve the issue, a 'qualified troubleshooter for specialised equipment' (hereinafter referred to as a qualified troubleshooter) at the workshop will assist. This qualification was internal and consisted of three levels. The group leader (*kumichō*), the subgroup leader (*hanchō*) and four operators are the only qualified members of this workgroup at any level. In addition to these workers, three other operators are not qualified but can handle minor equipment malfunctions.

When a qualified troubleshooter was unable to repair a defective item, maintenance personnel took care of it. In some cases, maintenance personnel respond immediately when equipment malfunctions are found. For example, water can accumulate in a machine that injects liquid into radiators.

10.3.2.3 Dealing with Changes in Production Volume

Dealing with an increase or decrease in car production volume of 10% or more results in a change in work cycle time, which alters the allocation of operations to each operator. When production volume increases, the work cycle time decreases, which reduces the workload of each operator while increasing the number of workgroup members. These fluctuations in production volume occurred at least twice a year, and the group and subgroup leaders responded accordingly.

10.3.2.4 Dealing with Changes in Production Methods

Dealing with changes in production methods is unclear in the workshop with a short work cycle time, however, the case study of a workshop with a slightly longer work cycle time of 8.5 min[5] shows some of this dealing.

In workshops with slightly longer work cycle times, new cars powered by electricity and petrol were assembled, and new production methods

[5] The work cycle time was 17 min until May 1998 and then 8.5 min.

were employed. In many workshops with short work cycles, operators worked along a moving line. However, in workshops with slightly longer work cycle times, the assembly lines remained stationary during work, and operators pressed the bell when the operation was complete.

The trim line, for example, had nine operators stationed there. Because that line was disconnected from other lines, such as the chassis line, the trim line moved only when all nine operators pressed the bell. The line to assemble new car models using this new production method was recently launched in December 1997. Seven months before the start-up date, operators with experience in other assembly shops were gathered. Thus, the nine operators were not newcomers, temporary workers (***kikankō***[6]), or even support workers (*ōensha*[7]), but rather experienced operators from other assembly shops.

10.3.2.5 Features of the Division of Labour

The following is a summary of the labour division for non-standard operations in the assembly workshop.

First, among quality defects and equipment malfunctions, assembly line operators deal with simple issues during working hours. For example, problems such as improper bolt tightening or a door that does not come off a conveyor lifter.

Second, other members of the workgroup handle problems that the operator cannot handle, such as not being able to check a gap within the work time or repair a defective piece of equipment. The other members are a reworker and a qualified troubleshooter. The qualified troubleshooter in a workgroup consists of the group leader, subgroup leader and several other members.

Third, maintenance personnel handle equipment defects or specific equipment malfunctions that the qualified troubleshooter cannot fix.

Fourth, the group leader and subgroup leader handle increases and decreases in production volume, not the operators.

[6] Kikankō is a blue-collar worker who signs a fixed-term employment contract with Toyota to work on the production line for up to 2 years and 11 months. Furthermore, there is a system in place to promote temporary workers to regular employees. According to Toyota's 'temporary worker recruitment' website, 1,023 people were promoted from temporary workers to regular employees between FY 2018 and FY 2022 (https://www.t-kikan.jp/support.html, accessed 9 April 2024).

[7] Ōensha are workers who move from a workshop with an excess of workers to a workshop with a shortage for a set period, typically within the same company.

Fifth, production engineers, manufacturing engineers and supervisors in workshops typically lead changes in production methods and the preparation of new products. However, experienced operators are also involved.

In short, operators are responsible for dealing with simple issues such as quality defects and equipment malfunction. The solution to simple problems is operation that does not require any qualifications. However, inexperienced operators cannot perform operation requiring qualifications and unprecedented operation. Operations requiring qualifications and unprecedented operations are handled by a variety of workers other than inexperienced operators, depending on the nature of the problem or change in question. In other words, the non-standard operations (operation requiring no qualifications, operation requiring qualifications and unprecedented operation) has a hierarchical division of labour structure: inexperienced operator → reworker, experienced operator or subgroup leader → group leader → maintenance worker or engineer.

10.4 Japanese Electrical and Precision Equipment Assembly Workshop: Cell Production at Company N

As mentioned in Chapter 8, in the late 1990s, Company N transitioned its assembly system for electrical and precision equipment (Product X) from traditional assembly lines to the cell production system, in which products were assembled in many assembly cells. The two basic types of assembly cells were the mini-assembly line (Type 1) and assembly by a single operator (Type 2), with the latter divided into the rabbit chasing method (Type 2a) and the stationary method (Type 2b).

Product X's main assembly processes included mounting, inspection and packing. In addition to these three main processes, the assembly cell workgroup's standard operation included logistics for assembled components and packaging materials. Furthermore, the workgroup was in charge of non-standard operations (non-routine operations), such as defect detection and rework. This section examines the division of labour in these operations using several assembly cell examples. The following descriptions, as well as Chapter 8, are based on information gathered from interviews with Company N's general manager of the manufacturing department, section managers and others between November 2000 and December 2003.

10.4.1 Division of Labour in Standard Operations and Its Characteristics

10.4.1.1 Four Assembly Cell Examples

Example 1. Mini-Assembly Line, Type 1 (as of March 2002):
A total of 657 small-sized products were produced per day. The work cycle time was approximately 44 s. The workgroup had 12 members: one cell leader (group leader), one material handler and ten assembly operators. The cell leader was a regular Company N employee, but the rest of the group worked for the contracting company.

Example 2. Mini-Assembly Line, Type 1 (as of March 2003):
One hundred and sixty-eight products are produced for office use per day. The number of components was approximately 70, the average assembly time per product was about 40 min, and the work cycle time was approximately 171 s (2.86 min). The workgroup had 14 members, and their responsibilities were as follows.

- One person serves as the cell leader (off-line and reworker).
- Reworker (and support for those who queue): one person.
- Material handler: 2 people.
- Assembly operators: 10 individuals.

  Mounting (6 processes): 6 people.
  Inspection (3 processes: electrical safety test, visual and instrumental image evaluation and appearance check): 3 individuals
  Packing (1 process): one person.

Example 3. Assembly by a Single Operator, Type 2b (as of March 2002):
Small-sized products were typically produced in 560 units per day. The product had 64 components. The average assembly time per unit was calculated to be approximately 8.6 min: [480 min × 60 s/ (560 units × 10 operators) = Approx. 514 s]. However, due to producing models with various specifications (product variants), the average assembly time per

unit was approximately 10 min. The workgroup consisted of 12 regular employees: one cell leader (and reworker), one material handler and ten assembly operators.

Each side of the assembly cell had five assembly stations, for a total of ten, with an inspection and packing station in the centre. Each of the ten assembly operators was responsible for mounting, inspecting and packing operations. The specific working procedures were as follows.

(1) Transfer the cover (body) to the mounting station, (2) mount at the station, (3) proceed to the centre and inspect (electrical safety test, visual and instrumental image evaluation, and appearance check), (4) pack and (5) place the finished product on a pallet.

Example 4. Assembly by a Single Operator, Type 2a (as of March 2003):
Twenty-two small office products were produced per day. The product had about 80 components. The average assembly time per unit was about 45 min. The workgroup included three regular employees, one of whom was the cell leader. One of the three was in charge of packing and logistics, while the other two were in charge of mounting and inspection. When production was low, one worker was responsible for it.

10.4.1.2 Features of the Division of Labour

The following characteristics of the division of labour in standard operation can be identified from the four assembly cell examples provided above.

First, in the assembly cell, the standard operation performed by a single operator was expanded in comparison with the traditional assembly line. That is, the traditional assembly line had a work cycle time of 20–30 s, whereas the mini-assembly line (Type 1) had a work cycle time of 40 s to less than 3 min, and assembly by a single operator (Type 2) took 10–45 min.

Second, the standard operation's content varied qualitatively across assembly cell types. In Type 2, an operator was in charge of a series of product mounting, inspection and packing, or simply mounting and inspecting. In other words, this standard operation was a unified process for producing finished products (including inspection and packaging).

In contrast, Type 1 did not always achieve the cohesiveness of the individual operator's work. However, the workgroup of about ten people was in charge of a variety of operations, including not only mounting complete products but also inspecting and packing. As a result, the workgroup performed standard operations in a consistent manner. This was a significant difference from the traditional assembly line. Individual workgroups did not perform a coherent series of mounting, inspection and packing operations on a traditional assembly line because multiple workgroups were typically located on a single line.

Third, the characteristics of the workforce responsible for standard operation varied depending on the type of assembly cells. As shown in Examples 3 and 4, Type 2 operators were only regular workers with long-term employment contracts with Company N.

Meanwhile, in Type 1, the operation was frequently performed by the contractor's employees, who often worked in Company N's assembly shop for a short period due to the contract's short duration. Company N was able to use contractor's employees because Type 1 had a shorter work training time than Type 2. In Example 1 (Type 1), 11 operators (excluding the cell leader) worked for the contracting company. In four Type 1 s other than the assembly cell examples, approximately 30 of the 40 operators were contracting company's employees (as of March 2003).

10.4.2 Division of Labour in Non-standard Operations and Its Characteristics

In the assembly cell, workgroups (cell leaders and operators) were involved in more than just standard operations like mounting, inspection and packing. The workgroups also started sharing some of the non-standard operations and administrative tasks that they had rarely handled on the traditional assembly line.

10.4.2.1 Dealing with Operational Errors, Quality Defects and Equipment Malfunctions

On the traditional assembly line, support personnel were replacing absent workers, reworking, troubleshooting and checking. However, in the assembly cell, absent workers were replaced and their task was reworked by cell leaders, whereas operators were in charge of checking. Checking was the process by which the back-end operator verified the work of the front-end process, and this work became part of the operator's standard operation. Some of the operators could rework, analyse defects and manage data.

Every day, problems arose in the assembly cell. These issues fall into three categories: work delays caused by operational errors, product quality defects and equipment malfunctions. The following was the treatment for these issues.

Dealing with operational errors: Operational errors occurred when workers dropped screws or other parts or gnawed on screw heads while working. At Company N, a system was in place to handle operational errors. For example, if a screw fell, the operator was required to replace it with a new one. Furthermore, operational errors are a common cause of work delays. Therefore, a percentage of the allowance (5%) was guaranteed to compensate for the delay within the standard operating time frame.

Dealing with quality defects: If the product did not meet the specified specifications after product inspection, it was defective, and the majority of quality flaws were due to defective components. The mounting and inspection process and the outgoing inspection (sampling inspection) performed prior to packing were the primary processes in which such quality defects were found. The operator was in charge of finding flaws in the mounting and inspection processes, whereas the QA (Quality Assurance) section of the general planning department was in charge of the outgoing inspection.

'Wrong shape of components', 'no good images', 'poor adjustment' and 'no increase in electrical potential' were some of the quality defects found by the operator. In these cases, the following rules were applied to deal with quality flaws.

1. If the operator believes something is 'wrong', he or she will turn on the red light and contact the cell leader.
2. The cell leader assesses the situation and decides whether he or she can solve the problem on his or her own or whether to contact the technical staff (in charge of the process)[8] in the quality staff area of the production department.

[8] The main assembly shop for product X belonged to Production Department 1 (which had about 200 employees), and the quality staff area was staffed by eight people: about half of the eight were primarily in charge of processes to solve technical problems, and the rest were primarily responsible for the production of jigs and tools.

3. If the cell leader cannot solve the problem himself or herself, he or she contacts the technical staff through the subsection manager, who will handle the situation.
4. If technical staff cannot solve the problem, they contact an engineer (production engineer) in the technical section of the engineering department via the subsection manager, and the technical staff and the engineer collaborate to solve the problem. The technical section of the engineering department consisted of technical Sect. 1, which specialises in electrical and mechanical engineering, and the components QA technical section, which specialises in components.

In short, the operator found quality defects but did not investigate the cause or implement corrective measures. Cell leaders, technical staff and engineers collaborated to determine the cause and implement countermeasures.

Dealing with equipment malfunctions: The assembly cell's main equipment, excluding the workbench (which also serves as a component shelf), was voltage and image inspection machines, as well as assembly tools (e.g. screwdrivers). If there was an issue with the image inspection machine, the operator would notify the cell leader. Typically, the cell leader was unable to respond, so the technical staff (in charge of jigs and tools) from the quality staff area was summoned, and they responded. In addition, if an assembly tool failed, the operator had to contact the cell leader to replace it.

As with quality defects, the operators found equipment malfunctions but did not investigate the cause or take corrective action. The cell leader and technical staff were responsible for determining the cause and countermeasures.

10.4.2.2 Assembling and Improving the Workbench, Improvement of Assembly Cell Layout

Line staff and leaders made all improvements to the assembly line layout and component shelves on a traditional assembly line. However, in the assembly cell, cell leaders and operators began to assemble workbenches (which also served as component shelves) and suggested improvements to the workbenches and assembly cell layouts. Concretely, the subsection manager created prototypes of the assembly cell and workbench layouts, which were improved with feedback from cell leaders and operators, and the workgroups improved them through weekly improvement practice meetings and extra work. Furthermore, during the preparation for production (start-up) of a new product, cell leaders and operators would occasionally consider the layout of the assembly cell.

10.4.2.3 Responsibility and Authority to Ensure Production Volume

On traditional assembly lines, the subordinate manager had the responsibility and authority to ensure daily production within scheduled working hours and to approve overtime work when production could not be secured. However, in the assembly cell, cell leaders were appointed by the subsection managers, and these responsibilities and authority were delegated to them. Cell leaders were basically line-offs, with one in each assembly cell. The cell leader's qualifications included being at least a fifth-level position in Company N's job classification system and the ability to rework.

10.4.2.4 Features of the Division of Labour

As previously stated, operators in the assembly cell were involved in both non-standard and standard operations. However, the cell leader was in charge of some administrative tasks (responsibility and authority for ensuring production volume). The operators were primarily involved in non-standard operations such as dealing with operational errors, quality defects, equipment malfunctions, assembling workbenches and making suggestions for improving workbenches and the layout of the assembly cell. The following are the characteristics of the division of labour in these non-standard operations.

First, operators and workgroups were responsible for dealing with simple problems such as operational errors and work delays.

Second, the operator and other employees divided labour to deal with difficult problems such as causes and countermeasures for quality defects and equipment malfunctions. In other words, the operator found the problems, but the cell leaders, subgroup managers and technical staff of the production department were in charge of determining the causes and solutions. Furthermore, the subsection managers created prototypes for the assembly and improvement of work benches, as well as the improvement of the assembly cell layout, which were later improved by the cell leaders and operators.

Third, for difficult problems that could not be solved by technical staff (manufacturing engineers), many of which were unprecedented operations, manufacturing engineers and engineers in the engineering department (production engineers) collaborated closely to solve the problems. Production engineers did not always view this collaboration positively. They recognised that they needed to work together because the manufacturing engineers were insufficiently strong. However, as explained below, this cooperative relationship is an intriguing form of labour division. This is due to the belief that production and manufacturing engineers have established a complementary relationship.

Production engineers received specialised education at universities and technical colleges, and were divided into detailed specialities such as mechanical and electrical, resulting in a deep specialisation but a limited range of expertise.

Meanwhile manufacturing engineers had extensive experience in production engineering as well as excellent assembly skills. They also had a system in place to learn for a set period in the engineering department, as well as people with engineering knowledge and experience working in assembly workshops. As a result, the manufacturing engineers were familiar with assembly procedures, and when a quality defect occurred, they were able to quickly identify 'something is wrong with this component'. They were unable to deal with technically advanced problems due to a lack of expertise in mechanical, electrical and other areas.

Although production engineers have unavoidably collaborated with manufacturing engineers, as a result, problems such as quality defects have been handled efficiently by having both parties work together to solve problems.

10.5 Conclusion

The cases discussed in this chapter do not provide a comprehensive analysis of the division of labour structure in assembly workshops because the data obtained from surveys and other sources are limited. There is little information available about the non-standard operations involving that workgroup, particularly since the Uddevalla plant closed down after a few years. Therefore, the conclusions that follow are tentative.

First, as the assembly system shifted from traditional assembly lines to assembly lines with ACP, cell production, or parallel production, the fragmentation and subdivision of standard operations performed by individual operators or workgroups was eliminated, while their work content expanded. Particularly, in cell production (Type 2: assembly by a single operator) and Uddevalla's parallel production, individual operators did not perform subdivided partial operations, but one or more operators assembled the entire product. In other words, these changes in the assembly system have suppressed the division of labour in standard operations and shifted it in a way that restores the functional coherence or wholeness of their work contents.

Second, restoring functional cohesion or wholeness in standard operations affects the division of labour in the assembly workshop in two directions.

One is to expand the possibility of utilising peripheral labour force, such as temporary workers and employees of the contracting company, when standard operation is limited to partial work performed within a very short work cycle time. This is because the fragmentation of work is eliminated and the work becomes more coherent, facilitating operators' work knowledge than before. This is the case with Toyota's ACP and cell production Type 1 (mini-assembly line).

When utilising such peripheral labour force, a 'separated system' (Koike, 1996) is developed in which standard and non-standard operations (primarily troubleshooting and improvement during mass production and production preparation) are shared by different workers. This separated system includes not only the division of labour among assembly operators, maintenance workers, supervisors and engineers, but also the expansion of the division of labour within assembly operators, i.e. the separation of peripheral workers, such as temporary workers and contracting company employees, from subgroup leaders, experienced workers and qualified troubleshooters.

Another characteristic of this division of labour is that, as seen in the cases of Toyota and Company N, non-routine operations other than relatively simple operations such as defect detection are handled primarily by group leaders and subgroup leaders (including cell leaders and reworkers), qualified troubleshooters and maintenance workers, whereas unprecedented operations are handled primarily by engineers. To perform non-routine operations effectively, an ordinary operator must be promoted to the position of subgroup leader or group leader. In other words, the division of labour in non-standard operations is hierarchical, which is a good way to train mid-level technicians and supervisors in the assembly workshop.

The other is to increase the possibility of delegating non-standard operations to individual operators or workgroups in a system where standard operations have long work cycle times and a single or a few operators assemble the entire product.

This is due to operators having a better understanding of the overall assembly of the product and improving their work competencies, as well as technical autonomy being ensured through the parallel assembly of complete products by individual operators or workgroups made up of several operators.

If this possibility becomes a reality, job integration will increase while the division of labour will be further limited. Administrative duties will include reallocating operations within workgroups, appointing and replacing group leaders and deciding whether to shorten or extend working hours. Parallel production at the Uddevalla plant and Type 2 cell production (assembly by a single operator) demonstrate the germination of this situation. Especially at the Uddevalla plant, all members of the workgroup had the opportunity to learn specialised skills such as teaching, maintenance, production engineering and quality control. In other words, in this assembly system, the division of labour in non-standard operation is shallow, and it is a system that focuses on improving the operator's overall competence rather than training mid-level technicians and supervisors in the assembly workshop.

Restraining the division of labour and promoting job integration imply the humanisation of work. Given the severe fluctuations in the product market, what is essential to expand the feasibility of this idea? It is up to those working in labour-intensive, high-volume assembly industries to reconsider their belief that a thorough division of labour increases efficiency. In other words, it is a matter of detecting waste in the division of labour.

In this regard, it is critical to consider the time loss analysis described in Chapter 2, as well as the study of the implementation of alternative assembly systems based on this analysis. However, it is difficult to change long-held beliefs about the division of labour in companies. As a result, collaboration with experts outside the company, such as researchers who are free of such beliefs, is essential.

References

Asao, U., Saruta, M., Nohara, H., Fujita, E., & Yamashita, M. (1999). *Shakai Kankyō no Henka to Jidōsha Seisan Shisutemu: Toyota Shisutemu wa Kawatta noka (Changes in the social circumstances and automobile production systems: Has the Toyota system changed?)*. Hōritsu Bunka Sha (in Japanese).

Blomgren, H., & Karlson, B. (1995). Assembly skills, process engineering and engineering design: An example from the Volvo Uddevalla Plant. In Sandberg, Å. (Ed.), *Enriching production: Perspective on Volvo's Uddevalla Plant as an alternative to lean production*, pp. 127–135. Avebury.

Boyer, R., & Freyssenet, M. (2002). *The productive models: The conditions of profitability*. Palgrave Macmillan.

Ellegård, K. (1995). The trajectory of Volvo's production system ideas: Paving the way for a reflective production system. *Sanken Ronshū (Industrial and Business Review)*, (14), 1–14. Sapporo University.

Engström, T., Jonsson, D., & Medbo, L. (1999). Developments in assembly system design: The Volvo experience. In Lung, Y., Chanaron, J., Fujimoto,T., & Raff, D. (Eds.), *Coping with variety: Flexible productive systems for product variety in the auto industry*, pp. 192–223. Ashgate.

Engström, T., Jonsson, D., & Medbo, L. (2001). *Assembly work structuring based on restructuring and transformation of product information*. Comparisons of management practices between Japan and Sweden. Handelshögskolan vid Göteborgs Universitet. Göteborg (later translated into Japanese and published in an academic Japanese journal: *Journal of Humanities and Social Sciences*, (12), 279–309. Nagoya City University).

Koike, K. (1996). *The economics of work in Japan*. LTCB International Library Foundation.

Koike, K., Chūma, H., & Ōta, S. (2001). *Monozukuri no Ginō: Jidōsha Sangyō no Shokuba de (Manufacturing skills: From workshops in the automotive industry)*. Tōyō Keizai Shinpō Sha (in Japanese).

Nakaoka, T., Asao, U., Tamura, Y., & Fujita, E. (2007). The division of labor to deal with "changes and problems" on manufacturing shopfloors. *Tōhō Gakushi (Journal of Toho Gakuen), 36*(2), 1–70. Aichi Toho University & Toho Gakuen Junior College.

Tamura, Y. (2003). *Borubo Seisan Shisutemu no Hatten to Tenkan: Fōdo kara Uddevara e (Development and conversion of Volvo production systems: From Ford to Uddevalla)*. Taga Shuppan (in Japanese).

CHAPTER 11

The Analytical Comparison of Work-Focused Improvement

Hikari Nohara

11.1 Problems and Solutions at Toyota and Volvo Uddevalla

11.1.1 Different Targets to Solve the Same Problem

Both Volvo and Toyota faced identical challenges. The first issue entailed was a labour shortage caused by the increased demand for workers and the young workers' reluctance to engage in disconnected and repetitive work. The second issue was related to efficiency resulting from labour shortage. Volvo grappled with these problems since the 1970s, while Toyota had since the late 1980s.

Both companies used a work-focused orientation to solve the issues, which differs from the traditional Western, Ford-type equipment-focused improvement, i.e. an automation-oriented approach, used during the

H. Nohara
Hiroshima University, Higashihiroshima, Japan

H. Nohara (✉)
Nagano University, Ueda, Japan
e-mail: nohara@nagano.ac.jp

H. Nohara and L. Medbo (eds.), *Technology Convergence and System Divergence*, https://doi.org/10.1007/978-981-96-1910-8_11

labour crisis in the 1960s and 1970s. This shift was possible because of the presence of manufacturing engineers, who enabled work-focused improvements (see 7.2.2).

Both Uddevalla and Toyota pursued work-focused technical methods due to their similar work-focused orientation. However, the respective managements chose radically different target because of their diverse social and intellectual backgrounds.

11.1.2 Uddevalla's Solution

A crucial aspect of cognitive capability lies in humans' ability to comprehend their actions by relating them to contextual meaning (see 11.2.1.). The theory of the holistic learning strategy emphasises that understanding details and their relationships works efficiently only when supported by a broader knowledge of the whole context (overview). In this context, managers and collaborators at Volvo Uddevalla accepted the holistic learning strategy and were consciously interested in leveraging their cognitive capability (see Chapter 4). Therefore, they attempted to integrate operators' segmented and fragmented work.

Furthermore, this concept of integrated work involves combining both conceptual and manual work. Such work can only be realised through the inner dialogue between workers and materials, mediated by using tools and machines in the labour process (see 11.2.1.). Each worker's inner dialogue operates at a unique pace based on individual differences, significantly contrasting with the machine-paced assembly line. A worker's discretion (autonomy) is essential for determining work pace. Therefore, there was a need to eliminate the machine-paced assembly line.

Instead of an assembly line, Volvo Uddevalla constructed a parallel product flow where products remained stationary at workstations during the work process. Buffers were necessary to absorb fluctuations in work caused by an operator's autonomy and other production fluctuations. The buffer function in the Uddevalla Production System (UPS) included high system capacity (see 11.1.3), indirect work such as subassembly and material handling, and a free (empty) working position. These buffer functions contributed to reduced losses, improved efficiency and minimised waste in group work.

Workers could generate a series of operations step by step only when engaged in an inner dialogue in labour process. However, for quick and accurate inner dialogue, workers required an overview of the labour process from its inception to the completion of the whole product. Such an overview could only be gained through integrated long-cycle work, which differs from the short-cycle work on the assembly line. Quick and accurate inner dialogue was supported by assembly-oriented grouping of parts and providing part-naming. Furthermore, a total kit system was implemented, in which all parts and components needed for assembling one automobile were arranged together. This was also supported by the parallel product flow, rather than a serial flow line, that facilitated the introduction of holistic learning strategy.

In retrospect, why was such a radical reform realised? In this context, while the top management did not have a concrete plan for the reform, the top management, i.e. Gyllenhammar, played a crucial role in establishing the Uddevalla plant. Under his leadership, the reformers successfully overcame several obstacles while introducing the new production system. In addition, the union supported the establishment of the Uddevalla plant.

11.1.3 The Characteristic Solution in the Case of Toyota

Japanese managers did not recognise that operators could achieve efficient and qualitatively distinct integrated long-cycle work by implementing a holistic learning strategy to display human cognitive capability. They strongly believed in the efficiency of short-cycle work, and their goal of improving was thus limited. Why did they adhere to this approach? In this context, such preference for short-cycle work may strongly relate to the Japanese mindset. Most Japanese are indifferent to the whole, but they are very interested in the details and good at partial improvement, in contrast to breakthroughs. This trait is common in the Japanese 'habits of the heart' (see Beller et al., 1965). For example, Japanese excel in writing essays that focus on details but are less adept at writing big novels. While Japanese philosophy excels in attention to detail, it lacks the systematic approach as proposed by philosophers such as F. Hegel (see Kato, 2002).

Let us delve deeper into this topic. As well known, K. Koike found and conceptualised the so-called multi-skilled worker in Japanese shop floors, which contrast with a single-skilled worker commonly found on Ford-type assembly lines (see Koike, 1996). From a comparative research perspective, his conceptualisation is significant because it clarifies the origin of competitive power within Japanese-manufacturing sites.

In retrospect, this comparison reminds us of Nilsson's characterisation of 'three kinds of knowledges' (see Ellegård et al., 1992): detail, relation and overview. Let us paraphrase Koike's concepts, i.e. multi-skilling and single-skilling using Nilsson's three kinds of knowledge. An assembly line operator at a Ford-type plant performs a single repetitive operation, gaining comprehensive knowledge about only one operation. On the other hand, a Japanese operator consecutively rotates their operations between the present task to the next. Thus, they gain extensive knowledge of the relationships among several operations because of frequent job rotations. While we agree with Koike's excellent characterisation, he does not address knowledge of overview. Undoubtedly, Japanese operators on the assembly line lack knowledge of overview, as reflected by Koike's conceptualisation. However, if the operator is interested in the overview, he cannot help but recognise the lack of knowledge of overview on the Japanese shop floor because Nilsson states that there are three types of knowledge.

In the case of Toyota, they seriously examined the possibility of using a standing-still station method for assembly work instead of a conveyor belt line from the late 1980s to mid-1990s. At the same time, they struggled to maintain the assembly line while overcoming the repetitive, segmented and fragmented assembly work. Finally, they chose the latter.

On further reflection, what is the most radical paradigm shift of UPS? It was not the elimination of the conveyor belt line, although it is most obvious from plant visitors' perspective. The shift from additive learning to holistic learning was the most radical change for UPS. This change in learning strategy from memory-oriented to context-oriented operations based on human cognition allowed operators to work efficiently in extensively long-cycle work (see Chapters 2 and 3). However, it is nearly impossible to understand the significance of the change in learning strategy when people lack understanding or interest in the knowledge of overview. Therefore, a production system based on knowledge of overview cannot be introduced if people remain indifferent to this perspective.

The historical trajectory shows that the 1980s was characterised by assembly line work that was just an accumulation of fragmented and unrelated operations. Operators found no contextually meaningful connection between the series of operations, i.e. assembly sequence. Thus, they did not recognise where they were and what they were doing on the way to complete the product, resulting in young workers' aversion towards their work.

Therefore, the very unique reforms were introduced at Toyota (see 7.3). They recognised that if operators could regain some contextually meaningful connections between the series of operations, they would become more understandable. As discussed in 7.3, the term 'contextually meaningful connections' indicates that operators were able to recognise where they were and what they were doing at any given moment in the process to finish a product. This was possible only when they introduced the rule that the order of assembly sequence was based on component functions and ease of assembly.

Several procedures were implemented to improve work processes. Among these, the most epoch-making invention was the introduction of the Autonomous Complete Process (ACP). In ACP, each component's function is completed within a single workgroup, ensuring that each operator's series of operations is contextually meaningful (see 7.3.2 and see also Chapter 10). Comparing ACP with UPS shows that ACP is not a particularly radical reform, UPS is a breakthrough from an assembly work perspective. However, it is impossible to find an intermediate invention such as ACP without manufacturing engineers with deep knowledge of assembly work. In general, it is interesting that Toyota always chooses extremely cautious incremental improvement over breakthrough. The following are a few examples. Regarding the product, they prefer RAV 4, which was the first off-road vehicle that was also available for on-road use. They also prefer hybrid vehicles to full-scale electric vehicles. Regarding the production process, they prefer small on-line robots that co-work with operators to fully automated lines. Regarding the assembly operation, they maintain an assembly line while pursuing maximum recovery of contextually meaningful connections among a series of operations. When examining such intermediate inventions, we observe that Toyota maintains a management philosophy favouring incremental improvement over radical change.

11.2 An Analytical Comparison of Work-Focused Means Between Toyota and Uddevalla

In this sub chapter, we discuss in detail the commonalities and differences between the work-focused technical means adopted by the new stage of TPS and UPS.

11.2.1 *Contextually Meaningful Connections in a Series of Operations*

11.2.1.1 What are contextually meaningful connections?

We closely examine the work characteristics. Integrated work that combines conception and execution includes contextually meaningful connections between a series of operations. As stated earlier, the aim of work-focused improvement was meaningful contextual work in the new stage of TPS. Thus, this connection was a common characteristic of assembly work in new TPS (NTPS) and UPS.

A re-examination shows that a contextually meaningful connection is not limited to a functional connection among components. When deciding the sequence of an operator's series of operations, the proximity of parts in the vicinity, the right-left and upper-lower symmetry of parts in the vicinity and the ease of operations are all considered (see T. Engström et al., 2002). Therefore, the term 'contextually meaningful connection' corresponds to all these factors. Once an operator is aware of these connections in advance, they can justify the reason of their choice of operation.

11.2.1.2 The Precondition of Contextually Meaningful Connections: The Way to Regroup Parts and Components

Parts are typically grouped in two ways. One is a 'design-oriented product structure', in which a whole product (automobile) is divided into many sections and eventually into each piece of a component, which is arranged along the line. Here, parts and components are grouped only from the perspective of a step-by-step hierarchical disassembly, which starts from the architecture of a complete automobile. The other method is 'assembly-oriented product structure', in which parts and components are regrouped according to contextually meaningful assembly sequence and organised in a kit box (see Chapter 2).

Considering the traditional serial flow assembly line, the parts were grouped according to the convenience of design departments and sent to the shop floor along the line. Then, the parts were regrouped according to the convenience of line balancing and arranged along the line. However, both UPS and NTPS adopted different methods. As a first step, both Toyota and Uddevalla grouped parts according to the convenience of the design departments.

However, these parts and components were regrouped based on the assembly-oriented product structure (see Chapter 2) at the Uddevalla plant. This involved plantwide regrouping parts and components. In the case of Toyota's ACP, parts and components were sent to each shop floor without changing the principle of grouping them. However, parts and components were regrouped and arranged on each shop floor based on new assembly-oriented grouping principles in which the functional relation of each operation and ease of assembly were seriously considered.

In summary, the key difference is that Toyota used a shop floor-wide distribution of parts and components according to the same principle as that of UPS, while UPS regrouped parts and components according to assembly-oriented principles of grouping plantwide. This was not done before Toyota introduced ACP. In UPS, the principle of grouping of parts changed radically from a design-oriented to an assembly-oriented product structure plantwide. Thus, they introduced a full-scale kit system in which all parts and components of a whole car were set.

Because of this radical change in how parts were arranged, an operator could find the part for every motion at every moment. The operator, thus, found a reciprocal instruction relationship in which an operation identified a corresponding part and a part identified a corresponding operation (product-process duality Engström & Jonsson, 1993). Specifically, the spatial arrangement of parts now showed the temporal order of the worker's assembly sequence.

11.2.1.3 The Role of Contextually Meaningful Connections

The significance of contextually meaningful connections differed between new TPS and UPS. Toyota long used line balancing as the only criterion for standardising work. Thus, each operator's work was highly segmented and fragmented, and they could not recognise which part of the automobile was being assembled. Therefore, they developed an aversion to performing such work, leading to decreased efficiency.

Toyota introduced ACP to address this issue. The ACP was, first, to complete the function of a component within a workgroup and, second, to give individual workers a series of operations with maximum contextually meaningful connections.

However, an operator still memorised these motions through repetitive training based on a worksheet. They implemented assembly sequences by memory, not through inner dialogue. However, ACP made it much easier to memorise the operations because of the meaningful assembly sequence.

Hence, the role of ACP was to help an operator memorise the sequence. In contrast, in UPS, the realisation of the connection between parts and a meaningful whole resulted from both a holistic learning strategy and the technical contrivances of integrating parts into larger work modules.

We now consider the role of contextually meaningful connections in UPS by deeply examining the inner dialogue. This is not the inner dialogue between self and another self without any bodily action during the thinking process (see Arendt, 1971, 2003). As previously shown, an operator faces a response from materials mediated by tools and machines whenever they perform an operation. The subsequent operation is the response to such a reaction. In this regard, a dialogue is exchanged between the operator and the materials' response. This is termed 'inner dialogue' (in labour processes) in this chapter.[1]

Even at UPS, an operation involved standardised industrial work. Therefore, an operator already knew the response because they had an overview and had frequently repeated such inner dialogues. An operator knew the work pattern of the assembly sequence, i.e. 'intra-group work pattern', which was the Swedish co-author's way of articulating this division of labour within the small parallel workgroups at the Volvo Uddevalla plant (see Engström & Medbo, 1994). The operator eventually received lengthy, repetitive training as well.

However, the outline of the response learned by repetitive training was not a real response. The real response resulted from adjusting to the reaction in the ongoing process. In this sense, inner dialogue is not the internal dialogue in an operator's mind but one between an operator and the kitted materials mediated by tools and machines in the ongoing labour process.

Work repetition can be decreased only by extending the work-cycle time. However, long work cycles without changes in the learning strategy will confuse the operator, thereby decreasing the work speed and quality.

To overcome this dilemma, we consider cognitive capability. By recognising contextually meaningful connections, along with training in using

[1] The author, Nohara's term 'inner dialogue' is the same as 'inner monologue' defined by L. Nilsson. It is his original concept. However, Nohara believes the term 'inner dialogue' expresses its content more accurately than 'inner monologue'. Therefore, Nohara suggested a change from 'inner monologue' to 'inner dialogue'. However, Nohara has not yet received his final approval for the change before his illness.

a clear work-related inner dialogue, an opportunity was created for individual workers to expand the content in assembly work without compromising on work speed and quality. Therefore, there was a fundamental reduction in repetitions at work.

In summary, in the case of NTPS contextually meaningful connections contributed to aiding memory without inner dialogue. On the other hand, they were an indispensable prerequisite for inner dialogue in UPS. This inner dialogue enriched an operator's work.

11.2.2 Assembly Sequence and Contextually Meaningful Connections

So far we have examined the relationship between work content and the grouping of parts. We now consider the actual work content. When working with a short-cycle time work, line balancing is the only principle used for standardisation for making standardised work (see 7.3.1). Thus, line balancing divides each operation into fragments to completely fill the cycle time without any lag or waiting between activities. Regarding standardised work, operators can memorise motions and parts even in random order; however, the volume and accuracy of memory is limited in such cases.

We observe a clear contrast in the case of UPS. When an operator recognises contextually meaningful connections in both the sequential order of motions and arrangement of parts, he/she can operate smoothly and accurately mediated by the inner dialogue based on assembly-oriented product structure. Moreover, these operations are guided by non-verbal communication between co-workers and other feedback mechanisms, such as partially empty-kitting fixtures and co-workers' ongoing work.

We now discuss standardised assembly sequences (SAS, Hyojyun Kojyun in Japanese) in New TPS (NTPS). Each vehicle consists of 108 units. Moreover, the order of assembly (termed as SAS) among these 108 units was also standardised to all Toyota plants worldwide (see 7.3.3).

Before initiating the ACP programme, the assembly work was standardised according to precise areas of the vehicle of a specific model at a particular plant, which is the traditional concept of standardisation of assembly line work. However, in the mid-1990s, Toyota shifted to a new level of standardisation. As stated earlier, this meant that the assembly sequence among these 108 units was standardised for all models and all Toyota plants. Regarding SAS, a worker always performed the same work, regardless of the vehicle model and the plant location, implying that the quality of assembly work was independent of individual worker's competence and attitude to work.

Unlike this tendency to exclude inner dialogue in assembly work in NTPS, UPS chose the opposite path, making inner dialogue the core method of assembly work. To reduce differences in work performance, the company developed several means for work process standardisation, including work modules and intra-group work patterns, and developed standardised questions for inner dialogue. These corresponded to three distinct phases of production processes: knowledge of materials and structuring, assembly, and inspection and adjustment.

In summary, both NTPS and UPS aimed to re-introduce contextually meaningful connections between a series of operations in the labour process; however, the roles differed significantly in each company. In NTPS, the results of work-focused improvement based on contextually meaningful connections were realised as an objective means distinct from individual competence, such as ACP, a grouping of parts and components and SAS. In UPS, however, contextually meaningful connections between a series of operations were an indispensable condition of inner dialogue. Work was performed based on those connections.

11.3 Work as the Result of the Work-Focused Reform: Toyota and Uddevalla

11.3.1 Conception and Execution in a Modern Factory

Consider a craftsman's work as a reference. As mentioned earlier (7.4.1), there are three kinds of conceptualisations concerning a craftsman's work. First, they conceptualise the kind of product to be produced. Next, the craftsman conceptualises the production process of making the product and, finally, they conceptualise how to work based on the preceding two conceptualisations. As the work progresses, detailed adjustments are made based on inner dialogue mediated by tools between the worker and the materials. Moreover, these conceptualisations and corresponding executions are implemented by an individual craftsman. This is termed as integrated work, in which conception and execution are integrated with an individual.

In such a case, we find three kinds of integrated work. If the focus is on product conception, the process of finishing product is corresponding execution. If the focus is on production processes, realising would be the corresponding execution. Finally, if the focus is on how to work, a rank-and-file worker's daily operations are the corresponding executions.

These descriptions of craftsman's work are the theories for analysing a modern factory, where operators are excluded not only from conceptions of a product and production methods but also from work methods, which are determined by an industrial engineer's time and motion study based on Taylorism. An operator is given a segmented piece of the operation through an instruction of how to work. Above the management and engineers conduct three different conceptions. Therefore, in the case of a rank-and-file operator on the line, there is no room for conception.

In addition, the belt conveyor was introduced in early twentieth century. In this case, line balancing was the only one criterion for organising each operator's work, resulting in not only segmented but also fragmented work. Therefore, the operator did not understand what they were doing while working.

This lasted from the early twentieth century to the 1960s and 1970s crises period. Mainstream Western managements' response to this crisis was to develop factory automation. In contrast, in the event of the subsequent labour shortage, only Toyota and Uddevalla, Volvo, adopted the work-focused reform. As previously described, Toyota has restored a contextually meaningful connection in an operator's work. As a result, an operator could now understand what their work processes are. In the case of Uddevalla, as indicated earlier, an operator regained an integrated work where conception and execution were combined in terms of working method (see Sect. 11.2.1.3), in addition to restoring a contextually meaningful connection. Finally, because of the work-focused reform, what perspective do we form from an operator's work as a whole? This is addressed in the following sections.

11.3.2 An Operator's Work in the Case of Toyota

11.3.2.1 Standardised Work at a Toyota Plant

In TPS, standardised work (SW) includes only execution. An operator must focus only on the ongoing operation while working. However, after introducing ACP, SW assumed a new characteristic. Specifically, an operator could now understand why another operation succeeded a particular operation, based on the contextually meaningful connection. Even in such a case, the operator could perform the job based solely on memory. However, when examining contextually meaningful connections between a series of operations, the worker found it easier to perform operations because they emerged generatively from memory. Additionally, each operator's total volume of memory could increase. The introduction of ACP facilitated repetitive short-cycle SW for all operators.

Therefore, SW itself does not include any conceptual work in both TPS and NTPS, the latter of which implies ACP. In this sense, there is no intellectual skill. However, there are some intellectual skills that are included in an operator's daily work in the field outside SW in Japanese automotive factories, especially in the case of Toyota. Such intellectual skills are built in the process of production system development and function to rationalise the system, i.e. increase the efficiency. Toyota's strategy regarding SW is the idea of the simpler, the better. Workers' mental potential should be included in KAIZEN activities. Therefore, they do not use a strategy to enrich SW itself.

An operator demonstrates three kinds of intellectual skills. First is KAIZEN (improvement) activities, which means activities after daily SWs to improve daily work, which involves equipment and the work environment. A long continuation of KAIZEN activity produces many excellent and loyal operators. Second, such operators are also involved in the process by which a draft of SW is modified and confirmed. Third, operators engage in responding to frequent irregularities in the manufacturing process (see Chapter 10).

11.3.2.2 The Process of Revisions of the Draft of SW

Next, we examine how work is standardised at Toyota. The first draft of a SW is prepared by the production engineering department, similar to Western auto companies. However, at Toyota, the next step is to organise the trial team, which includes manufacturing engineers, a group leader, a team leader and qualified veteran operators. This trial team repeatedly tries to perform operations based on the SW draft. As a result, the team finds points to be modified and creates a revised draft. Finally, the new revised SW is established after general manager's approval in the corresponding section. Moreover, even in the regular production phase, if an operator finds that the problem needs to be solved, they can again revise the SW after approval from the general manager. However, once the SW is established, even if the problem is found, this established work must be strictly maintained before the manager's new approval.

This implies that not only a team leader and a group leader (who are promoted from rank-and-file operators in the case of the Japanese inner promotion system) but also some selected veteran operators are involved in the work of conception. Such involvement by operators is not found in Western factories, except for Volvo, Uddevalla. However, only few selected operators are involved in such conception work. The management does not expand this involvement policy to all operators.

11.3.3 Total Comparison of an Operator's Work Between Toyota and Uddevalla

At Toyota, the duration of SW ranges in most cases from one to three minutes. However, at Uddevalla, it is from 100 to 500 minutes and 110 minutes in many cases, which is significantly longer than at Toyota (see Chapter 2). SW is the domain of management that should be strictly adhered to by an operator at Toyota; however, this is just a suggestion for an operator at Uddevalla. While working in SW at Toyota, an operator has no room to think about how to work. They are ordered to simply concentrate on parts and the components at hand. In contrast, at Uddevalla, pre-developed questions are given to an operator (see Chapter 3), who in turn must find the corresponding answers. This empowers the operator to think about the most appropriate operation at the particular moment as an answer to the question. Thus, the conception, i.e. the thinking, is in-built in the SW structure at Uddevalla.

However, this thinking activity is not about new things; rather, it is a recalling activity, implying that an operator recalls suitable operations (answers) based on experience when the question is raised. Moreover, this thinking activity is not about thinking of the meaning of an action (an operation) but about how to operate to make a good product. Thus, it is a means-oriented thinking.

Toyota's SW excludes KAIZEN activity and a response to irregularity; however, they are fairly included in Uddevalla's SW. Between Toyota and Uddevalla, it is common for an operator to participate in KAIZEN activity. However, Toyota's KAIZEN activity is characterised by the movement that organises and mobilises all operators. Consequently, KAIZEN activity becomes an indispensable part of an operator's everyday life at Toyota.

In sum, thinking (conceptualising) is embedded in Uddevalla's SW with regard to a formulated question and the corresponding answer, response to irregularity and other indirect work. Each operator is expected to participate in such thinking because everyone is involved in SW. In contrast, at Toyota, thinking is excluded from SW but implemented outside SW in daily work. This includes KAIZEN activity, response to irregularity and a revision of SW. All operators are organised and involved in KAIZEN; however, only selected operators and not all are involved in the response to irregularity and revision of SW.

11.4 Convergence and Divergence: The Different Social Roles of Work-Focused Methods

We now summarise the commonalities and differences of work-focused technical means used in NTPS and UPS. As described earlier, there were four work-focused technical means: re-introduction of contextually meaningful connections between a series of operations, grouping of parts based on contextually meaningful connections, kit systems based on new grouping of parts and buffer mechanisms. Each method aimed at solving the same problem, i.e. labour shortage; however, the concrete targets differed. NTPS aimed to re-introduce contextually meaningful connections to reduce work fragmentation and segmentation; however, UPS aimed at re-introducing integrated work to partially overcome the division of conception and execution, objecting the principle of Taylorism.

As both companies shared the same view, namely, work-focused improvement, they adopted similar concrete methods. However, the lean production theory, the key term proposed in the book *The Machine that Changed the World*, places TPS and UPS at opposite poles. Nevertheless, a comparison of the NTPS with the UPS show the contrary. Both companies followed the same four work-focused improvement approaches to solve the same problem. Moreover, both used the same or similar means.

However, the functions were different because each company had a different target. First, the first three methods share the common characteristic of involving contextually meaningful connections. Second, these contextually meaningful connections functioned differently in NTPS and UPS. NTPS aimed to ensure a certain level of efficiency and work quality, regardless of the worker's competence and attentiveness and therefore ignored any individual differences and considered each worker as a standardised operator in the short cycle of SW. Inner dialogue was not a concern and thus excluded. Consequently, these contextually meaningful connections functioned as a means for creating an objective method for achieving efficiency and work quality, independent of the individual.

In contrast, at UPS, contextually meaningful connections functioned as a means to promote quick and smooth inner dialogue. In summary, in NTPS, it was realised through an objective tool, while in UPS, it was reproduced in inner dialogue.

NTPS' grouping of work was based on contextually meaningful connections. Kit systems used these means to separate work from inner dialogue and standardise work. Conversely, UPS' new grouping of work and kit systems promoted inner dialogue, in which formulated and standardised questions in the inner dialogue steered a worker's mental orientation during expanded assembly work. This integration of mental and manual work led to efficient vocational training in the assembly industries in Sweden.

Buffers were the only one of the four work-focused methods with common technical functions. Both NTPS and UPS used buffers to introduce flexibility into the production process and absorb production volume fluctuations. However, the social function differed. The Toyota management regarded inherent flexibility as a necessary evil and it was strictly controlled to avoid abuse. UPS' integration of flexibility has a social impact that allowed workers' and workgroup autonomy.

11.5 Epilogue: Gifts to the Contemporary Industrial World

What gifts do the experiences at NTPS and UPS bring to the contemporary industrial world? In the case of NTPS, SAS will be a powerful means in any plant, regardless of an operator's characteristics and competence. In addition, both Toyota and Uddevalla, Volvo, followed the same strategy, i.e. work-focused improvement. The process of implementing this strategy was to accumulate knowledge about designing the man/machine interface, which is, for example, to contribute to how to design man/computerised-machine interface. In this regard, accumulated knowledge gained from the trials of work-focused improvement will guide future solutions.

Especially in the case of UPS, the experience gained will help in using the knowledge of cognitive science to improve human performance at human/computerised-machine interface. Holistic learning strategy, the integrated work based on inner dialogue and administrative autonomy and contextually meaningful operations will be useful in every area of industry and even beyond the industrial world. Therefore, the technical contrivances because of labour shortages have socially different consequences when there is no labour shortage.

In retrospect, the production system as a whole differed significantly between NTPS and UPS. Why was it different? To address this issue, we consider the following factors. What kind of worker image do engineers and middle managers have as a precondition for work-focused improvements? What kind of perceptions do they have of their worker and workplace as a result of these improvements? Such images have profound influence on the differences in production systems.

Both technical and societal factors should be considered when we examine the differences in production systems. Thus, we can only clarify the reality if we take a socio-technical approach as it can explain technology convergence and production system divergence between Toyota and Uddevalla, Volvo.

References

Arendt, H. (1971). *The life of the mind*. Harcourt Brace & Company.

Arendt, H. (2003). *Responsibility and judgement*. Schocken Books.

Beller, R., et al. (1965). *Habits of the heart: Individualism and commitment in American Life*. University of California Press.

Ellegård, K., Engström, T., & Nilsson, L. (1992). *Reforming industrial work: Principles and realities*. Tryckeri Balder AB.

Engström, T., & Jonsson, D. (1993). *Principles of perception and structuring of long cycle time assembly work*. Paper presented to 26th International Symposium on Automotive Technology and Automation, 13–17 September 1993.

Engström, T., Jonsson, D., & Medbo, L. (2002). Assembly work structuring based on restructuring and transformation of product information. Comparisons of management practices between Japan and Sweden. *Journal of Humanities and Social Sciences (Nagoya City University), 12*, 279–309.

Engström, T., & Medbo, L. (1994). Intra-group work patterns in final assembly of motor vehicles. *International Journal of Operations & Production Management, 14*(3), 101–113.

Kato, S. (2002). *A History of Japanese Literature: The First Thousand Years*. Tokyo: Kodansya.

Koike, K. (1996). *The economics of work in Japan*. LTCB International Library Foundation.

CHAPTER 12

The Summary and Points of Discussion for the Future

Lars Medbo and Hikari Nohara

12.1 The Summary and Points of Discussion, of Each Chapter

Chapter 2

Summary

This anthology chapter presents the practical and theoretical background of the advanced and unorthodox Volvo Uddevalla plant and several similar experiences within the Swedish automotive industry, including various initiatives to abolish assembly lines and assembly line work. Experiences

L. Medbo
Department of Technology Management and Economics, Chalmers University of Technology, Gothenburg, Sweden
e-mail: lars.medbo@chalmers.se

H. Nohara
Hiroshima University, Higashihiroshima, Japan

H. Nohara (✉)
Nagano University, Ueda, Japan
e-mail: nohara@nagano.ac.jp

H. Nohara and L. Medbo (eds.), *Technology Convergence and System Divergence*, https://doi.org/10.1007/978-981-96-1910-8_12

and insights gained during several decades of research and development work by others and the author predate the focused projection process.

The latest, most advanced and unorthodox initiatives created comparable, more flexible and efficient assembly work, combined with improved product quality (compared with the same period) for assembly lines and assembly line work for the same automobile model. Simultaneously, increased work satisfaction and improved ergonomic conditions were also gained. Such achievements were backed up by a similarly constructed assembly system for heavy truck chassis (the assembly docks at the Volvo Tuve plant).

The author and his research colleagues were heavily engaged during the assembly plant's projection process and short life; thus, they possess a deep practical and theoretical understanding of the dilemmas. This understanding, developed over years of research and experimentation, will be shared in this anthology chapter.

Apart from the details above, this anthology chapter comprises several complex descriptions of various administrative practices and routines used centrally in Gothenburg and other Volvo automobile companies, as well as locally in Uddevalla. These presentations require the reader to understand the magnitude of changes needed for the extensive assembly work cycles in which an advanced material feeding technique supported the small parallel workgroups.

Several unique features were practiced on the shop floor. One of the most scientifically spectacular features was how the holistic learning approach was purposely engineered with the help of a senior research competency. This approach is explained thoroughly in the anthology chapter.[1]

Discussion

(1) Is a complete cross- and multi-disciplinary research area available for alternatives to assembly lines and assembly line work?

[1] Two senior research competencies have evaluated what happened during the Volvo Uddevalla plant projection process from design theoretical and time geographical perspectives. Initiatives are enjoyable; however, these two initiatives are theoretical and methodological, proving something within their particular field of science. Nonetheless, in some essential respects, several crucial aspects are missed. Is there perhaps a more complete cross- and multi-disciplinary research area hiding somewhere in this anthology chapter's complex and intriguing details?

Or is this a gradually dying speciality used previously in small, relatively insignificant countries featuring a nowadays insufficiently knowledgeable practitioners and an aged small cadre of researchers?

(2) What exactly are the ambitions brought forward in the footnote above and to what degrees does each contribution to this anthology fulfil this? Do the authors agree and present a coherent anthology that hopefully avoids unfortunately self-appointed exegetical interpretations of the two recently closed Swedish unorthodox assembly systems (i.e. do they contribute to spreading questionable information worldwide)?

Chapter 3

Summary

This anthology chapter results from three decades of research and development that preceded the projection process for the Volvo Uddevalla plant. This partly inventive work was done at the Department of Pedagogics at Gothenburg University. Nilsson dedicated his entire professional life to this institution, while the other author, Engström, devoted his career to the Chalmers University of Technology for decades. Both authors were actively involved in the mentioned projection process.

Nilsson was engaged because the automotive company needed experience or insights regarding a new assembly work apart from a massive knowledge of short-cycle assembly work. A 'partly unexpected idea' appeared that knowledge from other sectors of society could provide constructive new principles for education and shop floor learning principles, thus engaging senior research competency in vocational learning.

At the initiation of the projection process, such principles were unavailable in the 'conservative' Swedish automotive industry; thus, qualitatively different assembly work was sought. Consequently, such initiatives were non-existent within the established company knowledge, saturated by knowledge and routines from short-cycle assembly work.

The final principles and practices gradually developed for the assembly plant were combined with advanced and unorthodox group work. This approach finally proved more efficient and flexible, resulting in higher product quality than the already established assembly line work; however, this assembly plant also underlined several other merits. These included the advantage of combining (1) the recruitment strategy chosen with

(2) the gradually developed education and shop floor learning principles (a temporary education and shop floor learning workshop predated the successively completed assembly plant).

Local recruitment that mirrored the characteristics of Swedish society (age, gender, etc.) ensured that several non-traditional categories of potential employees applied for work at the Volvo Uddevalla, including civil servants as teachers. As a result, many of these (primarily blue-collar employees) were later recruited internally together with the more traditional applicants for the more advanced work positions (often as white-collar employees).[2] Such measures differ significantly from recent occurrences in the Japanese automotive industry.[3]

Discussion

(1) Is knowledge of other social science disciplines available and applicable to the 'today's conservative' automotive industry? If this is the case, what potential positive effects are possible to gain?
(2) The recent partly one-directional transfer of experiences and insight from 'today's conservative' automotive industry may be questioned? Do the experiences and insights so far need to be crystallised further for future application?

Chapter 4

Summary
This chapter discusses the Swedish trajectory of developing the Uddevalla plant after the 1980s. The Kalmar and Uddevalla plants both showed the evolutionary process of production reform to the breakthrough of Ford. To break the tie of the Ford-type traditional flow assembly line, the Kalmar plant changed from a conveyor system serially to individual

[2] A completed education and a minimum age of 18 years were prerequisites for full-time employment in Uddevalla.

[3] International recruitment campaigns for male assembly workers were conducted in Japan. After that, far-reaching political chances of labour laws allowed part-time employment with far less favourable benefits. In this case, a more segregated labour market emerged featuring several generally accepted and intriguing mechanisms that the Swedish automotive industry does not always fully grasp.

separation moving by AGV (automated guided vehicle) and introduced stationary assembly in 20 minutes via a dock. In addition, to the Kalmar experiences, the Uddevalla plant abolished the conveyor system. Consequently, each car body stayed at one or two stations within the team area during the assembly process. The paper focuses on the Uddevalla plant compared with the Toyota production system (TPS), paying attention to the combination of technical, organisational and cognitional aspects.

This paper started with the following questions. In what ways do the Kalmar and Uddevalla plants differ? How was the Uddevalla plant able to achieve assembly in 120 minutes? Why could operators in the Uddevalla plant assemble a complete car without serial assembly line? The paper showed that the Uddevalla plant developed a unique assembly organisation and cognitional method to understand production without an assembly line. Specifically, the assembly process design was organised by four assembly groups, set from a combination of body and parts grouped in work modules. Each operator received a holistic image of a car as an object assembled and a hierarchical understanding of the assembly process in advance; therefore, it was possible to assemble a car supported by a kit parts supply to enhance cognitive sequence. This paper clarified the critical role and position of the cognitive field for the Uddevalla production system (UPS).

The paper compared UPS with TPS, revealing some interesting points. TPS use cognitive capabilities as a system, the same as UPS but with different functions. TPS use cognition in two fields, one for building organisational common understanding and the other for tool and method concepts.

Uddevalla plant was essential to understanding how production design and human activities are organised and how these combined are relevant in the IT age.

Discussion

(1) The 'neo-craftsmanship production', named and spread worldwide by Womack (famous as the father of Lean production), is important to re-evaluate. The most important point of UPS is to make a standard format for assembling. This formatting is comprised of an assembly subject and the human cognition domain. Craftsmanship differs between craftspeople. Therefore, it is for individual

methods, not organised ones. TPS uses different type of formatting formation between two domains depending deeply on production system.

(2) This paper showed that activities in this formatting field (especially the relationship between production systems and human action, and assembling objects and human cognition) would be an important research point for the digital age as human activities would be more individualised. The cognitive formation in this paper should be developed in more detail, especially in organisational activities, such as how an organisation builds cognitive formation to organise employees to push production and activities forward.

(3) The critical rationale and significance of studying UPS lie in the strategic interplay between human activities and cognition in practical applications. UPS strategically integrates human action and cognition across both technical and organisational domains. This investigation is grounded in the socio-technical system (STS) framework; however, STS does not traditionally encompass cognition as a primary subject. By examining the application of cognition at UPS, we aim to clarify how objects are perceived, contextualised and interpreted. This approach allows us to present cognitive methodologies as a potential model in workplace settings.

In the contemporary future, the advancement of information and communication technology (ICT) is anticipated to expand the scope of work worldwide, rendering individual tasks increasingly complex and isolating within organisational structures. Under these developing conditions, while ICT personalises work for each individual, a critical need arises for collaboration and mutual understanding to execute tasks effectively. This situation necessitates the development of appropriate cognitive frameworks and the preparation of cognition for action. The emphasis on forming a shared cognitive understanding becomes a common 'language' platform to communicate and think for assembling actions among operators, individuals and organisational dimensions, including projects, teams and divisions. As mentioned, the study of UPS offers critical insights into the role of cognition in technological advancement. These insights provide a framework for understanding how human organisations can adapt to emerging technological innovations to achieve heightened efficiency from individual and organisational perspectives.

Chapter 5

Summary

The Volvo Uddevalla factory represented a new way of producing cars in a world where assembly lines had been the norm since the Ford Model T. The factory was based on experiences from several production systems designed within the Swedish automotive industry. It was seen as the future of the automotive industry in achieving sustainable industrial work; thus, it is essential to understand the operators' work environment and content to develop essential prerequisites for successful production. Self-determination theory can be used to understand operators' work environments and how they impact operational performance and health. Therefore, this study explored operators' social and technical work environments at the Volvo Uddevalla plant and how the organisation considers operators' need for autonomy, competence and relatedness.

The research data are based on interviews with 17 Volvo Uddevalla factory assemblers in the spring of 1993. Each interview was video recorded and lasted one and a half to two hours. These interviews were transcribed and coded, providing a base for the analysis and findings.

The findings show that the design with parallel flow and long work cycles created significant opportunities for high performance and good working conditions. Consequently, the production system enabled operators' basic psychological needs to be fulfilled in several aspects: autonomy through the fact that the teams essentially organised themselves, competence through kitting, overview and working in pairs and relatedness through organisation in teams. However, management planned, controlled and led the production system as if it were a traditional assembly line; thus, its potential was not utilised. With few exceptions, the managers lacked coaching skills and deep knowledge regarding the process and the leadership was traditional, focusing on achieving goals but not on how objectives could be achieved. There was limited focus on learning and development; for example, operators' skills to develop standardised work were not utilised, leading operators to distrust the standardised work.

Discussion

(1) The design with parallel product flow, extensive work cycles and content generates the potential for high performance and good

working conditions. However, the productivity and quality potentials were not utilised due to how the production planning and control were conducted, meaning that the takt (pace) was set based on the most complex product variants, leading to long waiting times for operators (i.e. non-value-adding time). Furthermore, the product was designed to be assembled at a traditional assembly line, not in a job shop with limited space for component racks and tools. Consequently, management at Volvo Uddevalla planned the workshops in the same way as if they were an assembly line. They did not understand that working in pairs with group work patterns is efficient because there are no system losses (no waiting for other workers).

(2) The production team organisation, with seven operators per team, provided opportunities for operators to have good working conditions. Generally, the assemblers appreciated the work, stating that they found it pleasant and they developed their work together with their colleagues. They often formed strong friendships with their co-workers. Managers play an essential role in forming the social system; however, with a few exceptions, at Volvo Uddevalla, they lacked coaching skills and deep knowledge regarding the process. Instead, the leadership was traditional, focusing on achieving goals but not on how objectives could be achieved; thus, there was a limited focus on learning and development. For example, operators' skills to develop standardised work were not utilised; hence, operators largely distrusted the standardised work developed by production engineers.

(3) The technical system with a parallel flow pattern, kits for material feeding and work in relatively small teams enabled the fulfilment of basic psychological needs in several aspects. These include autonomy, as the teams essentially organised themselves, competence through kitting, overview and working in pairs and relatedness through organisation in teams. With leadership based on the production system's design and needs, well-being and performance would have improved significantly.

Chapter 6

Summary

The TPS is supported by Toyota's personnel management/labour-management relations. The two pillars of the TPS are the 'just-in-time' and the 'Ninnben no tuita Jidoka' (Automation with a Human Element). Managed education is compulsory and high school education supports the system in the community.

The 'Ford production system' is famous for the 'assembly line' of automobile production and is characterised by the thorough use of physical labour based on simple and repetitive labour. In the case of Toyota, it goes one step further. Even on the same line, labour is characterised by thoroughly utilising the workers' physical ability and intellectual one for the purpose of management.

This process requires 'human resource development' through in-house education that is, the development of so-called Toyota men willing to continue working for Toyota's corporate objectives. This approach has been implemented through individual, small-group and all-employee labour management.

Toyota's personnel and labour management is moving towards flexibility (i.e. the use of non-regular workers), 'de-working hours' and motivation management centred on economic and social stimuli (i.e. no disclosure of base increases and wage decisions based solely on personnel evaluations). This structure leaves no room for the demands and claims of individual workers. It is hard to deny that this disregard for the individual and individuality has led to a spate of occupational accidents, deaths from overwork and suicide.

Toyota avoids labour-management conflict, and class extinction has emerged within the company. Toyota's development fosters managed competition and immobilises wage inequality and discrimination; thus, the chances of realising a free and egalitarian 'welfare state' like Sweden's are slim.

Discussion

(1) The labour and living conditions of Toyota and its subcontractors do not come up in discussions on TPS and lean production systems (LPS), such as work-related accidents and deaths from overwork and the increasing number of non-regular workers.

Understanding how to utilise the capabilities of Toyota workers in the Toyota way requires that we keep in mind the personnel, labour management, industrial relations and the production system. Toyota's personnel, labour management and industrial relations (which manage Toyota's regular and irregular workers in a unified manner while dividing them into individuals, small groups and as a whole) cannot be ignored. Unfortunately, human resources management and industrial relations between Toyota and Volvo have not been compared; this is an issue for future research.

(2) In areas where Toyota plants are located, roads are built for Toyota's parts transportation (JIT), which is highly inconvenient for people, especially children and older adults. Roads and urban development must also be considered when considering the TPS.
(3) The UPS parallel production system is a shop floor production system. Conversely, Toyota includes the entire process, from the delivery of parts from suppliers at various levels to production and sales, which JITs link. Moreover, at Toyota, workers are linked through the 'human Kanban' system, from recruitment to retirement, including non-regular workers and even the attraction and repulsion of workers from subcontractors. Employment management is adopted so that Toyota will not waste money.
(4) Unlike Volvo, Toyota has continued to build plants and a total production system reform may not be compatible with Toyota's policy of gradual 'kaizen'.

Chapter 7

Summary

Chapter 7 focuses on the production system itself. FPS was transformed into TPS because of the constraints in post-war Japan. The constraints give birth to such transformation; however, most people regard TPS as the most rational system worldwide.

While people were absorbed into TPS, Toyota established a new stage of standardisation, which included autonomous complete process (ACP) and standardised assembly sequence (SAS). Because of these new concepts, the work content of SW and sequential order of assembly became the same in Toyota's plants worldwide. This structure gives

Toyota a significant advantage from the viewpoint of efficiency and quality in global competition; however, people have overlooked its importance.

In the case of TPS, conceptual work was still entirely excluded from standardised work (SW); however, it is included in Kaisen, in which all operators were involved. Furthermore, some selected operators were included even in the revision of SW and in the activities of responses to irregularities, both of which were conceptual works.

Second, Chapter 7 focuses on TPS's influences on society. The excellence of TPS includes a pull system, levelling production, visual management, the principle of solving problems immediately, the concept standard (i.e. a crystallisation of all wisdom found on the manufacturing site) and the presence of a manufacturing engineer. These excellences show us the ultimate modern rationality from the viewpoint of waste elimination.

However, several dangers exist in such modern rationality. The concept standard does not imply whether the criterion is appropriate—the strong orientation towards standardisation results in the unification of how to think. Enforced concentration on a production system brings about indifference among people to a world beyond a production system, depriving people of citizenship in civil society.

Discussion

(1) The relationship between Lean Production (LP) and TPS.

As far as Japanese co-authors know, TPS is far from allowing workers autonomy. The LP concept is a generalisation of TPS; however, the Swedish LP seems friendlier to workers. If so, Japanese LP might be transformed into Swedish LP influenced by Swedish tradition, which differs from Japanese tradition(see 12.3.1). Such differences may tell us that the same concept functions differently between countries. For example, while Nohara conducted research at NUMMI at the beginning of the 1990s, all employees, including the president, wore the same uniform with the name tag on the chest. It seemed to Nohara to be the symbol of workers' integration into the organisation but most workers were happy to wear the same uniform because it symbolised equal partnership at the plant. This situation suggests that the uniforms function differently for Japanese critics and ordinary workers at US factories (see 12.3.2).

(2) PS's rationality vs. its dangers to society.

The author describes TPS as the ultimate modern rationalisation from the viewpoint of waste elimination. However, whether it works well without the vertical division of labour and hierarchical bureaucratic organisation may be an open question. Conversely, once we widen our horizon beyond the production system from the viewpoint of citizenship, the very rational production system may bring about uniformity of thinking, which is the entrance into totalitarianism and indifference to citizenship and public affairs (see 7.5.2.3).

Chapter 8

Summary

Chapter 8 clarified the fundamental attributes of the cell production system (CPS) and the societal and technological conditions for transitioning from traditional production systems to CPS.

When comparing CPS to traditional production systems, CPS stands out due to its four fundamental characteristics: quick response to changes in the market, significant elimination of 'Muda', rigorous control of efficiency, and the autonomy and completeness of work. These basic characteristics have improved efficiency by boosting productivity, enhancing quality, augmenting flexibility, reducing floor space and elevating the work morale of operators. Conversely, CPS has also implemented specific enhancements to make work more humane. This is because it has granted workgroups or individual operators the autonomy and completeness of work.

There are three types of assembly cells. The basic types are mini-assembly line (Type1) and assembly by a single operator (Type2), with Type2 further divided into the rabbit chasing method (Type 2a) and the stationary method (Type 2b). When comparing these assembly cells based on their level of autonomy and completeness of work, Type 2b is characterised by a high level, Type 2a falls in the intermediate range and Type1 is considered low.

The quantity of assembly cells and combinations of types fluctuated due to modifications in products and production volumes. Many assembly cells were classified as Type1, which were short assembly lines. The workgroup on this new assembly line experienced greater autonomy and completeness than the traditional assembly line. However, the individual operators' work was fragmented and lacked autonomy or completeness.

Furthermore, five specific social and technical factors were necessary and facilitated the transition from traditional production systems to CPS: the emergence of issues inherent in traditional production systems, labour-intensive nature of the assembly process, extensive decision-making authority of management, expertise of manufacturing engineers in production processes and the unique characteristics of the product.

Discussion

(1) Work efficiency and Humanisation.

The CPS successfully achieved simultaneous improvements in work efficiency and humanisation. However, it should be noted that there were more Type1s, which prioritise improving efficiency rather than humanising work. This was probably influenced by factors, such as the extensive discretionary power of plant management and the absence of pressure in the local labour market. Therefore, in the future, if workers and trade unions have a greater voice at the enterprise and industry level, or if labour shortages become more acute, the CPS will shift to a greater emphasis on the humanisation of work. If the CPS shifts in this direction, the efficiency (resource-saving character) of the CPS will also evolve and enhance the sustainability of the industry.

(2) Prospects for CPS diffusion.

Among the five factors of CPS diffusion, two particularly important conditions are the emergence of issues inherent in traditional production systems and the labour-intensive nature of the assembly process. The former is due to changes in the product market. It is uncertain whether the trends in product markets, such as product diversification, shorter product life spans and increased price competition, will persist in assembly industries like electrical and precision machinery. However, product diversification would be irreversible. Similarly, whichever the advancement of artificial intelligence will alter the labour-intensive assembly process remains uncertain. Assuming product diversification, it will be difficult to alter the labour-intensive nature of the assembly process. Therefore, it can be anticipated that the transition from traditional production systems to CPS will progress in situations where the five social and technical conditions or comparable conditions, are present.

Chapter 9

Summary

There are three levels of knowledge: details, relationships and overview. Even if knowledge of relations between details is expanded and accumulated, knowledge of relations never spontaneously reaches the overview level. We can achieve an overview only when we focus explicitly on how to reach this level, which is known as a holistic learning strategy.

In a traditional Ford-type assembly line, an operator's time was fully occupied by a single segmented and fragmented task. In contrast, once a Toyota assembly operator becomes used to a specific operation, they shift from the current operation to one of the next-door operations. Such shifting occurs often during a worker's work life, known as job rotation. In this case, an operator learns work based on an 'additive learning strategy'. There is knowledge of detail and the relationship between details; however, there is no place for knowledge of overview.

Through this job rotation, the worker becomes an unskilled multifunctionalist; however, the worker can only handle several unskilled tasks, each requiring no expertise.

In the case of UPS, an operator could adequately handle a current operation because they clearly understood the target and maintained knowledge of the overview. They also clearly understood the connections between a series of operations due to maintaining the knowledge of relations.

UPS created the first reform since the introduction of the Ford production system, which paid attention to operations and the operator regarding human cognitive capability. UPS operators use a holistic learning strategy, gaining an overview of an automobile's total architecture and assembling process; therefore, an operator exercises an inner dialogue using their initiative without a manager's order.

Discussion

(1) As the holistic learning strategy (Nilsson, 2001) and its materialisation (Engström et al., 2001) tell us, we cannot spontaneously reach

the overview without a strategy for how to reach it. In other words, when we only focus on our daily routine, we never get an overview; thus, we must have our own strategy for reaching the overview. In the case of automobile assembly, the object of overview is the physical entity, i.e. an automobile body. Therefore, Nilsson and Engström found a unique way to reach the overview: first, they disassemble complete automobiles and then arrange disassembled components and parts on the floor according to 'assembly-oriented product structure'. In the case of social system analysis, the object of overview is not a physical entity but the social system, which is not easily recognised. Therefore, although our daily routine and concerns are locked in the subordinate sub-system, we must find the original way to get an overview of the social system. Otherwise, we cannot defend the destiny of a total social system against catastrophe, such as totalitarianism(see 7.5.2 and 12.3.4).

(2) As the last remarks of the above summary suggest, in the case of UPS, the number of managerial ladders of the production organisation could be less than the other type of production organisation. If an individual wants to maintain a considerable number on the managerial ladder, they may prefer TPS or FPS to UPS. In this regard, the preference for a production system may be influenced by the organisation's preference between a highly bureaucratic organisation and a more horizontal decision-making system.

Chapter 10

Summary

Chapter 10 clarified two points. First, as the assembly system shifted from traditional assembly lines to assembly lines with Toyota's ACP, Company N's cell production and Uddevalla's parallel production, the fragmentation and subdivision of standard operations performed by individual operators or workgroups were eliminated, while their work content expanded. In other words, these changes suppressed the division of labour in standard operations and shifted it in a way that restored the functional coherence or completeness of their work contents. Second, restoring functional cohesion or wholeness in standard operations affects the division of labour in two directions.

One is to expand the possibility of utilising a peripheral labour force, such as temporary workers and employees of the contracting company when standard operations are limited to partial work performed within a very short work cycle. In this situation, the fragmentation of work is eliminated and the work becomes more coherent, facilitating operators' work knowledge than before. This case occurs with ACP and cell production Type 1 (mini-assembly line).

The other is to increase the possibility of delegating non-standard operations to individual operators or workgroups in a system where standard operations have long work cycle times and a single or a few operators assemble the entire product. In this situation, operators have a better understanding of the overall assembly of the product and improve their work competencies. Technical autonomy is also ensured through the parallel assembly of complete products by individual operators or workgroups of several operators. If this possibility becomes a reality, job integration will increase while the division of labour will be further limited. Parallel and Type 2 cell production (assembly by a single operator) demonstrate this situation.

Discussion: Thorough Division of Labour vs. Humanisation

Restraining the division of labour and promoting job integration imply the humanisation of work. Given the severe fluctuations in the product market, what is essential to expand the feasibility of this idea? Those working in labour-intensive, high-volume assembly industries must reconsider their belief that a thorough division of labour increases efficiency. In other words, detecting waste in the division of labour is a matter. In this regard, it is critical to consider the time loss analysis described in Chapter 2 and the study of implementing alternative assembly systems based on this analysis. However, changing long-held beliefs about the division of labour in companies is difficult. As a result, collaboration with experts outside the company, such as researchers who are free of such beliefs, is essential.

Chapter 11

Summary

Toyota and Volvo faced the same labour shortage problem. Regarding how to solve it, both had the same work-focused orientation. Therefore, they used similar or identical technical means to solve the problem. As a result, we found technical convergence between their experimentations; however, they had different targets realised in the production system.

Toyota aimed to reduce fragmentation and segmentation of work to realise more comprehensive work, meaning that work recovered the contextually meaningful connection among the series of operations in hand. In contrast, Uddevalla's target was to realise integrated work to overcome the division between conception and execution. Therefore, the production system became different, representing system divergence.

In both cases, there were four work-focused technical means: reintroduction of contextually meaningful connections between a series of operations, grouping of parts based on contextually meaningful connections, kit systems based on new grouping of parts and buffer mechanisms. Each means aimed to solve the same problem, i.e. a labour shortage; however, the concrete targets differed.

Both companies shared the same view (i.e. work-focused improvement), so they employed the same or similar means to solve the same problem; however, the functions varied because each company had a different target. The New Toyota Production System (NTPS) aimed to ensure a certain level of efficiency and work quality regardless of the worker's competence and attentiveness. Therefore, it wanted to disregard individual differences as much as possible and regard each worker as a standardised operator; hence, inner dialogues were not a concern and were excluded. As a result, these contextually meaningful connections created an objective way to achieve efficiency and work quality independent of the individual.

In contrast, at UPS, contextually meaningful connections functioned to promote quick and smooth inner dialogue. In summary, it was realised by an objective tool in NTPS while reproduced in the inner dialogue in UPS.

Discussion

When we understand humans' inherent cognitive capability and overview, inner dialogue works well on the labour process, making the extended cycle work possible. In other words, longer cycle work improves the segmented division of labour to a more expanded range of work.

Generally, once locked in the segmented division of labour, we lose concern for the total system. This indifference to the whole invites crisis. Overcoming this indifference to the whole requires widening the range of the segmented division of labour. Furthermore, the segmented division of labour is the indispensable condition of people's indifference to the whole and it makes possible the maintenance of a divided line between ruling and ruled classes.

Therefore, to address people's concerns, people must obtain an overview. In the case of auto assembly, the object of overview is the automobile body itself. Therefore, the way to get an overview is as follows. We disassemble the whole body and arrange all components and parts on the floor according to contextually meaningful connection, i.e. assembly-oriented product structure. In this case, all components and parts are tangible physical entities. In contrast, in the case of an organisation and social system, these are not physical entities, so we must develop our own way of obtaining an overview (see 12.3.4).

Note: The above summaries and discussions (12.1) are written by each chapters' authors respectively.

12.2 Condensed Structure of This Project

How can this book be understood regarding the contributions of the different chapters? Where do the analysis and discussions of the various contributions lead us?

The traditional assembly line has been questioned based on the same problems in Japan and Sweden: labour shortages and unattractive jobs. Similar technical solutions were introduced in Japan and Sweden through the NTPS, CPS and UPS. The various chapters show that organisation and social systems developed in different directions due to disparities in culture and experiences and how the situation developed in the two countries. These disparities led to significant differences in production systems.

Toyota further developed its previously successful production system (TPS) in terms of efficiency. For its part, Volvo had both efficiency problems and problems with the working environment in its traditional line-based production system and it made radical changes in its production system. These changes meant the need for completely new ways of organising operations. New learning strategies and alternative material supply methods were needed to enable long work cycles; however, they did not focus on changing leadership for UPS's new type of operation. After developing UPS over a few years, various solutions and experiments reached a production system with good performance and working environment compared to Volvo's traditional line-based production system.

For Toyota, the gradual change meant great operational stability through far-reaching standardisation of product and production processes to facilitate overview and understanding of production. The standardisation and one-sided problem setting given by the management have affected the operators' human conditions and led to uniformity in thinking and disregarding individuality. NTPS has meant high performance.

At the same time, CPS was developed in Japan as an alternative to achieve flexibility and efficiency. CPS meant production systems with improved work humanisation, primarily through autonomy and work cycle time, depending on the flow layout.

Figure 12.1 shows four distinct topics where UPS and NTPS differ (see bidirectional arrows). The topics are context in work, humanisation of work, leadership and performance.

As discussed in this chapter, in Sweden, the implementation of Toyota-inspired lean production was largely influenced by previous experiences such as UPS.

12.3 Randomly Selected Discussions for the Future

Readers will find many discussion points for the future in the above two sub-sections.

Now, the editors will propose four topics in the following sub-sections. Although editors do not insist that they are most important, the editors believe that these topics are, at least, worth considering.

12.3.1 Swedish Lean Production Compared with the Toyota Production System

Chapter 7 discusses the development of Toyota's production system until today's New Toyota Prodcutio System(NTPS), focusing on how the production system affects the employees of today's Japanese society.

The TPS has gained global spread through lean production since it was coined in the 90 s. Lean production is globally dominant as a basis for designing production systems. In different countries, lean production based on TPS has had different realisations and meanings since the

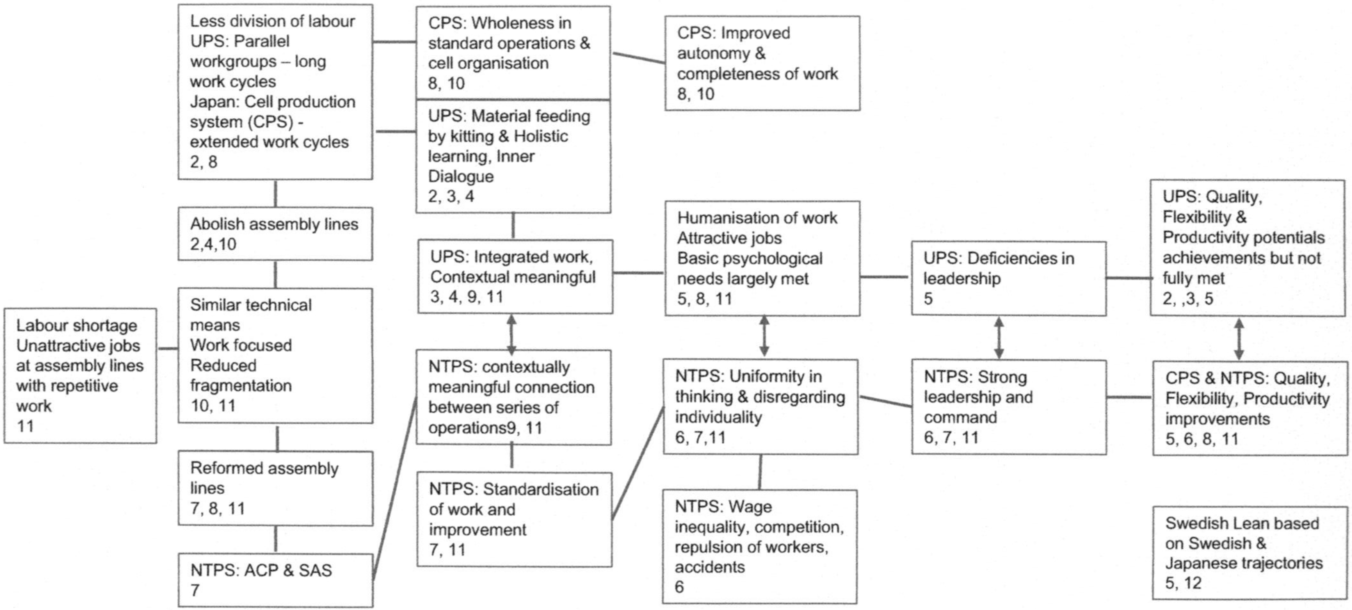

Fig. 12.1 Schematisation of results and discussions in the book's chapters and how they relate *Note* The numbers indicate chapters

90 s. Different backgrounds in the form of culture and experiences in production have contributed to different trajectories for lean production in various countries.

In Sweden, the development of what can be described as Swedish Lean production system(LPS) differs significantly from the development in Japan from TPS to NTPS, as described in Chapter 7. Swedish lean production means lean production, as is generally practiced in Sweden. Of course, there is a widespread difference in how companies claiming to be lean conduct their business; however, we try to describe how we perceive lean production to have developed in Sweden and its meaning. This view largely aligns with the Swedish national programme 'The Production Leap' ('Produktionslyftet'), which supports companies' development of LPS (ref. https://www.produktionslyftet.se/).

As this book shows, the Uddevalla factory resulted from experiences from Swedish companies, research and society, resulting in a unique car production system. In a Swedish context viewed in retrospect, this situation is not strange but results from gradually developing production systems. For example, see Sect. 2.2 where the development within Volvo, Scania and Saab is described.

With the global introduction of LPS and the subsequent closure of the Uddevalla factory in the 90 s, the type of Swedish production system that the Uddevalla factory represented fell out of favour. Most of these production systems were converted to what was then considered LPS. When we now look at what LPS means in Swedish industry, we believe that the steps in production system development that the Uddevalla factory brought about, as well as other Swedish production experiences and culture, contributed to strongly influencing the implementation of LPS in today's Swedish industry. Cooperation and an informal approach to the relationship between people in Swedish society are distinctive. In workplaces, collaboration in groups and ease of communication between hierarchical levels in companies is common.

Here, we will relate Swedish LPS to NTPS, as discussed in Chapter 7. First, we must look more closely at what Swedish LPS means and how its meaning has changed over time. From the beginning, the perception of LPS was based on the IMVP project and the book, *The Machine That Changed the World* (Womack et al., 1990) introduced 'lean', where effects were related to working methods and tools at Toyota. When lean implementations in Swedish companies did not become sustainable, it was understood through experience that aspects other than tools were essential for success. The understanding of the meaning of Swedish LPS has changed since LPS production was introduced in 1990 in the West. Three phases can be distinguished; initially, Swedish LPS was a matter of different tools observed by researchers in the IMVP programme and other Western studies in Japan. Then came a second phase that focused on the value stream. Unlike the Taylor functional organisational approach, Swedish LPS focuses primarily on meeting customer needs quickly. In the third phase, the focus has shifted to seeing LPS as a business strategy, meaning that a company develops its own production model based on its conditions. Thus, one does not copy Toyota but takes inspiration from Toyota and other companies. Today, two distinguishing features of a Swedish lean business strategy are the focus on value stream and continuous improvement through the personnel working in the company.

Such a business strategy requires the company's organisational structure to change to reward the value stream. An essential area is the relationship between engineering and production. Unlike Toyota, Swedish companies are often organised, so technical departments with engineers are often separated from production. These engineers are responsible for developing and deciding how production processes should occur. There is a particular change towards organising engineers to support production through Swedish LPS, but there is a long way to go within many Swedish companies.

With the focus on improving the value stream within Swedish LPS, abolishing functional organisation is essential. In a functional organisation, the business is structured based on different areas of expertise to achieve economies of scale in production and learning/knowledge; thus, the ability to reach effective value streams in the business is not in focus. The value stream mapping methodology is used frequently. The improvement involves creating a flow that is as continuous as possible and controlled based on demand, i.e. pull. Usually, a continuous flow cannot be reached, and the different flows (which make up the value stream) are synchronised with some pull control. In such a value stream, employees create and machines create value and tasks that represent waste are reduced. It is also essential that production lead times decrease, flexibility increases, quality increases, floorspace needs decrease and production overview improves. Unlike the NTPS line assembly organisation, the focus is thus not unilaterally focused on addressing variation in work pace.

The focus on value stream, an overview and relationship to the whole is provided and the employees understand which whole and which value is created. A clear improvement can be seen concerning the whole and overview of small and medium-sized enterprises (SMEs) that implement LPS. Accordingly, the basic psychological needs of relatedness are met (discussed in Chapter 5). Of course, achieving these effects in large production systems with line assembly and short-cycle times is difficult.

In Swedish LPS, the team producing/assembling, including the team leader, modifies the engineer's draft and suggests and updates standardised work. Unlike NTPS, there is a need within Swedish LPS for operators to use intellectual skills to develop standardised work. Thus, there is a need to satisfy the basic psychological needs of competence. Manufacturing engineers will regularly review the 'new' standardised work and the group leader (front-line manager) will formally decide to update SW.

Swedish LPS companies are often smaller (because they are the most) and the production leap is aimed at SME companies. In these companies, the cycle times are significantly longer than at Toyota's assembly lines, meaning that the standardised work is less detailed; thus, internal dialogue and dialogue between team members is common. If the work does not occur on an assembly line, the requirements for standardised work will differ and the job will not be as fragmented as in NTPS.

Within Swedish LPS, continuous improvement and learning are important, starting from the shop floor, where the operators in the work team are in control; the real knowledge of the work and the processes is considered to exist within the team on the shop floor. In that way, the work becomes more interesting for the operators. Within NTPS, a large company's centralised engineering resources develop better jobs using ACP and SAS. Therefore, are the opportunities to develop and improve too limited for the improvement work to be rewarding for the operators?

Entrusting deviations to be handled mainly by manufacturing engineers, supervisors, maintenance and workers in NTPS differs from Swedish LPS.

Through a short daily management meeting, the work team (the group leader usually participates) deals with all deviations from the previous day/work shift; they decide what can be solved within the team (responsible are appointed) and what must be solved outside the team by engineers, i.e. maintenance (group leader responsibility). Deviations resolved within the team are done directly by the person appointed (simple problems), while other deviations, together with improvement suggestions, are processed and developed within the team's improvement team (KAIZEN team). Therefore, there is a significant focus on the operators in the work team, who take responsibility for deviations handled within the team or being brought up in the organisation for the team to improve. A methodology is often available to evaluate which improvement proposals are to be implemented.

Thus, in these respects, the degree of autonomy is higher in Swedish LPS than in NTPS; however, in Swedish companies, company management is often weak in setting target conditions and goals using appropriate methods (e.g. Hoshin Kanri). Nonetheless, Swedish company management can develop goals for the shop floor other than eliminating waste and levelling production, which NTPS dominates (Chapter 7) and formulate them to act as a focus for the team.

For operators to take an active part in and be responsible for deviation management and improvements, the method must be taught and used by operators and team leaders. Such methods include A3, PDCA and value stream mapping. Considerable training is required to successfully and continuously use the methods to solve problems and develop the company. The different methods and tools used are developed gradually within the company by employees and managers. An example is the design of visualisation in daily management and improvement meetings.

Compared with NTPS, problem-solving is more of a team activity than an individual activity. The role of leaders in coaching is central to training and applying the methods. The methods become standardised (institutionalised); therefore, these are not methods for explicit engineers.

These methods and tools are used within Swedish LPS and by Toyota, but within Swedish LPS, the application differs from that of NTPS, where 'ends are already given before starting the thinking process' (Chapter 7). Instead, the aim is to develop and improve the operations (business), work and working conditions by utilising the employees' knowledge and experience. Whether using predetermined methods means that thinking processes are limited or facilitated can still be discussed.

Examining individual autonomy regarding the collective to solve problems in the team likely requires methods in both cases. In the team, success is achieved through the experiences and skills of the team members. In the team, problems are solved through the common method, and the operations are developed in consensus; thus, a certain measure of autonomy through the team is achieved in Swedish LPS. Thus, all three basic human psychological needs (Chapter 5) have relatively large possibilities to be reached in Swedish LPS.

Furthermore, in Swedish LPS, as in NTPS, targets are set with the company's operations that workers strive to achieve; thus, in some sense, it is about problem-solving and improving operations to reach the targets.

From a humanistic perspective, our perception of Swedish LPS compared to NTPS is significantly more positive in Sweden than in Japan. Swedish LPS means that the employees in the companies and on the shop floor contribute to developing and improving the business, suggesting that companies develop their operations through their employees and organise based on the value stream. Having had a division into a thinking and an executing organisation, it is now the case that blue-collar workers perform specific white-collar tasks. Overall, the hierarchical distance in the company is reduced. Production teams can take on new roles through teaching and training methods and implementing tools to solve problems and improve operations. It has often been the case that white-collar workers (i.e. engineers) have neglected developing production systems and only focused on daily operations. Thus, introducing methods does not negatively impact autonomy and thinking, which is the case with Toyota in Japan. This situation can be seen as our inspiration from TPS contributing to increased competitiveness and better, more interesting and humane jobs in the Swedish context.

12.3.2 Social Preference That Makes the Same Concept Function Differently

In the following, one of the editors (Nohara) will describe how the same concept functions differently in different contexts. The other editor (Medbo) refers to the lean production concept in 12.3.1. Regarding introducing lean production from Japan into Sweden, he does not necessarily say that a vertical hierarchy became more horizontal in Sweden than in other countries; instead, he says 'ease of communication between hierarchical levels is common' in Sweden. In this regard, he suggests that even if a formal institution is the same, its function could differ between cultures because its culture influences the institution.

His second remark is also very interesting, as he indicates that people take inspiration from Toyota, but they do not copy Toyota. Therefore, they may use different tools that are more appropriate to their society or system based on their inspiration. Alternatively, they may make the same concept function differently based on their inspiration.

The following is an example in a different field. Human beings are understood to have three fundamental rights in the modern era: liberty, equality and fraternity. Furthermore, most Japanese people are very conformity-oriented and not individualistic. Therefore, in the trajectory of modernisation, these fundamental rights mean something else in modern Japan. Liberty has not been fully understood and is unpopular among ordinary, non-individualistic people. Equality is easily accepted as conformity among non-individualistic people (see Kato, 1985).

As another example, at the beginning of the 1990s, Nohara interviewed 99 operators (supervisors included) at NUMMI(Toyota's actual first transplant in USA). The interviews included one specific question: 'What do you think about the company uniform?' In those days, uniforms with name tags were worn by all company members, i.e. NUMMI, including the president himself. It was announced as the symbol of equal partnership and was very popular among Japanese mass media and researchers. However, Nohara believed that mandatory company uniforms would make individualistic American workers unhappy.

In Japan, school uniforms were very common, and students in most high schools had strict rules for wearing uniforms. From most intellectual viewpoints, uniforms were regarded as the symbol of militaristic integration and obedience to the organisation to which students belonged; however, Nohara incorrectly believed that individualistic American workers hated uniforms. In the United States, blue-collar workers worry about their perceived social status as second-class citizens compared with white-collar workers. Surprisingly, in the 99 interviews, all respondents except one viewed the uniform policy as positive.

Nohara was impressed that the same tool or concept could function differently in different contexts. In retrospect, even in this book's context, techniques such as 'buffer' and 'contextually meaningful connection among series of operations' function differently in different contexts, like in TPS and UPS (see Chapter 11).

12.3.3 Social Preference That Makes Different Technology Selections Solve the Same Problem

12.3.3.1 Social Character of Technology Selection

The selection of technology in a particular era of a certain society is often explained as a result of inevitable technological development in human history; however, this is only the case when considering technology selection as the most rational choice compared to other possible alternatives. This way of thinking is similar to One Best Way by F.W. Taylor. Furthermore, such technological determinism is one of the popular bases for the historical view of the inevitable path of human history in which people take a path only towards a specific direction, even if there are some deviations. At the same time, we can also say that technology determinism is one variation of the historical view of inevitability.

However, it is not an inevitable result of technology development that determines the choice of technology and how it is utilised. Rather, the social selection of technology determines the desirable technology or the desirable way to use it; clearly, the technology selection reflects its social character.

We should now review our research on manufacturing technologies in this book. FPS was transformed into TPS in post-war Japan as a unique version of FPS. Facing the same labour shortage problem, we witnessed two branches of technology development. UPS emerged in the middle of other traditional FPS in Sweden. Another was NTPS transformed from traditional TPS. NTPS and UPS both focused on work reform, which differs significantly from equipment-oriented reform seen in most car manufacturers in the Western world. They both developed four similar or identical technical means for work reform. Nevertheless, such technical means pursued different directions and roles (see Chapter 11).

Such divergence comes from differences in social and technical conditions between Uddevalla and Toyota. In addition, we uncover an argument about the choice of technology we took and how it was utilised. We now examine this argument on the social selection of technology.

12.3.3.2 Selecting Manufacturing Technology and the Pursuit of Different Social Orientation

(1) **The Development of the Same Partial Technology**

Next, we compare the development of technology between TPS and UPS. As discussed in Chapter 11, the development of the total production system itself took a different path between Toyota and Uddevalla; however, both acquired four similar or identical partial technologies, each was a component of two different production systems. That is to say, four technological means were introduced, such as contextually meaningful connection among series of operations, reconsideration of the importance of buffer mechanism, reorganisation of parts grouping and the introduction of a kit system. When we focus on each partial technology, we see it has the same characteristics in both systems.

Therefore, it is possible that each management pursued different production systems but developed similar or identical partial technology. Furthermore, it shows that mutual technical learning among these technologies is possible even if the production system differs.

(2) Different Social Functions of the Same Partial Technology

A production system can be regarded as a work organisation. We must pay careful attention to the fact that selecting a particular manufacturing technology does not necessarily result in selecting a certain type of work organisation. It is a matter of social selection for people to choose a work organisation based on particular partial manufacturing technologies. For instance, the buffer function was taken seriously in NTPS and UPS; however, the social function differed significantly. In Toyota and Uddevalla, buffers were used to increase the efficiency of the technical system. In Uddevalla, the buffers also allowed the operators some autonomy. To facilitate the social role, the buffer function in Uddevalla included traditional buffers of automobiles and low operator density (high system capacity), as in Toyota; however, Uddevalla also included subassemblies, free worker positions and indirect work tasks (non-assembly). These measures also aim to increase the efficiency of the technical system.

The following is another example of a contextually meaningful connection among a series of operations. What happened in fragmented and segmented operations once ACP was introduced at Toyota? The degree of fragmentation of the series of operations decreased significantly and the accuracy of the segmented operations increased. As a result, the operator's sense of unbearable operations decreased. Later, in the 2000s, transplants abroad became common. At this time, a standardised assembly sequence (SAS) based on ACP pursued the high quality and quantity of a series of operations independent of an operator's skill and loyalty to the company. In contrast, the recovery of contextually meaningful connections among a series of operations realised the conceptional work in terms of recalling within standardised work in UPS.

(3) Social Preference for Technology

Social selection determines what kind of technology is chosen and how it is utilised, but it is not necessarily a result of the inevitable path of natural selection or technological development in human history. Such awareness could be the tool to criticise the view of social evolution based on technological determinism. It is up to people to change or not change their society.

Furthermore, let us study this social selection of technology differently. A specific manufacturing technology is found according to an orientation towards a particular work organisation; however, said technology contributes to realising the anticipated types of work organisation and can also contribute to developing different types of work organisation. It is up to people how to use the developed technology.

12.3.3.3 The Realisation of the Production System as a Result of Social Preference

(1) The Divergence Between Different Production Systems

As established, NTPS and UPS are the products of the labour crisis, which includes labour shortage and worker evasion against fragmented simplified work. Such products were the result of work-focused reform but not the result of equipment-focused reform. Because manufacturing engineers thoroughly knew manufacturing techniques and operations, they could realise work-focused reform against automation-oriented reform, which was the mainstream trend. In this regard, the presence of manufacturing engineers was necessary for work-oriented reform. However, who selected such work-focused reform rather than automation-oriented reform? The top management, an informal group of partial middle managers and partial workers on the manufacturing site, manufacturing engineers and vocational training specialists who worked for universities in the case of UPS pushed this work-focused reform. At Toyota, middle managers, manufacturing engineers were key persons who proposed such a work-focused reform and the top management accepted this proposal from the manufacturing site. They have pursued automation-oriented reform until the end of the 80 s at the Tahara plant, the frontrunner plant at that time. Nevertheless, they changed course in the middle of equipment-focused reform. Such a conscious changeover of manufacturing strategy was realised only because of such key people.

These work reforms from traditional FPS went as far as abolishing belt conveyors at UPS. Toyota moved towards the abolishment at the beginning of the 90 s; however, they ultimately decided to pursue work reform to the limit while maintaining a belt conveyor and Tayloristic work management principle.

Why did those divergences at assembly work occur? One of the reasons was the various social orientation of the management. Specifically, how seriously did they take the issue of de-humanisation of work at the manufacturing site? To what extent did they believe conceptual work can be realised on the manufacturing site without decreasing efficiency? In other words, to what extent did they trust an operator's potential to develop conceptual work? To what extent did they accept operators' autonomy, namely their discretion? The difference in the management's social orientation composed of such elements resulted in different production systems, namely NTPS and UPS, even when both plants faced identical labour crises and adopted the same work-focused reform and similar or identical technical means.

(2) Social Preference for Technology at the Divergence on the Evolution Path of the Production System

First, the total production system expresses the management's social orientation. Since a specific type of production system is tied to a particular social orientation, it could not be a tool for realising other different social orientations, even if it consists of similar or identical elements. That is to say, UPS is the product of the social orientation that pursues both efficiency and the humanisation of work. It results from the social selection preference of technology but is not necessarily due to technology evolution. In retrospect, the fact that UPS once withdrew its business resulted from abandoning social orientation and pursuing more humanised work. It was not the result of defeat in efficiency competition. UPS's retreat resulted from social preference, so its revival was a matter of social preference.

Second, even under the same social orientation, transplantation of the entire production system from one country to another is challenging due to the different social supporting conditions and restrictions against the production system. When FPS was transplanted into Japan, it was transformed into a variation of FPS, later called TPS. The challenge was mainly the shortage of capital and engineers and a much smaller domestic market. Furthermore, TPS also transformed into NTPS because of the change in the social environment, namely the young labour shortage between the late 1980s and the early 1990s (Chapter 7).

The comparison between the two systems revealed the issue of social preference for technology. Choosing a system is a matter of social preference, not the inevitable end that technology evolution traces.

12.3.3.4 Conclusion

We found three points regarding each partial manufacturing technology as follows. First, technology could be developed, transplanted and learned by different societies even if social orientations or social conditions vary among these societies. Second, a technology could achieve alternative functions depending on the social orientation. Third, since the entire work organisation, namely the production system, results from a social orientation, the same production system could not be selected if the social orientation differs. In sum, whether it is a partial technology or technology as an entire system, the social orientation of society or an intermediate organisation (such as a firm) determines what role the technology plays in the social function. In this regard, social preference for technology selection is critical.

Therefore, our conclusion now leads us to the theme of technology selection in industrial history. Whether it is technology to develop a product or manufacturing technology, we should re-examine the trajectory of technology evolution, including what Noble (1984) calls 'the road not taken' in a broader historical and social context.

12.3.4 Why Is Overview so Important? How Do We Obtain It?

12.3.4.1 Knowledge of Overview and Uddevalla

One of the editors (Nohara) has long wanted to grasp the logic behind UPS practice and the overview concept is essential for this purpose. The role of concept can be summarised as an overview based on the writings of and discussions with Nilsson, Engström and Medbo. Nohara's understanding of Uddevalla is not necessarily shared among co-authors, but he believes it is worth discussing.

(1) **Natural Work**

First, we look back at Nilsson's original concept, i.e. 'natural work'. In this case, a single person or team composed of a few people makes one complete product. While working, they mobilise their human potentialities, including physical and mental (including willingness, sense and reflection). These human potentialities are interrelated and function simultaneously; therefore, it is unnatural to divide them into their elements or prevent them from functioning simultaneously.

When examining the 'natural work' process in any case, we find three activities, i.e. planning, doing and control and adjustment. These activities compose work as its elements but are interrelated and overlapping. While workers engage in these three activities, they fully mobilise their human potential, as mentioned above.

This situation is the starting point of our analysis based on Nilsson's 'natural work'.

From the viewpoint of 'natural work', work division among planning, doing and control and adjustment does not function well and creates problems because it is not natural, which is typical in the case of Taylorism.

As 'natural work', the labour process is where three activities, i.e. planning, doing and control and adjustment, are done repeatedly. As a result, a worker develops their process towards the end product. While workers engage in their labour process, they repeat their inner dialogue between their human potentialities and tools and materials. That is to say, in their labour process, they raise questions concerning planning, doing and control and adjustment and find answers by themselves, which is the inner dialogue in the labour process.

The 'natural work' concept does not express nostalgia for the lost craftsmanship. The characteristic of 'natural work' is now reproduced daily in the young generation. The pattern of division of labour differs from age to age and in different societies. In this regard, the division of labour is not a 'natural' product but an artificial one. Conversely, while the young generation is not mature, they are not yet incorporated into any social system ruled by division of labour, characterised as the combination of what Nilsson calls 'work distribution' and 'work division'.

However, when they become adults, they will be incorporated into some social system ruled by the division of labour as pillars of their occupations. Human potentialities and work elements (planning, doing and control and adjustment) are interrelated and simultaneous. Therefore, 'natural work' is not nostalgia but continuously reproduced reality. On the contrary, work distribution and division are very artificial products in modern society.

(2) **Three Kinds of Knowledge**

The second point, which Nilsson emphasises, concerns the knowledge about the physical object. In this regard, he insists that there are three kinds of knowledge: knowledge of detail, knowledge of relations (between details) and knowledge of overview (of the physical object as a whole). Once a worker experiences the knowledge of different kinds of detail, he/she can learn about the relations between old and new details. Therefore, only through job rotation can he/she increases their his/her knowledge of relations. In this regard, it is not necessary to find any particular strategy. Daily routines and job rotations make it possible to gain knowledge of relations; however, the individual never understands the overview without a conscious strategy for obtaining it.

(3) **Disassembly and Re-grouping of Parts and Components**

Obtaining the overview when facing a complicated and extensive physical object like an automobile is a problem for engineering science. By nature, humans have one inherent cognitive capability and one inherent cognitive characteristic. Engström et al. (2001) regard it as their starting point for discussion based on achievements in Gestalt psychology, cognitive science and generative grammar.

Gestalt psychology means the capability to recognise the shape as a whole (overview) when viewing a physical object. For example, when we look at the shape of a dog created with dotted lines, we recognise it as a dog, not as just a collection of dots. Human beings have the inherent cognitive ability to recognise the shape as a whole (overview), as Nilsson and Engström emphasise.

Nohara adds one point to their observation: people can grasp the overview only when the shape as a whole comes into view. Using the dotted-lined dog example, it is hard to recognise the dots as a dog when only the ear or tail is visible.

This example is the same as in the case of short-cycle work on an assembly line, in which many obstacles interrupt this inherent cognitive capability for the assembler to obtain an overview of the automobile. Therefore, Engström et al. (2001) struggled before determining how to get the overview, i.e. the answer which they found was to disassemble a complete car and arrange components on the floor. This approach was so-called reverse engineering; however, the arrangement was complex. They reorganised the arrangement and changed how components and parts were grouped from design-oriented to assembly-oriented (assembly-oriented product structure). Arranging parts and components on the floor was done according to this new grouping, i.e. assembly-oriented product structure. It was a process by which the author disassembled and tested (in a dialogue with practitioners and researchers) different groupings of the components to find an appropriate and useful assembly-oriented product structure. Disassembly and assembly-oriented product structures were simultaneously established; it was difficult because no one knew the assembly-oriented product structure in advance. This approach demonstrated how to get an overview of the structure of a complete car and an overview of the assembling process from the viewpoint of an assembler.

(4) **Overview from the Assembler's Viewpoint**

Nohara explains why an assembler's viewpoint is referred to in this context. An assembler's viewpoint differs from that of others (e.g. a pedestrian on the street) indifferent to automobiles. From the viewpoint of such a pedestrian, a vague picture is enough as an overview because they have nothing to do with assembling an automobile. Conversely, from the viewpoint of an assembler, the overview should include content that supports their assembly operation. Therefore, in the case of an assembler, the concept of overview should include two pictures. One picture shows the whole architecture of an automobile composed of various relationships between components. The another picture shows the whole assembly process. When an operator gets these two pictures as the overview, they also obtain the surveyability of a series of operations. Of course, these two pictures could be one because of 'product/process duality', based on 'the ground/figure effect' of Gestalt psychology. This example represents the overview specified for an automobile assembler. That is to say, an overview differs per viewpoint.

The overview made it possible for a worker to find the end of their labour process and the right path towards the end through their inner dialogue on their labour process. In other words, this meant they could avoid the deviation from the right path.

(5) The Meaning that an Assembler Finds in the Structure and Series of Operations

As stated, human beings have, by nature, one inherent cognitive capability and one inherent cognitive characteristic. The latter means that by nature, human beings understand things according to their meaning which they find by themselves. What is the meaning of assembling an automobile? In this regard, Engström et al. (2001) found five relatedness factors: the parts-whole relationship, the functional relationship between components, the similarity between parts or components, the proximity between components in the architecture of an automobile and the symmetry between components in the architecture. These five factors compose what Nohara calls contextually meaningful connections in the architecture of an automobile and among a series of operations (Chapters 7 and 11). These five factors are beneficial for an assembler to imagine the subsequent operation to the present one and to prepare a physical position/posture appropriate to the next operation.

(6) Lessons from Uddevalla Concerning Knowledge of Overview

(1) Criticism of the division of labour between conception and execution:

  (i) Human potentialities are interrelated and simultaneously function.
  (ii) Work consists of planning, doing, and control and adjustment; these elements are interrelated and overlap.

(2) Human inherent cognitive capability and characteristics:

  (i) People can grasp the overall shape only when the shape as a whole comes into view.

(ii) Human beings, by nature, understand things according to their meaning which they find.

(3) Points concerning overview:

(i) Without a clear and conscious strategy, we never obtain an overview.
(ii) An overview differs by the viewpoint.
(iii) In the case of extremely long-cycle work, inner dialogue leads operations. In this case, if an operator lacks an overview of the whole structure and process, their inner dialogue cannot properly lead operations.

12.3.4.2 Intellectual Skill Without an Overview: Japanese Shopfloors

As described in Chapter 1, differences between Japanese and Western plants have been debated for several decades concerning the comparative study of the excellence between several production systems. In this regard, Kazuo Koike compared an assembly worker's operation on the line between Japanese and Western automobile plants.[4] He characterised Japanese assembly operators in auto plants as multi-skilling, which was very different from single-skilled workers in Western auto assembly plants. According to his thesis, Western assemblers only perform one task while working on the same shop floor. In contrast, Japanese assembly workers in an auto plant rotate their jobs frequently, which they call job rotation. As a result, Japanese auto-assemblers become multi-skilled, which differs significantly from single-skilled workers in Western plants. Koike said this was the very origin of the excellence of Japanese shopfloors. He also insisted that Japanese multi-skilled workers had some intellectual expertise because they had acquired the knowledge of relations among several simple tasks (Koike, 1996).

Regarding intellectual skills, much discussion has emerged among the Japanese parties concerned; however, this point is not the focus of this chapter. The point was that anyone including Nohara himself never refered to the lack of the knowledge of overview. In retrospect, most people in Japan are indifferent to getting an overview of a whole on

[4] Masahiko Aoki generalised Koike's study in his comparative institutional analysis. As a result, Koike's thesis of comparison of work between Japan and the Western world has become well-known.

any occasion. Therefore, it might be unthinkable among managers and researchers that the lack of an overview was the problem in the case of a blue-collar worker's job. Such a phenomenon seems to express thorough indifference to an overview among Japanese people.

12.3.4.3 Overview and How to Recognise Society as a Whole

(1) **Work Distribution and Work Division**

We commonly assume that the development of the division of labour (work distribution and work division) is unavoidable if people pursue higher productivity, namely a materially wealthy life. However, UPS shows that long-cycle work in a parallel workstation without a belt conveyor is more productive than short-cycle work in traditional assembly lines (Chapter 2). In this case, extremely long-cycle work includes two connotations: abolishing a belt conveyor (liberation from machine-paced work), and the degrees of work distribution and work division are less than short-cycle work.

According to Nilsson, human potentialities consist of mental and physical potentialities. The former includes willingness, sense and reflection. Such element of human potentiality is interrelated and overlappedly functions. As described earlier, work consists of planning, doing and control and adjustment. Each work element is also interrelated and simultaneously functions. Accordingly, artificial work division obstructs human beings from actualising human potentialities, at least to some extent.

Furthermore, it is also important that manufacturing techniques be set as being compatible with actualising human potentialities. The parallel flow pattern provides the conditions for significantly lower system losses than in a serial traditional line flow/conveyor belt. The latter becomes the technical obstacle to fully utilising human potential.

In sum, common sense should be re-examined that assembly work on the line and highly developed division of labour (work distribution and work division) in a modern industrialised plant are unavoidable for the 'wealth of people'.

(2) Why Is an Overview so Important?

In the case of extremely long-cycle work, inner dialogue leads operations. In this case, if an operator lacks an overview of the whole structure and process, an operator's inner dialogue cannot properly lead operations.

Now, we investigate why knowledge of an overview is so important. When we focus only on detail, we can not prevent the collapse of the whole (Chapter 7). This situation is especially prevalent in the contemporary world because the division of labour (work distribution and work division) is highly developed, and most people involved in the greatly segmented division of labour have become indifferent to the whole. In this regard, what Nilsson calls 'natural work' is worth remembering. Generally, without an overview of a whole system, our jobs in a particular sub-system cannot be adequately conducted; we cannot judge whether each action within our sub-system is appropriate from the viewpoint of total optimisation but not partial optimisation. It is especially worth remembering in Japanese society, where most people are indifferent to the whole.

(3) How Do We Get an Overview of Our Society?

We can learn about relations through job rotation, as Koike points out. However, we never obtain an overview without a clear, conscious strategy. Therefore, obtaining an overview is the problem. At the beginning of the Uddevalla project, Engström et al.(2001) found a way to get an overview: disassembly and re-grouping of parts and components, as already described. However, can this approach be applied to society, i.e. a total social system composed of many sub-systems? In this regard, society is undoubtedly an existing entity, but it is not tangible. What is the method to get an overview of the society compared with the disassembly and re-grouping of parts and components?

Although obtaining such an overview is challenging, Nohara posits the following approach. Anyone living in a particular society has already learned some tangible facts about their society. These are only fragments of things that occurred in their society. People gather with each fragment and discuss it with each other to get an overview of their society. Fragments are not the whole, but everyone has different fragments that reflect one aspect of our society. Therefore, willing and frequent discussions based on these different fragments might be the only way to get an overview of our society.

(4) **Breakthrough and Overview**

Why are Japanese companies good at incremental improvement but not good at breakthroughs? The Japanese way of thinking, namely indifference to the knowledge of overview, is strongly related to this point. Without knowledge of an overview of the product/process (only knowledge of partial relations and partial details), it is challenging to find the entrance of the fundamental innovation of the product/process or even image the fundamental innovation itself.

12.4 The Complement: Why Did Toyota Survive in the 1990s and 2000s and Why Did Uddevalla Close in 1989 to 1993?

Toyota and Uddevalla were concerned with work-focused improvement and developed their corresponding work-focused means. These were the same or similar; however, the means functioned differently because of the diverging aims of the two companies. The target of Toyota was to regain contextually meaningful connections between a series of operations. In contrast, the target of Uddevalla was eventually the reintroduction of integrated work, although the target was rather vague at the start-up of the Uddevalla project, like more humanised work. The two companies took different trajectories because of the difference in the target they wished to realise as an operator's competence; thus, it was not because of Toyota production system with greater technical rationality that Toyota survived and Uddevalla did not.

To explain the reason for this difference, we focus on the social choice of the technical system each company prefers. At that time, no company had a social concern about developing workers who were engaged in integrated work. This situation was the background behind why Toyota survived and Uddevalla closed.

Regarding the social concern to workers' competence, let us compare the level of competence between workers performing short-cycle work with about two or three weeks of training and workers performing integrated long-cycle work with about longer weeks of training. In Uddevalla, a worker needed one month of training to start working as a freshman assembler in running production.

Of course, on average, the latter worker's competence will be much higher than the former's concerning standardised work. Therefore, a relatively longer training period is not difficult for managers when long-term employment becomes more common than short-term employment. As a result, the management's and public's preference for implementing technology like the Uddevalla model will likely be revitalised.

We have discussed the background of Uddevalla's plant closure, shown above. We next examine the official statement about the closure. In the final decision, Volvo announced that the closure was not due to low productivity. Volvo claimed that the three assembly plants in Sweden (Göteborg, Kalmar and Uddevalla) were equally good, but the cars had to be assembled in one plant in Sweden with a combined body shop, painting shop and assembling shop. Neither Kalmar from the 1970s nor Uddevalla from the 1980s had a body or painting shop.

Another reason behind the plant closure was Volvo's financial crisis at the beginning of the 1990s, which resulted in very low demand and production volumes in the Volvo primary plants in Göteborg and Gent (Belgium). This situation was why Volvo 'had to' reduce their production fixed costs, meaning they closed down their two small factories in Kalmar and Uddevalla.

In 1988, Volvo decided that its expansion had to be abroad and Volvo needed to try to cooperate with other carmakers because of the expected high costs for the future development of engines and platforms.

The Renault Company, which Volvo was closely linked to in the 1990s, did not like production systems without belt conveyors. This aversion stemmed not from the productivity point of view but from the viewpoint of lower (as they believed) control of workers.

The main point is to underscore that the Uddevalla plant was not closed for productivity reasons. In reality, the productivity was better in Uddevalla than in two other Volvo plants in Sweden and the flexibility and the learning costs were much lower.

References

Aoki, M. (1988). *Information, incentives, and bargaining in the Japanese economy*. Cambridge University Press.

Engström, T., Jonsson, D. och Medbo, D. (2001). Assembly work structuring based on restructuring and transformation of product information: Comparions of management practices between Japan and Sweden, *Journal of Humanities and Social Sciences* (Nagoya City University), *12*: 279–309. See more the references of chapters 2 and 3.

Kato, S. (1985). Liberty and/or equality. *Sekai, 470*, 31–38.

Koike, K. (1996). *The economics of work in Japan*. LTCB International Library Foundation.

Nilsson, L. (2001). Kumitaterodo no Alternative to Gakusyu Senryaku (Alternative on assembly work and its learning strategy), *Rodo Horitu Jyunpo, 1510, 1512, 1514, 1516.*

Noble, D. (1984). *Forces of production*. Alfred A. Knopf.

Womack, J., Jones, D., & Roos, D. (1990). *The machine that changed the world*. Palgrave Macmillan.

Index

A

Additive learning, 301, 337, 344, 378, 404
Additive perspective, 131, 168
Advanced assembly work instructions, 54, 68, 72, 88, 96, 111, 112, 114, 116, 118, 127, 146
Advanced material feeding technique, 22, 36, 39, 61, 67, 105, 111, 127, 134, 146, 158, 159, 392
Appointment tests, 271
Assembly by a single operator (Type 2), 325, 327, 333, 335, 337, 338, 363, 365, 371, 372, 402, 406
Assembly cells, 325–333, 336–338, 363, 365–369, 402
  assembly by a single operator (Type 2), 325, 327, 333, 335, 337, 338, 363, 365, 371, 372, 402, 406
  mini-assembly line (Type 1), 325, 330, 333, 335, 337, 338, 363, 365, 371, 402, 406
  rabbit chasing method (Type 2a), 325, 327, 330, 338, 363, 402
  stationary method (Type 2b), 325, 327, 333, 338, 363, 402
Assembly functions, 208, 209, 211, 215
Assembly-oriented product structure, 25, 59, 63, 69, 70, 72, 75, 77, 79–84, 91, 108, 110, 113, 136, 165–171, 179, 181, 182, 294, 298, 345, 346, 380, 381, 383, 405, 408, 424
Autonomous Complete Process (ACP), 16, 294, 296–298, 357, 359, 370, 379, 381, 384, 400
Autonomous workgroups, 31, 43, 49, 50, 61
Autonomy, 46–48, 235, 241, 258, 278, 282, 376, 398, 402, 409, 414, 418, 420
  administrative autonomy, 48, 49, 76, 89, 357, 389
  and completeness of work, 330, 332, 338, 402

H. Nohara and L. Medbo (eds.), *Technology Convergence and System Divergence*, https://doi.org/10.1007/978-981-96-1910-8

technical autonomy, 47–49, 51, 52, 74, 76, 89, 90, 117, 225, 354, 357, 372, 406

B

Balance, 12, 77, 90, 147, 148, 155, 198, 214, 248–256, 259, 261, 282
Basic psychological needs (BPN), 15, 240, 242, 243, 256, 258, 261, 397, 398, 413
Breakthrough, 13, 159, 344, 377, 379, 429

C

Cell leader, 329, 364–369
Cell production systems (CPS), 10, 16, 323–336, 338, 363, 402, 403, 408, 409
Central assembly process information (the PKIs), 23, 25, 36, 164
Central information system (the KDP), 23, 25, 33–36, 54, 61, 131, 160, 163, 164, 166, 175, 179, 181
Central product information, 23, 27, 36, 55, 159, 172
Co-Determination Act (MBL), 199
Cognitional standard, 204
Cognitive capability, 4, 15, 294, 310, 346, 376, 377, 382, 404, 407, 423–425
Cognitive form, 224, 227, 229
Cognitive format and formation, 206, 211, 222, 224–226, 235, 396
Cognitive perspective, 219–221
Common platform, 5
Competence, 11, 49, 93, 106, 136, 139, 146, 240, 243, 248, 249, 252–255, 259, 261, 278, 297, 298, 304, 306, 317, 319, 346, 354, 355, 372, 384, 388, 389, 397, 398, 407, 413, 429, 430
Contextually meaningful connection, 17, 293–302, 310, 378–385, 388, 389, 407, 408, 418, 425, 429
Continuous improvement, 270, 276, 412, 413
Contractor's employees, 336, 366
Craftsman's work, 384, 385

D

Dealing with changes, 350, 351, 361
Dealing with problems, 351
Design-oriented product structure, 28, 30, 55–63, 65, 66, 70, 73, 74, 76, 131, 132, 166, 380
Development Agreement (UVA: Utvecklingavtalet), 199
Disassembly, 23, 25, 38, 59, 65, 69, 71, 79, 82, 84, 85, 162, 172, 207, 208, 380, 424, 428
Division of conception and execution, 10, 285, 293, 388

E

Educational concepts, 143, 144, 146–148, 153
Education and training management, 276
Efficiency, 1, 3, 7, 9–11, 17, 21, 43, 44, 46, 48, 103, 135, 158, 182, 208, 214, 217–219, 230, 233, 259, 269, 287, 301, 305, 308, 309, 311, 312, 315, 324, 330, 332, 338, 350, 372, 375–377, 381, 386, 388, 396, 401–403, 406–409, 418, 420
Elimination of waste, 303, 311
Engineer

manufacturing engineer(technical staff), 3, 25, 50, 55, 56, 64, 66, 67, 88, 114–116, 147, 163, 179, 180, 229–231, 288–290, 301, 307, 308, 335, 336, 339, 363, 370, 376, 379, 386, 401, 403, 413, 419
production engineer, 230, 231, 246, 247, 251, 253, 259, 289, 292, 336, 363, 368, 370, 398
Equality, 416
Equipment-focused improvement, 375
Expanded assembly work, 74, 126, 128, 156, 157, 164, 167, 177, 389
Experimental workshop, 23, 25, 35, 37–42, 54, 56, 62, 63, 65, 68, 71, 75, 79, 80, 84, 86, 127, 135, 139, 152, 159–161, 164, 169, 171–174
Extensively long assembly sequences, 54, 67, 68, 71, 110–113
Extremely short assembly sequences, 64–67
Extremely short assembly sequences in standardised forms, 64

F

5S, 232, 277
Five Whys, 232, 303
Fixed-term workers, 272, 279
Flexible Manpower Line policy, 279
Ford Production System (FPS), 3, 5, 16, 269, 285, 287–290, 292, 293, 346, 399, 400, 404, 405, 417, 420, 421
Functional equivalent, 5
Function Group Register, 30, 58, 59, 74, 132

G

Government-led Spring Struggle, 276, 281

H

Hamster wheel, 319, 320
Holistic learning, 4, 13, 22, 24, 25, 58, 68, 69, 74, 86, 93, 107, 108, 113, 206, 207, 301, 345, 346, 376–378, 382, 389, 392, 404
Holistic perspective, 167, 168, 207, 219
Human conditions, 409
Humanisation, 3, 7, 10, 11, 43, 46, 158, 182, 338, 372, 403, 406, 409, 420
Humanised work, 10, 420, 429
Human kanban system, 271, 282
Human resource development, 15, 270, 276–278, 399

I

Incremental improvement, 344, 379, 429
Indifference to a world beyond a production system, 317, 401
Indirect work, 248, 354, 355, 376, 387, 418
Inherent flexibility, 49, 90, 117, 389
Inner dialogue, 225, 293, 298, 299, 301, 344–346, 376, 377, 382–384, 388, 389, 404, 407, 422, 425, 426, 428
Inner monologue, 74, 147, 154, 156, 225, 382
Integrated work, 17, 149, 344, 376, 380, 384, 385, 388, 389, 407, 429
Intellectual skill, 302, 386, 413, 426
Intra-workgroup pattern, 53, 108–111, 114, 171

J

Japanese management, 199, 288, 290, 344
Job complexity, 241, 252
Job design, 240, 250
Job design theory (JDT), 15, 240
Job integration, 356, 372, 406

K

KAIZEN, 199, 233, 289, 290, 301–309, 311–318, 386–388
Kalmar plant, 92, 100, 127, 196–198, 200, 201, 234, 394
Knowledge, three levels of, 343, 345, 404

L

Labour-intensive nature of assembly process, 333, 334, 339, 403
Labour-Management Declaration, 270
Labour-management talks, 280, 281
Leadership, 15, 242, 249, 253, 256, 258, 261, 397, 398, 408, 409
Lean production system (LPS), 195, 196, 202, 215, 216, 228, 229, 233–235, 399, 411, 412
Learning strategy, 4, 11, 13, 128, 301, 343, 344, 376, 378, 382
Levelling production, 303, 306, 311, 312, 314, 401, 414
Local 'limited regular employees', 272
Loss
  balance loss, 44, 45, 331, 332, 359
  handling loss, 44, 45, 105, 331
  system loss, 44, 45, 89, 93, 94, 153, 257, 398, 427
  time loss, 37, 43–48, 102, 105, 117, 331, 372, 406
  variant loss, 44, 45

M

Machine-paced, 1, 11, 44, 48, 76, 86, 102, 376
Machine-paced product flow, 48, 90
Man/machine interface, 389
Management education, 269, 270, 277, 278, 281
Material kit (kitting), 22, 39–42, 60, 67, 70, 71, 73–75, 77, 80, 91, 92, 94–96, 99, 104, 105, 107, 110, 117, 133, 134, 136, 139, 159, 160, 164, 165, 177, 245, 247, 248, 251, 255, 353
Mechanistic approach to perceiving the product, 144, 145, 162, 164, 167, 180
Mechanistic learning approach, 93
Memory, 54, 136, 145, 152, 209, 294, 302, 344, 381, 383, 385, 386
Modernity, 16, 92, 286, 312, 316
Most advanced and unorthodox assembly systems, 22, 34, 53, 118, 119
Most refined versions of Japanese assembly lines, 85, 88, 89, 119
Motivation, 235, 240, 242, 243, 249, 279
Motivation management, 278, 399

N

Nationwide Karoshi Hotline, 273
Natural work, 60, 126, 140–142, 144, 422, 423, 428
New Factories project, 196–199
New language, 210
New Toyota Production System (NTPS), 297, 344
Non-machine-paced product flow, 48, 90
Non-traditional assembly systems, 6–8, 128, 135, 153, 180

O

Ohno line, 229, 230
Operations
  non-routine operations, 350, 363, 371
  non-standard operations, 350, 352, 354, 355, 360, 362, 363, 366, 369–371, 406
  operations requiring no qualifications, 350, 351
  operations requiring qualifications, 350, 363
  standard operations (routine operations), 16, 204, 310, 350, 351, 357, 363, 365, 366, 369, 371, 372, 405, 406
  unprecedented operations, 363, 370, 371
Operator concentration, 46, 92–94
Operators' inevitable stochastic variation, 44, 89
Organic approach to perceiving the product, 126, 144, 145, 161, 162, 165, 166
Organisational culture, 242, 256
Organisational design, 223, 242, 246
Organisational learning, 247, 256
Overview, 21, 36, 91, 94, 105, 110, 125, 141, 144, 161, 165, 168, 170, 206, 212, 213, 252, 261, 301, 318, 324, 343–346, 352, 353, 376, 378, 382, 397, 398, 404, 405, 407–409, 412, 413, 421, 423–429

P

Parallel production, 16, 370–372, 400, 405
Partial technology, 418, 421
Performance and performance-based pay, 275
Peripheral labour force, 371, 406
Petty bureaucrats, 313, 314
Physical context, 245
Plurality, 305, 311, 316
Problem setting, 303, 304, 311–314, 316, 409
Problem solving, 241, 303, 304, 307, 311–313, 315, 316
Product architecture and product variation, 27, 35, 36, 53–55, 58, 60, 61, 71–73, 77, 80, 84, 96, 102, 129, 131, 159, 165, 170, 180
Product change orders, 63, 67, 98, 99, 103, 147, 171
Product characteristics, 327, 335, 337
Product information, 23, 27, 28, 33, 36, 38, 41, 54–58, 64, 66–68, 85, 99, 108, 112, 114, 218, 227
Production teams, 15, 245, 248, 253, 254, 256, 258, 398, 415
Product variant codes, 55, 56, 63–67, 98
Progressively organised material and information processing, 97, 98
Psychological safety, 242
Pull system, 306, 401

Q

QC circles, 273, 276, 277, 279, 282
Qualified troubleshooter, 351, 361, 362, 371

R

Recalling, 387, 418
Recruitment management, 271, 272
Re-grouping of parts and components, 428
Relatedness, 240, 261, 397, 398, 413, 425
Reporting, communication and consultation, 278

Reworker, 360, 362–365, 371

S
Same concept, 401, 415
Scientific management, 6, 119, 129, 137, 141, 300
Self-determination theory (SDT), 15, 240, 241, 397
Set Parts Supply (SPS), 218, 295
Smallest work elements in the PKIs, 64, 66, 67
Social function, 389, 418, 421
Social preference, 415, 416, 419–421
Socio-technical approach, 8, 287, 390
Socio-technical research, 7, 8, 21, 31, 127, 129, 134, 158, 182
Socio-technical systems theory, 15, 240
Standardisation, 163, 225, 293, 297, 298, 301, 304, 305, 309, 311, 312, 315, 316, 383, 384, 400, 401, 409
Standardised assembly sequence (SAS), 295–299, 383, 384, 389, 400, 413, 418
Standardised questions, 298, 384, 389
Suggestion system, 270, 276, 279, 282
Surplus competencies, 106, 107, 120, 129, 137, 145, 150
Swedish industry, 6–8, 411
Swedish lean production system (LPS), 411–415
System divergence, 17, 390, 407

T
Team culture, 242, 254, 258, 261
Team leaders, 198, 231, 246–250, 254, 303, 314, 355, 357, 386, 413, 414
Technical intern trainees, 272
Technology convergence, 390
Thoughtlessness, 313
Tilting devices, 86, 87, 94, 110
Totalitarianism, 305, 312, 313, 315, 316, 320, 402, 405
Toyota Calendar, 273
Toyota Man, 270
Toyota Production System (TPS), 1–4, 10, 13–17, 195, 196, 202, 214–216, 227–230, 232–235, 269, 270, 273, 276, 285–288, 290, 291, 299, 302, 305, 306, 308, 312, 317, 331, 379–381, 385, 386, 395, 399–401, 405, 408, 409, 416, 417, 421
Toyota Way, 216, 276, 280, 400
Traditional learning, 301, 344
Traditional production systems, 16, 325–327, 329, 330, 332–335, 338, 339, 402–404
Transformation of production system, 6, 16, 285
Transplants, 2, 287, 292, 301, 418, 421

U
Uddevalla Production System (UPS), 1–4, 10, 12, 13, 17, 195–197, 200–205, 207, 209, 210, 215, 216, 218, 222–224, 226, 229, 231, 232, 234, 235, 287, 344–346, 376, 379–382, 384, 388–390, 395, 396, 404, 405, 408, 409, 417, 419, 420, 427
Uddevalla Project, 200, 428, 429
Unification how to think, 311, 401
Uniform, 305, 315, 316
Uniformity, 304, 305, 312, 313, 315, 316, 402, 409
Unique system of illustrations, 38, 75, 81

V
Value stream, 239, 412, 414, 415
Vertical division of labour, 10, 11, 402

W
Well-being, 15, 243, 261, 398
Work elements, 65, 66, 112, 354, 356–359, 422
Workers
  nonregular, 270–272, 274, 279, 281, 282, 336, 399, 400
  regular, 279, 336, 366
Work-focused improvement, 9, 10, 298, 376, 380, 384, 388, 389, 407, 429
Workgroup cohesion, 48, 49, 51, 74
Working-up, 51, 89, 100
Work modules, 93, 108, 110–112, 114, 211, 214, 216, 217, 219–221, 223, 226, 229, 298, 346, 382, 384, 395
Workshop manager, 244, 250, 253–256, 259, 261, 355, 357
Work situation, 15, 240, 241

The manufacturer's authorised representative in the EU is Springer Nature Customer Service Centre GmbH, Europaplatz 3, 69115 Heidelberg, Germany. If you have any concerns regarding our products, please contact ProductSafety@springernature.com

Printed and bound by CPI Group (UK) Ltd, Croydon, CR0 4YY

01/10/2026

02225924-0011